Disruptive Technologies for Sustainable Development

About the Conference

The intent of organizing the international conference is to provide tremendous opportunities to academics, researchers and practicing engineers to share the scientific knowledge, industrial experiences and research outcomes. The theme of conference is to address the recent societal issues and provide solutions to them through the application of Artificial Intelligence, Machine Learning and Deep Learning and big data. This conference covers the latest research and computing techniques applied to save Environment and Energy, Waste Management, Engineering Science and Technology. This also provides an opportunity to develop network and collaborations with international and national Universities and Industries.

Sustainable Goals

- Agriculture
- Quality Education
- Health Care
- Clean Water
- Energy Management
- Climate Modelling
- Green Environment

Technologies for Sustainable Development

- Machine Learning and Deep Learning
- Big Data Analytics
- Image Processing
- Nano Computing
- Mobile Communication
- Network Security
- 5G technologies
- Internet of Things

Proceedings of the International Conference on Application of AI, ML, DL, Big Data on Recent Societal Issues (ICARSI 2023)

Disruptive Technologies for Sustainable Development

Edited by

Dr. G. Nagappan

Professor and Head,
Department of Computer Science and Engineering,
Saveetha Engineering College

Dr. V. Uma Rani

Associate Professor,
Department of Computer Science and Engineering,
Saveetha Engineering College

CRC Press
Taylor & Francis Group
Boca Raton London New York

CRC Press is an imprint of the
Taylor & Francis Group, an **informa** business

First edition published 2024
by CRC Press
4 Park Square, Milton Park, Abingdon, Oxon, OX14 4RN

and by CRC Press
2385 NW Executive Center Drive, Suite 320, Boca Raton FL 33431

British Library Cataloguing-in-Publication Data
A catalogue record for this book is available from the British Library

ISBN: 978-1-032-55000-8 (hbk)
ISBN: 978-1-003-42847-3 (ebk)

DOI: 10.1201/9781003428473

Typeset in Times LT Std
by Aditiinfosystems

Disruptive Technologies for Sustainable Development – Dr. G. Nagappan et al. (eds)
© 2024 Taylor & Francis Group, London, ISBN 978-1-032-55000-8

Contents

Disruptive Technologies for Sustainable Development – Dr. G. Nagappan et al. (eds)
© 2024 Taylor & Francis Group, London, ISBN 978-1-032-55000-8

List of Figures

List of Tables

Disruptive Technologies for Sustainable Development – Dr. G. Nagappan et al. (eds)
© 2024 Taylor & Francis Group, London, ISBN 978-1-032-55000-8

Preface

We feel greatly honoured to have been assigned the job of organizing the AICTE Sponsored International Conference on Application of AI, ML, DL, Big Data on Recent Societal Issues (ICARSI'2023) on April 21 & April 22,2023 at Saveetha Engineering College. The international conference is a platform that brings together the brightest minds from across the globe to share their ideas and insights on the recent societal issues with Artificial intelligence, Machine Learning, Deep Learning, Big data and emerging technologies. With an aim to promote collaboration and foster innovation, this conference promises to be a melting pot of ideas and knowledge sharing.

We would like to extend our gratitude to the management for permitting us to organize the conference on "Application of AI, ML, DL, Big Data on Recent Societal Issues (ICARSI' 2023)" for conceptualizing and executing this event with utmost dedication and perseverance. We would also like to express our gratefulness to All India Council for Technical Education (AICTE) for sponsoring and approving this conference. We sincerely thank our beloved founder, president Dr. N. M. VEERAIYAN and our director Dr. S. RAJESH for their significant support to make this conference successful. We really need to convey our sincere thanks to the Principal Dr N.Duraipandian, Vice Principal Dr. R. Senthil Kumar and Dean Dr. V. Vijaya Chamundeeswari for their continuous support and guidance in organising the conference. We thank all our department staff, students and participants for making this conference successful.

ICARSI'2023 is a testament to the unwavering commitment of our academic community to push the boundaries of knowledge and pave the way for a better future. This conference covers the latest research and computing techniques applied to save environment and energy, waste management, image processing, machine learning and deep learning along with other engineering science and technology. This also provides an opportunity to develop network and collaborations with international and national Universities and Industries. Out of 229 papers submitted from various international and national universities and industries, around 67 have been accepted which is highly impressive. From the selected papers 54 out of 67 are registered and presented in conference. All papers which are presented and approved will be published in conference proceedings series on Disruptive technologies to achieve sustainable goals book series of Taylor and Francis (Scopus Indexed)

Disruptive Technologies for Sustainable Development – Dr. G. Nagappan et al. (eds)
 ISBN 978-1-032-55000-8

About the Institution

Saveetha Engineering College (SEC) was established in 2001, by the Founder President Dr N. M Veeriayan, a committed and dedicated Medical Professional. SEC has 13 UG courses & 7 PG courses including MBA, Doctoral Programs (Ph.D.) in five Departments. The college is autonomous under Anna University, accredited by NAAC with 'A' Grade. NBA accreditation for UG Courses in IT, CSE, ECE, EEE, Mechanical Engineering and has been certified for ISO 9001-2000 by BVQI. SEC has Five Research Centres recognized by Anna University, MOU with reputed academic institutions in India and abroad.

SEC established in the year 2001 by the Saveetha Medical and Educational Trust

SMET - 35 years of excellence in Education

Founded by Dr. N. M. Veeraiyan, a visionary philanthropist par excellence.

Accredited by NAAC with A Grade, Five UG Programs accredited by NBA,

Recognised by UGC with 2 (f), 12 (B) for research funding

Recognised as SIRO: Scientific and Industrial Research Organisation by DSIR, GoI

NIRF National Institutional Ranking Framework Innovation Rank 101-150 Band

Total strength of 5469 students in 15 UG , 8 PG including MBA , Doctoral programs in 6 Departments

About the Department

The Department of Computer Science and Engineering was established in the year 2001 with an intake of 60 students. It was increased to 120 students in the year 2005-2006.In 2017-2018, it is increased to 180. The department offers B.E Computer Science and Engineering, M.E Computer Science and Engineering programs. The department has a team of sincere, dedicated, competent and experienced faculty with commitment to serve the college in achieving its goals. The department has conducted many workshops, seminars, FDP, value added courses, conferences etc, at regular intervals with nationally and internationally renowned resource persons.

SEC FLEXILEARN

Learners can select Subject, Faculty, Schedule & Domain Electives (fully learner centric)

FLEXILEARN promotes interdisciplinary learning by creating opportunities to interact with peers from different programs. FLEXILEARN gives opportunities for gifted Learners to complete the credit requirements earlier and have longer duration for internships/ projects enhancing employability.

Centres of Excellence for Futuristic Learning

AI / ML, AR/ VR/ Game Development, Mobile Robotics, Cloud Computing, Industrial IoT, ARM Microcontroller System, Industrial Robotics, Centre for AI Tools, Blockchain, Drone Technology, Embedded Design Innovation Center, 3D Design & Printing, Renewable Energy, Big Data Analytics, Cybersecurity, Fully Automated Robotic SMT Assembly, Full Stack Development, Mobile App Development, Electric Vehicle, Nanotechnology, Mems Design Centre

UNIQUE FEATURES

LEARNING BY DOING - All subjects are integrated with 50% Practical through PBL - Project Based Learning & CLAB - Class in LAB system

LESS IS MORE - SEC follows Finland Model of Education

LIFELONG LEARNING through 5C's - Communication, Collaboration, Critical Thinking, Creativity & Caring.

SCOFT - Saveetha Consortium for Future Technologies to prepare the learners for INDUSTRY 5.0 enhancing Employability and Entrepreneurship.

HOLISTIC LEARNING - Transdisciplinary courses like Design Thinking, Problem Solving, Music, Theatre, Yoga, Dance, Arts, Philosophy, Psychology, Civics, Biology, Humanity, Sports etc., as part of Curriculum.

Blended Learning, Access to content and Assessment through Moodle LMS - Learning Management System for Interactive learning experience.

MASTERY IN LEARNING - Made possible by learning each course in smaller Modules.

Highlights

Active Entrepreneurship Cell & IPR Cell.

Placement Training from First Year onwards through Exclusive Training embedded within the Curriculum.

Courses conducted by Faculty from Industry to enhance Technical skills.

IELTS | Business English Certificate (BEC) Training for all in collaboration with the University of Cambridge, UK.

Infosys Campus Connect Program.

German, French, Japanese, Chinese Courses as part of Curriculum.

START-UP eco system through Incubation Centre.

Exclusive Training for Competitive Exams like GRE, GMAT, CAT, MAT, Toefl, IES / GATE etc.

ACADEMICS

30 Students per Class Room

1:13 Staff : Student Ratio

Faculty : 201 Ph.D; 96 pursuing Ph.D

96% Graduates Received Degree

Compulsory Internship / Paid Internship

6 Crore Merit Scholarship every year

AWARDS

Winners of Smart India Hackathon 2017, 18, 19, 20 & 21, conducted by the Ministry of Education, Govt. of India.

Chief Minister's Award for Excellence in e-governance - CII Connect Hackathon 2018.

EMC2 Academic Partner Award.

Best Academic Partner Award by ICTACT Academy, a Society formed by GoI, Govt. of TN & CII, for 5 Consecutive Years.

Stanford University, UK - Best Teaching Department Awards for EEE, Mechanical.

Best student Project Awards by Texas Instruments & Petrofac.

Aspiring Minds Excellence award by AMCAT in Producing Maximum number of Employable Students for 5 Years.

IEEE Best Student Branch Award for 2019.

Won IIIrd place in L&T Construction - CAD Challenge hosted by IIT Madras - Shaastra 2022.

PALS Award of Excellence in Participation from IIT-Madras for two consecutive years (2019-20 and 2020-21).

Disruptive Technologies for Sustainable Development – Dr. G. Nagappan et al. (eds)
© 2024 Taylor & Francis Group, London, ISBN 978-1-032-55000-8

Message from Founder President

Be the Best

International conference on "Application of AI, ML, DL, Big Data on Recent Societal Issues (ICARSI-2023) takes privilege in being a platform to address the various hassles in Computer and Technology, thereby providing ample opportunities for creating new products, services & applications. This conference is a step towards achieving our vision in becoming a world- class academic and research institution in order to produce human capital with first class mentality.

This conference will be a good starting point for Colleges and International Universities to interchange knowledge and skills in the area of recent technologies. We are looking forward to find new solutions in this area and forecast future trends in order to realize India's aspiration and to contribute to global needs. It is a great pleasure to welcome all delegates and participants to this conference. I would like to congratulate the Department of Computer Science and Engineering, for their commitment and superb drive in organizing this conference ICARSI'2023 as fourth time in this series with latest updates every time. I am very certain that this occasion will be able to provide a platform towards strengthening our relationships in knowledge sharing while at the same time provide the necessary thrust in joint research collaborations and product commercialization within the research society.

It is my aspiration that this conference will be a foundation for the growth of new ideas towards a better tomorrow. Last but not the least; I would also like to thank all the conference industry sponsors. With your continued support and interest in us, I am sure that the quest of making SEC a top-class college is not going to be impossible. Thank you.

Dr. N. M. Veeraiyan
Founder President

Disruptive Technologies for Sustainable Development – Dr. G. Nagappan et al. (eds)
© 2024 Taylor & Francis Group, London, ISBN 978-1-032-55000-8

Message from Director

It is my ardent belief that we will see INDUSTRY 5.0 very soon even before we get a grasp of INDUSTRY 4. We at SEC have invested today in the Emerging Technologies like AI/ML, Data Science, 3D Printing, VR/AR, Robotics, IoT, Cybersecurity, Block Chain, Quantum Computing to prepare our Learners for tomorrow with the help of our excellent industry academia partnerships, collaborations with foreign universities vast research base and above all continued patronage of our alumni.

Computer Science is the cornerstone of all Emerging Technologies. I thank the organizers for conducting this conference and setting a platform for exchange of ideas. I wish more such interactions and exchange of innovative ideas to happen in future between Professionals of Computer Science, other engineering disciplines and all other fields including Health and Humanities which is required to tap the potential of Future Technology for sustainable growth of our Universe.

I welcome one and all to participate in the International conference on "Application of AI, ML, DL, Big Data on Recent Societal Issues" ICARSI'2023, organized by Department of Computer Science and Engineering, Saveetha Engineering College.

I wish the conference a grand success. Thank You.

Dr. S. Rajesh
Director

Disruptive Technologies for Sustainable Development – Dr. G. Nagappan et al. (eds)
© 2024 Taylor & Francis Group, London, ISBN 978-1-032-55000-8

Message from Principal

It is my great pleasure to invite you all to the international conference on "Application of AI, ML, DL, Big Data on Recent Societal Issues", ICARSI'2023, which takes place in Saveetha Engineering College, Chennai on 21 & 22 April 2023.

The facets of Science, Technology and Engineering are changing very fast. Engineering and Technology are becoming increasingly inter and multi-disciplinary with needs for multi-institutional and multi-country participation. Major experimental facilities require very large material, human and intellectual resources. The various Conferences, training programs and workshops in this quarter here are focused on relevant areas such as advances in computer science. Education is what we imbibe from our books, our peers, our teachers and our surroundings. As we move towards our goals, we must always keep in mind that we owe more to the world than it owes us. Each of us can and must do his or her bit to help win the Universe that will nurture our future generation.

I thank AICTE, Taylor and Francis, all of the contributors, reviewers and experts for their categorical support in organizing this conference. I congratulate the Computer Science and Engineering Department- Head of the Department, faculty members, non-Teaching faculty members and Students for their tremendous efforts in organizing an event of this grandeur.

I congratulate all the researchers and students for their commitment for active participation and wish you all a grand success.

Dr. N. Duraipandian,
Principal

Disruptive Technologies for Sustainable Development – Dr. G. Nagappan et al. (eds)
© 2024 Taylor & Francis Group, London, ISBN 978-1-032-55000-8

Message from Vice Principal

I am immensely pleased to learn that the Department of Computer Science and Engineering is organizing the AICTE sponsored international conference on "Application of AI, ML, DL, Big Data on Recent Societal Issues", ICARSI-2023 on April 21-22,2023, through Hybrid mode.

Quality is never an accident; it is always the result of high intention, sincere effort, intelligent direction and skillful execution. It represents the wise choice of many alternatives. The thing is, continuity of strategic direction and continuous improvement in how you do things are consistent. They are mutually reinforcing. Science can amuse and fascinate us all, but engineering changes the world. The engineer has been, and is, a maker of history. Scientists study the world as it is; engineers create the world that has never been. The way to succeed is to double your failure rate. I hope this conference can make a change in the field of Computer Science and Engineering.

I wish the HOD, coordinators, faculty members and students of the Computer Science and Engineering department grand success in this event.

Dr. R. Senthil Kumar
Vice Principal

Disruptive Technologies for Sustainable Development – Dr. G. Nagappan et al. (eds)
© 2024 Taylor & Francis Group, London, ISBN 978-1-032-55000-8

Message from Dean

I am indeed happy that the Department of Computer Science and Engineering of SEC is going to organize the AICTE sponsored international conference on "Application of AI, ML, DL, Big Data on Recent Societal Issues", ICARSI-2023 on April 21-22,2023, through Hybrid mode.

Facing enormous pressure to stay competitive, manufacturers are turning to technology to digitize and automate operations. We look at the evolution of manufacturing and where technology is bringing it next, from human-powered exosuits to collaborative robots and beyond.

In response to global disruption, the industrial sector has changed dramatically since 2020. Industries are increasingly looking for new technologies, from advanced AI, ML, robotics in R&D labs to computer vision in warehouses, that increase productivity and remain competitive. The timelines and technologies will vary by sector, but nearly every manufacturing vertical will see sweeping change in the next decade. Digitization and automation of the factory are lucrative long-term investments because they decrease labour costs, increase uptime, and slash error rates. This conference discusses all the current trends in computer science engineering especially in AI, ML and Deep learning.

I convey my hearty greetings to the computer science and engineering Department organizers and laud SEC Management's efforts.

Dr. V. Vijaya Chamundeeswari
Dean

Disruptive Technologies for Sustainable Development – Dr. G. Nagappan et al. (eds)
© 2024 Taylor & Francis Group, London, ISBN 978-1-032-55000-8

Message from Conference Convener

Engineering is getting excited by the day. Emerging Technologies with fascinating possibilities are being explored. Disruptive technologies would offer enormous in the near future. Smart and sustainable computing systems are likely to change our lifestyle and become part of our daily life and not just the subject of seminar or talks.

It is a matter of great pleasure to proclaim that the Department of Computer Science and Engineering is organizing the international conference on "Application of AI, ML, DL, Big Data on Recent Societal Issues", ICARSI-2023. It is a matter of great pride that the organizers of the Conference have been successful in creating such a huge impact on the target participants in such a short span of time.

The overwhelming response received from all over the country in these conferences is simply amazing. It is heartening to acknowledge the fact that", ICARSI'2023 has been sponsored by All India Council for Technical Education (AICTE).

It is a great achievement on the part of the organizers to arrange the publication of the Proceedings of the conference in the form of an E-book to be published with Taylor and Francis. The publication of the selected papers of this conference in SCOPUS indexed Taylor and Francis conference proceedings would add another dimension to the overall impact of the outcome of this conference on a larger section of research community all over the world. It is also encouraging to note that delegates of international repute in the field including a foreign professor are going to deliver invited lectures and key-note address in this conference.

Information provided in various papers and reproduced in the proceedings is aimed at benefiting the Engineers, professionals and indirectly the society. It is expected that the purpose would be served in a satisfactory manner through in-depth discussion and interaction among the participants during the conference.

On behalf of the Department and also on my personal behalf, I take this opportunity to record my heartfelt appreciation and gratitude to all the authors, delegates, conference organizers and all others participants, for their untiring efforts and constant endeavour to make the conference scale new heights.

I wish ICARSI-2023 a Mega Success!!

Dr. G. Nagappan
Professor & Head

Disruptive Technologies for Sustainable Development – Dr. G. Nagappan et al. (eds)
© 2024 Taylor & Francis Group, London, ISBN 978-1-032-55000-8

Message from Organizing Secretary

The international conference ICARSI'2023 is a platform that brings together the brightest minds from across the globe to share their ideas and insights on the recent societal issues with Artificial intelligence, Machine Learning, Deep Learning, Big data and emerging technologies. With an aim to promote collaboration and foster innovation, this conference promises to be a melting pot of ideas and knowledge sharing. We would like to extend our gratitude to the management for permitting us to organize the conference on "Application of AI, ML, DL, Big Data on Recent Societal Issues (ICARSI' 2023)" for conceptualizing and executing this event with utmost dedication and perseverance.

We would also like to express our gratefulness to All India Council for Technical Education (AICTE) for sponsoring and approving this conference. ICARSI'2023 is a testament to the unwavering commitment of our academic community to push the boundaries of knowledge and pave the way for a better future. We are confident that this conference will provide a platform for all the participants to learn, network and create lasting partnerships. We wish all the participants a fruitful and enriching experience at the conference. Let us work together to create a better world through our collective knowledge and innovation.

I wish you the best experience.

Many thanks for attending the conference.

Dr V.UmaRani
Associate Professor/CSE

Disruptive Technologies for Sustainable Development – Dr. G. Nagappan et al. (eds)
© 2024 Taylor & Francis Group, London, ISBN 978-1-032-55000-8

Details of Programme Committee

International Advisory Members

Dr. Arivalagan Pugazhendhi, Ton Duc Thang University, Vietnam.
Dr. Eldon Rene, IHE-Delft, Netherlands.
Dr. Jorge Beltramini, Queensland University of Technology, Australia.
Dr. Eric van Hullebush, Diderot University, France.
Dr. Asadi Srinivasalu, Bluecrest University, Monrovia, Liberia.
Dr. Edward Danso Ansong, University of Ghana and CEO, Fredan Computers, Accra, Ghana.
Dr. Wael M.Alenazy, Associate Professor of Computer Systems, King Saud University, Saudi Arabia.
Dr. Abdullah Saleh Alqahtani, Associate Professor, CFY Deanship King Saud University, Saudi Arabia.
Dr. You Wing Leung, Hong Kong Bapist University, Kowloon Tong, Hong Kong.
Mr. Gus Bjorkeroth, Managing Director, EMEA, Cprime-California, Unites States.
Mr. Niranjan Shankar, Web Developer, Zurich Insurance Company Ltd, Dublin, Ireland.
Mr. Siraj Mohammed, Senior Technical Manager, Central Gropu, Bangkok, Thailand.

National Advisory Members

Dr. Vedaraman N, CLRI - CSIR, Adyar.
Dr. Sivasubramanian, NIT Calicut, Kerala.
Dr. J. Arun, Scientist – C, Sathyabama Institute of Science and Technology, Chennai.
Dr. P. Narayanasamy, Professor,PSG Tech,Coimbatore.
Dr. S. Chitrakala, Professor, College of Engineering (CEG), Anna University, Chennai.
Dr. S. Venkatesh, Associate Professor,Indian Institute of Information Technology, Allahabad.
Dr. V.Mahendran, Assistant Professor, Indian Institute of Technology, Tirupati.
Dr. Subha L, Faculty Scientist, AU-KBC Research Centre, Madras Institute of Technology, Chennai.
Dr. Vasantharaj Rajagopal,Chief Scientist, Madras Institute of Technology, Chennai.
Dr. Sameer Sahasrabudhe, Director, Educational Multimedia Research Centre(EMRC),Pune.
Dr. Yogendra Pal, Assistant Professor, NIIT University, Neemrana, Rajasthan.

Program Committee Members

Dr. S. Sasikumar, Professor, Saveetha Engineering College
Dr. M. Vijay Anand, Professor, Saveetha Engineering College

Dr. Anitha Julian, Professor, Saveetha Engineering College

Dr. E. Sujatha, Professor, Saveetha Engineering College

Dr. R. Kaladevi, Associate Professor, Saveetha Engineering College

Dr. R. Ramyadevi, Associate Professor, Saveetha Engineering College

Dr. V. Loganathan, Associate Professor, Saveetha Engineering College

Dr. V. UmaRani, Associate Professor, Saveetha Engineering College

Mr N.V Ravindhar, Assistant Professor, Saveetha Engineering College

Mrs. Preethi, Assistant Professor, Saveetha Engineering College

Mrs. Rekha, Assistant Professor, Saveetha Engineering College

Chief Guest

Mr. Sridhar Swaminathan, Associate Vice President (AI & Analytics), Cognizant Technologies

Mr. SriramVembu, Associate Vice President (AI & Analytics), Cognizant Technologies

Key Note Speakers

Dr. Lochandaka Ranathunga, Department of Information Technology, University of Moratuwa, Sri Lanka.

Dr. Asadi Srinivasulu, Professor & HOD, Head Research of Data Science, Data Science Research Laboratory, BlueCrest-Monrovia-1000.

Shreyas R, Associate Software Engineer, Fox Corporation, Cincinnati, Ohio, United States

Muruga Laxmanan, Information Security Officer, PCCW Solutions, Singapore.

Disruptive Technologies for Sustainable Development – Dr. G. Nagappan et al. (eds)
© 2024 Taylor & Francis Group, London, ISBN 978-1-032-55000-8

About the Editors

Dr G. Nagappan

He is currently working as Professor & Head of CSE department of Saveetha Engineering College, Chennai. He has 22 years of Academic experience. His Experience focuses on Teaching and Research with a global activity-based blended learning model with multicultural learners spanning across the globe. He has received his Doctorate from Faculty of I&C, Anna University Chennai, alumnus of Department of CSE, College of Engineering, Guindy, Anna University Chennai where he did his Engineering Post graduation in Systems Engineering and Operations Research. He is also alumnus of Mahendra Engineering College where he did his Bachelor of Engineering in C computer Science & Engineering under Periyar University, Salem.

He is outshining as PALS Innowah - Innovation Challenge Coordinator, Remote Center Coordinator, Aakash Project Coordinator and Spoken – Tutorials Project Manager cum FOSS Promoter working under NMEICT project of IIT Bombay and Kharagpur run by Ministry of Youth Affairs, Government of India. He has opulent experience in organizing 25 workshops for faculty & Students and submitting 16 android projects for Aakash Tablets. He has been project investigator for IITB - E-Yantra Robot competition & Aishu semi humanoid Robot for AVON Corporation, Mumbai and MSME Projects. He has also been Proud Mentor of the Finalist Team in Smart India Hackathon 2017 and Gujarat Rajkot Hackathon 2018.

His extensive experience across Engineering and Management discipline s include Course curriculum Development for Autonomous , Text book content development, Teaching course, Project Coordination, Project Supervision, Question paper setting, Answer script Evaluation, Keynote addressing, Website development, Conference and Workshop Organization, Management of projects such as Aakash, Spoken –tutorials and Remote Center, Doctoral Committee Member, Promotion of New technologies and Social Media which makes him a versatile highbrow. He has been creating innovative pedagogical techniques and analyzing its significance in Engineering Education with his experience in Wipro Mission 10X & IITM STLC. He has also been resource person for versatile topics in various Workshops, Guest Lectures and Conferences. He has received **'Bronze Partner'** award from INFOSYS for his involved contributions in CAMPUS CONNECT Program which involves both Foundation and Soft Skills programs. Received "Aram Achiever Award 2023" from Aram Seiyya Virumbu Foundation and OXFAA University on 28 May 2023. He has been one of the Top Finalist for **'Best Techno Faculty'** award in ICTACT – BRIDGE 2014 in appreciation of his exceptional utilization of Technology towards Education and Academic Superiority. He has received **'Jewel of Saveetha'** award for his stupendous contributions towards the development of new interdisciplinary projects involving all the constituent colleges of Saveetha University. He has received **'Best teacher'** award for last 8 consecutive years in appreciation of his outstanding excellence, commitment, technical expertise, motivation and hard work. He has received **'Best NSS Unit and Programme Officer'** award from Honorable Governor of Tamil Nadu, His Excellency Surjit Singh Barnala, for year 2006 indicating his high dedication, efficiency and outperformance amongst peers. He has published 16 Patents, 17 international journal papers and has presented his research papers in 6 international conferences. He has authored a Book Titled 'Database Management system' published by Anna University, Centre for Distance Education, Chennai-25 and another Book titled Programming and Data Structures published by Sams Publishers. He is Reviewer of five International Journals and has refereed few International Conferences , many symposiums and competitions . He is life member of ISTE, IACSIT & Transportation Research Group of India. He has also participated in innumerous seminars, workshops and conferences to nurture his knowledge.

Dr V. Uma Rani

She is currently working as Associate Professor in department of computer science and engineering, Chennai. She has received her Ph.D from the department of computer science, Bharathiar University, Coimbatore, Tamil Nadu, India. She has received her Master of Engineering and Bachelor of Engineering from Kalasalingam College of Engineering affiliated to Anna University, Chennai, Tamil Nadu, India. She has more than 20 years of teaching and research experience. She has published more than 40 papers in reputed international journals, conferences and 3 patterns in artificial intelligence domain. Her area of interest includes wireless sensor network, Artificial intelligence, Deep Learning and Metaheuristic optimization. Now she is working on metaheuristic optimization of projects on medical and agricultural domain. She has been reviewer of wireless personal communication journal, international journal of Engineering Science and Technology (IJESCT) and international Journal of Research and Analytical Reviews (IJRAR). She has been reviewer of various conferences like AICTE Sponsored IEEE international conference on Networks, Multimedia and Information Technology (NMITCON) organized by NMIT, Bengaluru, International Conference on Ambient Intelligence, Knowledge Informatics and Industrial Electronics (AIKIIE-2023),Karnataka, **IEEE International Conference on Data Science and Network Security (ICDSNS),** Tiptur and IEEE International Conference on Intelligent and Innovative Technologies in Computing, Electrical and Electronics, Bengaluru. She is a life member of international Association of Engineers (IAENG) and All India Council for Technical Education (AICTE). She has received best teacher award in the department of computer science and engineering, Jaya Engineering College during the year 2015.

Disruptive Technologies for Sustainable Development – Dr. G. Nagappan et al. (eds)
© 2024 Taylor & Francis Group, London, ISBN 978-1-032-55000-8

Handsign Language Translate to Hindi

1

V. Loganathan[1]
Associate Professor, Department of Computer Science and Engineering,
Saveetha Engineering College, Chennai

K. Hari Priya[2]
M.E Scholar , Department of Computer Science and Engineering,
Saveetha Engineering College, Chennai

Abstract: A sign language is a form of communique substantially usage by persons who are deaf. Signs are made with the hands rather than with words. Only persons with disabilities use sign language, which is a rare language. It is not used by many so they don't try to learn this language, and many can't understand the signs. To overcome this aspect based on AI technology, a new invention has been made by developing a sign language recognition conversion. This method mainly uses image processing to capture the sign in the form of live recorded video using a webcam, recognize the image, and convert it into text. Earlier used algorithms are using open-source datasets and techniques, which affect the efficiency of recognition. To overcome this, we have used the SVM+HOG algorithm for image processing. Then the recognized image is converted to text. The converted text is then subjected to TTS to convert it to a speech format so that people can hear and guess it more easily. This is also further developed by converting the text into any other possible language, which we require to find the meaning of the sign in all other languages to ease the communication process between people at any place in the world. Our model overcomes the deficiencies of previous models by increasing the efficiency of performance to more than 90% accuracy to improve the system.

Keywords: Sign Language, SVM+HOG algorithm, TTS, Pipeline Translation method

1. Introduction

Computer recognition and image processing of sign language is based on the sign gesture acquisition and extraction and continues until the conversion to text or speech generation. It can also convert the text to the language that we select, so it will be easier if the deaf and dumb people move to the United States. The sign languages between countries are also different from each other, so it makes it difficult for them to communicate, even for their daily needs. This scanner helps recognize and convert it to the language that we want in text or speech format, which is easy to communicate between people. This research is a combination of two research: sign language acquisition and conversion to text or speech, similar to the format of Google Translate. As a result of this method, disabled people will find it easier to communicate and live their lives. The processes used in this research are image processing, speech conversion, and text translation. The algorithms used in this research are the SVM and HOG algorithms for image processing, TTS, and ANN. TTS is a computer simulation of human text-to-speech using machine learning methods. The ANN is used for language translation.

[1]loganathan@saveetha.ac.in, [2]sonaharipriya@gmail.com

DOI: 10.1201/9781003428473-1

2. Existing System

In the existing system of sign language recognition, it is performed using a webcam or pictures, which are identified and converted to text. These concepts are achieved using neural networks and large amounts of data. To obtain more accuracy, more datasets must have to be trained, which takes more time and space.

Ilya Makarov (2019) has proposed a novel model that is based on the latest deep convolutional neural networks (CNN) and utilises real-time dactyl recognition (DR) sign language systems. Once they had their own dataset, they estimated dactyl recognition using Russian(RSL) sign language and their approach. A study based on Peruvian Subscribed Language Recognition (PSL), often known as the bracket issue, was completed by Roberto Nuri na-Jara in 2021. A comparison of four baseline classifiers in the Peruvian Sign Language Recognition (PSLR) task was made in order to determine and assess the dataset's efficacy, and the best result was 99.32% in the F1 measure's normal range.

Zhibo Wang (2020) This shows the designs for the new Deep SLR real-time SLR system. To capture the graphical features, two armbands outfitted with an IMU detector and multi-channel seem detectors attached to the forearms are permitted. To achieve accurate, scalable, and end-to-end nonstop SLR without sign segmentation, it suggests a grounded encoder-decoder model with a multi-channel convolutional neural network (CNN). It imposed heavily on a smartphone and gauged its efficacy through thorough assessments.

Ananya Roy (2018) has described techniques for understanding American Sign Language. The American Sign Language alphabets are compared with the datasets to provide the desired consequent outputs based on user needs. This recognition of American Sign Language is accomplished using OpenCV, which is frequently used to identify images using a camera.S. Pramada (2013) has suggested binary sign language datasets for recognition here, but it is user-friendly in that it can be converted to any form of sign language recognition, such as English or others, based on the user's needs with prior image processing. The output is converted to text or speech using the TTS or STS engine. It mainly focuses on the language translation.

Pankaj Chandankhede (2021) has developed this mainly based on language translation. This recognizes the handwritten texts from the books, and these handwritten texts are recognized based on image processing techniques by comparing the images with the given alphabets of datasets. The recognized output is initially converted into regional languages such as Marathi, English, and Gujarati. These handwritten texts are recognized using the OCR (Optical Character Recognition) technique, and for language translation, a translator using MATLAB has been used to achieve the desired output based on user needs.

These existing systems have many features that are used by the users. These features are separately available for the user's need. Therefore, it will be difficult for the user to handle each option one by one on a different website or app. So, to make it easier and more user-friendly, To make it more streamlined and suited to their particular needs and requirements, we integrated the components of the sign language recognition system and language translation conversion into a single portion.

3. Proposed Methodology

- Data collection
- Image acquisition
- Feature extraction
- Gesture recognition
- Text-to-speech
- Convert to another language

3.1 Data Collection

Data is collected for both the testing set and the training set. As the number of pure data increased, the efficiency and recognition of the system increase. The data in our research is collected by a real-time method of using a webcam to collect pictures on the spot from the user, and it gets stored in a training dataset folder. It is then tested with the given algorithm to live detect the results of the signs.

3.2 Image Acquisition

Here, we used image acquisition techniques to collect data in real time. In this image acquisition system, the first step to follow in this system is vision-based, that means it is used to acquire images at runtime via the web cam or camera. Additionally, the acquired images are then saved in a directory folder where the stoner will store and train all of the images. The saved and trained images will then be used to compare the recently captured image with the earlier images. So here we do not use the external image data set collection from an open source system because those are mostly related to American sign languages, which do not match our system, and we are here performing our tasks to create our own dataset using this technique, which could also increase the accuracy rate rather than using the open source data collection pictures for training and testing purposes.

3.3 Feature Extraction

The information is provided to the images via image segmentation. The images are first converted from color to grayscale by using our algorithmic steps. Thus, the unnecessary background data is eliminated, and the bit size of the featured image, which was used for the system's further training, is also decreased. The process of identifying or matching the attributes of the image with the image shot by the user to forecast indications is made more effective with the use of algorithms like Gaussian Blur and Median Blur.

3.4 Gesture Recognition

The gesture recognition stage is supposed to come next. The operation will also compare the featured extracted image character with the previously saved and trained data in our database once the previous steps have been completed. Once the stoner had set the threshold, the gestures were examined. The programme is started when it is recognised. As a result, it is anticipated that the input signal's text will be shown. The programme will run again from the beginning if the sign characteristic cannot be extracted.

It is possible using the SVG HOG algorithm. The histogram of acquainted slants , which usually consists of any series of ways that give an array of image features that basically represent the objects contained in the dataset. Later, the image features are used to identify the same features in other images. To homogenize the colour of an image, histogram equalization is applied. In another edge discovery , driver is launched to estimate the image slants. This way, the image silhouettes, shape, and texture information are attained. The grade is calculated for each colour channel, and the locally dominant grade is used to achieve invariance against colour. It generates descriptors. Slants or exposures of the edges at each cell pixel are accumulated in a 1-D histogram.

3.5 Text-to-Speech

A new technology called text-to-speech (TTS) reads digital textbooks aloud. It is sometimes referred to as "read audibly" technology. We made use of a Python text-to-speech conversion package. It is compatible and works both online and offline, unlike essential libraries. It is usable because we can simultaneously receive text and the sign language transcription in our system.

3.6 Convert to Another Language

Pipeline translation is an import library function in the Python programming language, which is already user-defined. Therefore, when these library functions are imported and invoked in the Python function, they help in translating the text from the regional language to the target language based on the user's needs and requirements. This system accepts text as a string or an array. However, the return type is a string if the input is a string. However, the return type is a list if the text is a list. This form of the Python library is used to provide better output with better efficiency, which can also be used in offline mode, and it does not affect the other modules of the research. It is easy to work with other parts without affecting the efficiency of the research.

4. Results and Analysis

The Results and Analysis section is mostly about the analysis of the performance of the system. Here, this system mostly uses OpenCV for image processing for sign recognition using hand gestures. Other older algorithms like ANN increase running time and complexity of the model. Therefore, we have used the SVG+HOG algorithm in our system to overcome this issue. SVG+HOG gives 97.5% accuracy of results compared to older systems, which makes the system give more accurate results using the image processing technique. This will also help in the prediction of signs at the first try,which makes the system less time-consuming and has more accuracy.

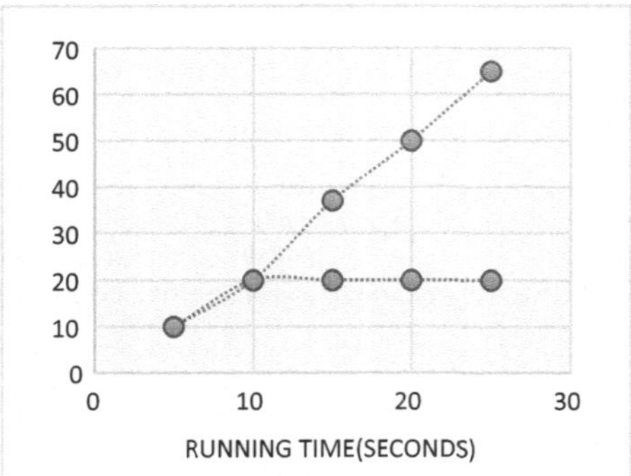

Fig. 1.1 Performance for SVG+HOG

Fig. 1.2 Recognition of sign

Based on the observed results during the running of the system, the performance measure was been calculated. After the use of algorithmic steps, the image is being processed using the SVG+HOG algorithm, which makes the image more clear and easy to recognize the features of the image, which increases the performance characteristics of the system. Based on this, our system has increased efficiency and accuracy, and the time and space complexity has been reduced, which were major necessities during the development of the system.

5. Conclusions

This idea will be helpful for disabled people who are unable to hear or speak. The communication of these people is much more difficult in real life with normal people, which makes them disconnected from the real world. This idea may help these people to ease communication, have a better connection with the real world, and live like all other people in society.

References

1. Wei-Chieh Chuang, Wen-Jyi Hwang, Tsung-Ming Tai, De-Rong Huang, and Yun-Jie Jhang.(2019).Continuous Finger Gesture Recognition Based on Flex Sensors.
2. M. Taskiran, M. Killioglu, and N. Kahraman.(2018) .A real-time system for recognition of American sign language by using deep learning, 41st International Conference on Telecommunications and Signal Processing (TSP). IEEE.
3. D. Chakraborty, D. Garg, A. Ghosh, and J. H. Chan.. (2018). Trigger detection system for american sign language using deep convolutional neural networks, 10th International Conference on Advances in Information Technology. ACM.
4. Oscar Koller, Necati Camgoz, Hermann Ney, and Richard Bowden. (2019) .Weakly supervised learning with multi-stream cnnlstm-hmms to discover sequential parallelism in sign language videos, IEEE Transactions on Pattern Analysis and Machine Intelligence.
5. X. Zhang and X. Li. (2019). Dynamic gesture recognition based on memp network, Future Internet.
6. A. Vaitkeviˇcius, M. Taroza, T. Blazauskas, R. Damasevicius, R. Maskeliunas,and M. Wo´zniak. (2019). Recognition of american sign language gesturesin a virtual reality using leap motion, Applied Sciences.
7. J. Schioppo, Z. Meyer, D. Fabiano, and S. Canavan. (2019). Sign language recognition: Learning american sign language in a virtual environment.
8. S. J. Qin and L. H. Chiang. (2019) Advances and opportunities in machine learning for process data analytics, Computer. Chemical Engineering
9. Ankit Ojha; Ayush Pandey; Shubham Maurya; Abhishek Thakur; Dr. Dayananda P . (2020) Sign Language to Text and Speech Translation in Real Time Using Convolutional Neural Network, International Journal of Engineering Research & Technology .
10. Dinh-Son Tran, Ngoc-Huynh Ho, Hyung-Jeong Yang, Eu-Tteum Baek , Soo-Hyung Kim and Gueesang Lee, (2020) Real-Time Hand Gesture Spotting and Recognition Using RGB-D Camera and 3D Convolutional Neural Network, Applied Science .
11. Ananya Roy, Dr, Sandhya Roy. (2018) Recognition of sign language using image processing, International Journal of Business Intelligence and Data Mining.
12. D.Mane , D.More, R.Dolas . (2013) sign language recognition using image processing, IOSR .
13. S Pramada, D Saylee, N Pranita. (2013) Intelligent sign language recognition using image processing ,IOSR.

Note: All the figures in this chapter were made by the author.

Disruptive Technologies for Sustainable Development – Dr. G. Nagappan et al. (eds)

Hybrid Cryptographic Algorithm to Improve Security in Cloud File Storage

2

M. Hema[1], K. Valarmathi[2], Naveen Sakthi J.[3], Saravanan S.[4], Suravarapu Sai Charan Reddy[5]
Information Technology, Easwari Engineering College, Chennai, India

Abstract: Cloud Computing is one of the most advanced and useful Technology. As cloud storage continues to grow in popularity, ensuring the safety of the confidential data stored there becomes increasingly important. While conventional encryption methods have their limitations, a new approach combining different encryption techniques can provide a stronger defense against cyber threats. This paper proposes the hybrid cryptographic algorithm in order to improve the cloud security by combining two existing robust cryptographic algorithms to form a new and superior version of hybrid cryptography. This innovative technique holds great potential to secure sensitive information in the cloud, promoting peace of mind for individuals and organizations alike. In our paper we will be using two encryption algorithms to achieve hybrid cryptography, RSA and Fernet.

Keywords: Noise picture, Encrypting, Key, Decryption, Cryptographic

1. Introduction

Data availability, integrity, and confidentiality are provided by cryptography. For those who are not permitted to disclose the information, it changes the data from its plain form to another scrambled form known as cipher text Encryption is achieved by using certain encrypting algorithm to covert the cipher text to plain text. Decryption is defined as the process of converting cypher text into plain text. Both encryption and decryption require the key as an additional input. Cryptography can be divided into symmetric-key and public-key cryptosystems based on the number of keys used. Combining two or more cryptosystems is referred to as hybrid cryptography. In this article, we've suggested conducting research on hybrid cryptographic algorithms used to protect cloud data. Two key components of cryptography are symmetric-key and asymmetric-key cryptosystems. Instead of using simple cyphers one at a time, the hybrid cryptographic approach aims to produce a more secure, better performing, and robust algorithm.

2. Literature Review

Data owners are still hesitant to store their data with a third party, despite the fact that cloud-based services offer a number of advantages. The main issues with outsourced data are non-repudiation, confidentiality, integrity, and privacy. Tong, Jie, Yihong Long (2021) proposed an ABE system but Several enhancements can be made to the aspect of security of data. Similarly, A.N, A.V.K et al. (2022) Proposed BPCS Steganography to overcome the existing drawbacks in steganography approaches but it is possible to acquire a greater number of shares, which would result in a much-improved level of safety for any future research. M. Y. Shabir, A. Iqbal et al. (2016) This paper proposes the use of classical Encryption Techniques in Cloud Computing but the

[1]hema.m@eec.srmrmp.edu.in, [2]valarmathi.k@eec.srmrmp.edu.in, [3]naveensakthijega@gmail.com, [4]saravana28101@gmail.com, [5]charansaisuravarapu@gmail.com

DOI: 10.1201/9781003428473-2

algorithm lacks many security features of the current advanced systems. Kahla, Mohammed Elhabib, et al (2021) This system uses MNF-G and El-Gamal Encryption to improve image encryption but There is a possibility that the image quality will alter the results. C. Qin, J. Hu, F. Li et al. (2022) Proposed a JPEG encryption scheme to improve size constraints but it is limited to only one format. S. KanagaSuba Raja, and M. Hema. (2019) proposed an algorithm to improve resource utilization in cloud. Hema M, and S. KanagaSuba Raja. (2023) this paper explores the utilization of underloaded servers in cloud.

3. Algorithms

3.1 RSA

The RSA calculation is used in asymmetric cryptography. The usage of two different keys, such as the Public Key and Confidential Key, is what is meant by the word "asymmetric." As hinted by the name, the Confidential Key is kept a secret, but the Public Key is made available to everyone. While the secret and the public key are used for decoding and encryption respectively. public and private keys are produced first. We will then store the keys—a set of public and private keys—in files. We will make a keys envelope in our project organizer so that we can place the keys into the records. The two sheets that make up the private and public keys for the Keys organizer will each contain one important key. We actually want to stack the keys, so that should come next. The keys are stacked once we open the records we produced before and return both the private and public keys. We will create a confirm strategy for the message check and pass in the message, the checkmark, and the key.

$$M < n$$

$$C = M^e(mode\ n) \qquad M = C^d(mod\ n)$$

Fig. 2.1 Encryption and decryption formula for RSA algorithm
(Where, C – Cypher Text, M – Plain Text, e,n – public key, d,n – private key)

3.2 Fernet

Without the key, a communication that has been encrypted with Fernet cannot be read or controlled. A symmetric verified cryptography implementation is called Fernet. The class of cryptography, fernet, and fernet provides services for both encoding and decoding (key). Data is first encrypted using the strategy encrypt before a new fernet key is generated using the class function produce key() (data). A secure communication is produced by this encryption, which cannot be read or changed without the key. In addition to having high security and authenticity assurances, it is base64-encoded and URL-safe. A "Fernet token" is the name given to it. The border of the message that you would want to obfuscate is thought of as the information that must be in bytes or else, we could have a "type error." Because an encrypted communication includes the time it was created in plaintext, a potential attacker will be able to determine when a message was made. The function encrypt at time(data, current time) precisely encrypts data while pausing.

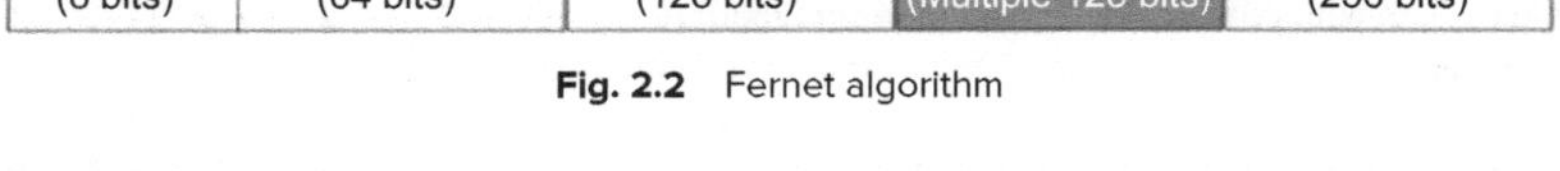

Fig. 2.2 Fernet algorithm

4. Proposed Methodology

The Proposed System uses RSA and FERNET Encryption Algorithms. For enhanced security the files uploaded to the cloud are encrypted in two-layers with the corresponding algorithms. RSA algorithm is a 512-bit algorithm which has two different keys for both encryption and decryption resulting in it being an asymmetric key algorithm. The interface consists of options to upload files. After uploading the file, clicking the send button will start saving the file to the cloud. If an error occurs during loading, it will either report the error to the user. When a user uploads a file, it triggers a chunking system that splits the file into three parts. Therefore, each part will be encrypted separately in two layers. Once this process is completed the file is stored in cloud as separate encrypted chunks. Once the user requests the file to be downloaded, the UI asks for two keys for decrypting the 2 layers. The Keys Which are used for encryption (Public and Private keys) needs to be reversed as this is asymmetric encryption technique.

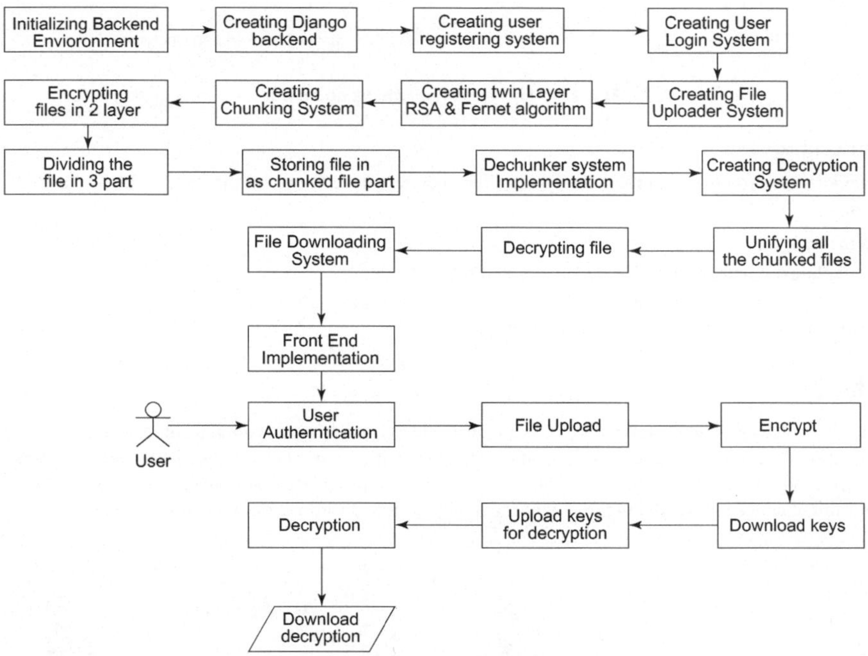

Fig. 2.3 Proposed system architecture

4.1 Registering User—Account Creation

The User First has to Register in the system to access the encryption system. The Registration process is required in order for the users to have a separate space where they may upload (encrypt) the files and later in order to download the encrypted files for themselves without any third party accessing the files. The stored files are encrypted and chunked. thus, even if the files are accessed by any attackers, they will not be able to decrypt them without the keys. Moreover, they will not be able to know which algorithms are used and in what combination or in what sequence these are used.

4.2 Logging in to the System

Upon Registration, the user will now have the ability to Log in/Log out of the server and now will have access to all the functionalities of the system. The Front end and the back end of our system is built in conjunction with both Python and the Django framework. The User Once logged in will have the access to upload files thus encrypting them, view the encrypted files associated with that account and to download the encrypted files.

4.3 Encryption Process

The User can now upload their confidential files to the server in order to store them in an encrypted manner. The Uploaded Files goes through the encryption and chunking process. The chunking process chunks the file into three equal parts. The Two Layer encryption process encrypts the files with both RSA and Fernet. The files are then stored as encrypted chunks in the server. Upon Completion of the Encryption process. The keys will be generated.

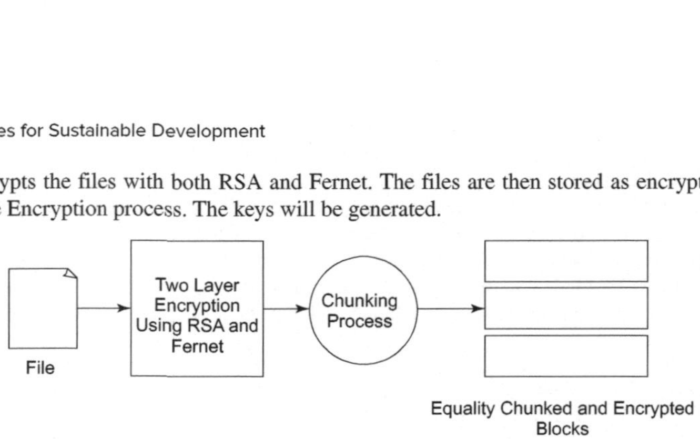

Fig. 2.4 Encryption and chunking process

4.4 Decryption Process

If the User decides/wants to download their Encrypted files, they will have to upload the keys. Once the user uploads the key, the system then,

1. Verifies the key
2. Unifies all chunked parts to form a single encrypted file
3. The Two Layer Encrypted file is now Decrypted to obtain the original file
4. And finally, the user can download the original file

5. Results

The Proposed system used with the combination of symmetric and asymmetric encryption techniques provides our system with enhanced encryption/decryption time compared to the existing algorithms. For our tests we have used file sizes ranging from 1 MB up to 8 MB to better understand the time difference. In the encryption tests, our hybrid system consistently outperforms the single algorithm counterpart, producing the highest delta in the 8 MB file size of up to ~14%.

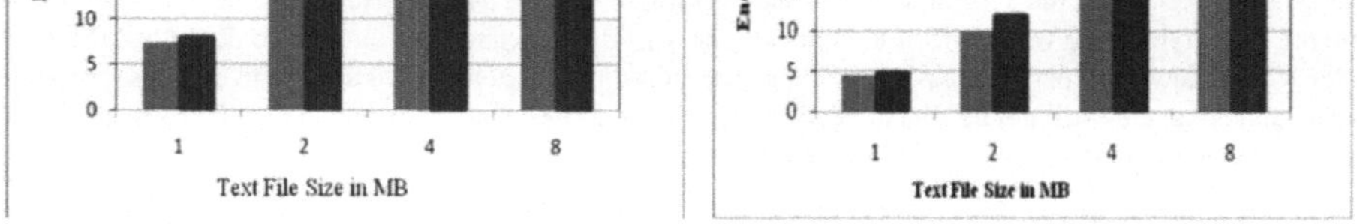

Fig. 2.5 Encryption and decryption time comparison of proposed system and existing system

Decryption Tests: the time taken for decrypting file sizes till 4 MB is higher than the encryption time, but it evens out at the 8 MB file size. The Proposed Hybrid System outperforms here too with an average delta of 13-14 % better performance in proposed system.

6. Conclusion

The goal of our project, which is to provide a secure and efficient algorithm is achieved with the help of our hybrid cryptographic algorithm. Achieving a better file safety with twin layer encryption and three-part chunking. The files thus stored in cloud/

servers are better protected and safeguarded to leeway the users from storing the files in a hassle free manner. Improved file encode time achieved compared to the existing AES algorithm with a maximum difference of up to 14%. An Average of 13-14% of improvement achieved in the file decryption process with the use of our proposed algorithm in comparison with the existing AES algorithm in the file size of 1,2,4 and 8 MB making it more efficient as the file size is increased. Future improvements can be made to the system to encrypt video files.

References

1. C. Qin, J. Hu, F. Li et al. (2022), "JPEG Image Encryption with Adaptive DC Coefficient Prediction and RS Pair Permutation," IEEE Transactions on Multimedia vol. 10 doi: 10.1109/TMM.2022.3148591.
2. S. KanagaSuba Raja, and M. Hema. (2019) "An Optimal Algorithm to Improve Resource Utilization in Cloud Data Centre." International Journal of Engineering and Advanced Technology (IJEAT) vol. 9 doi: 10.35940/ijeat.A9704.109119
3. M. Y. Shabir, A. Iqbal, Z. Mahmood et al. (2016) "Analysis of classical encryption techniques in cloud computing." Tsinghua Science and Technology vol. 21, no. 1, pp. 102–113, Feb. 2016, doi: 10.1109/TST.2016.7399287.
3. Tong, Jie, Yihong Long, and Quan Liu. (2021) "A File Encryption System Based on Attribute Based Encryption." 17th International Conference on Computational Intelligence and Security (CIS). IEEE, 2021. pp. 454–458, doi: 10.1109/CIS54983.2021.00100.
5. A. N, A. V. K and N. R (2022) "Sharing Confidential Images with Abbreviated Shares using Steganography and AES Algorithm." (CONIT) pp. 1–5, doi: 10.1109/CONIT55038.2022.9847932.
6. M. E. Kahla, M. Beggas, A. Laouid et al. (2021) "Asymmetric image encryption based on twin message fusion." 2021 International Conference on Artificial Intelligence for Cyber Security Systems and Privacy (AI-CSP). 2021, pp. 1–5, doi: 10.1109/AI-CSP52968.2021.9671161.
7. Hema M, and S. KanagaSuba Raja. (2023) "An Efficient Framework for Utilizing Underloaded Servers in Compute Cloud." COMPUTER SYSTEMS SCIENCE AND ENGINEERING 44.1 (2023): 143–156.

Note: All the figures in this chapter were made by the author.

Disruptive Technologies for Sustainable Development – Dr. G. Nagappan et al. (eds)
© 2024 Taylor & Francis Group, London, ISBN 978-1-032-55000-8

Twitter Based Sentiment Analysis to Predict Emotions using Machine Learning Algorithm

3

Panimalar S. P.[1]
Assistant Professor, Saveetha Engineering College,
Chennai, TamilNadu

Hemaroshini M.[2], Jayathraa V.[3], Reethika R.[4]
Final Year Student, Saveetha Engineering College,
Chennai, TamilNadu

Abstract: Twitter becomes the principle source of facts. The objective for our projectis raw records extracted fromtweets. To do that, we automate the process of extracting tweets and divide them into three categories: positive, negative and neutral. Sentiment analysis technique is extensively used in textual content evaluation. In precis, Twitter sentiment analysis the way that users of text evaluation the use of advancedtechniques to investigate thesentiment of a text (here, a tweet) in high quality, terrible and neutral shape. It is likewise acknowledged to research opinions and mainly to analyze conversation,review and exchanges (all within the form of tweets) for naval affairs, political evolution, and to evaluate public movement.

Keywords: Twitter sentiment analysis, Machine learning, Text mining

1. Introduction

In trendy global, social networks like Twitter, Facebook, Tumbler and so forth. Twitter is a microblogging platform that offers a wealth of facts that will be used for various applications of sentiment analysis, including forecasts, surveys. Sentiment evaluation is the method of evoking traits based on mind, feelings, and feelings. And post any critiques on social networks. Sense evaluation is the end result of the division of herbal language texts into those classes as fantastic, bad and impartial. The amount of information generated from social networking web sites is huge; This statistics is informal and won't provide any meaningful facts till it is evolved. Therefore,that allows you to employ the massive quantity of information, we perform inference analysis, that is, we extract functions from this information and make predictions. Opinion evaluation may be very essential in modern international, as human beings are usually stimulated via the mind and critiques of different human beings. Today, if someone wants to buy a product, vote, many others, then the first aspect he wants to understand is what other humans' reviews, reactions, and reviews are about that product, candidate, or film on social networks. Like Twitter, Facebook, Tumbler etc. Therefore, we want with a view to automatically generate a sentiment analysis machine from this huge amount of information. Python is a understandable, excessive-level, easy and dynamic programming language this is nice recognized for its natural language processing functionality the usage of NLTK (Natural Language Toolkit). NLTK is a Python library that offers a frame work for programming and analyzing facts. NLTK also gives a graphical demo to show distinct results or tendencies, and affords sample records for education and testing distinct classifiers.

[1]panimalarsp@saveetha.ac.in, [2] hemaroshini711@gmail.com, [3] jayathraa07@gmail.com, [4] reethureethika2002@gmail.com

DOI: 10.1201/9781003428473-3

2. Objective

The foremost intention is to connect with Twitter and find tweets containing specific key phrases after which rank high quality, terrible or neutral tweets. Opinion evaluation is ready extracting humans's opinions accurately from a massive wide variety of informal text evaluations and studying them. Mood training, i.e. Definitely, negatively or neutrally.

3. Literature Survey

"Opinion gaining knowledge of" Opinion mining is a wide region of natural language processing, text evaluation, and computational linguistics, which incorporates pc exploration of styles, critiques, and emotions expressed in textual content M. Comesaña, A. P. Soares, M.Perea, A. P. Piñeiro, I. Fraga, and A. Pinheiro in 2020, although opinion or mind-set is extra approximately feelings than reason, it's also colloquially called emotion, hence it is equal to the opinion of the collecting or sentient being purged. P. Bhoir, S. Kolt in 2015 theory mining has been said to have many applications, which include accounting, regulation, research, leisure, training, technology, politics, and engineering. In historical instances, many social networks gave internet clients the possibility to specific and share their mind and critiques Twitter B. Twitter is a famous real-time microblogging carrier that permits clients to percentage short statistics, called tweets, up to one hundred forty characters lengthy A. K. Jose, N. Bhatia, and S. Krishna in 2010. Customers write tweets to explicit their evaluations on diverse topics related to each day lifestyles. Twitter is a surely perfect platform for eliciting public opinion on particular problems Taboada, M., Brooke in 2011 a collection of tweets is used as the principle corpus for sentiment evaluation, which involves using opinion amassing or natural language processing M. Rambocas, and J. Gama in 2013 C. E-commerce Microblogging He stated that organizations generating such merchandise have started conducting microblogging surveys to find out the overall mind-set within the direction of the product, often these businesses study the reaction of customers and respond to customers on microblogs. internet world start dialogue D. Osimo, and F. Mureddu in 2010 found a fashion of increasing the wide variety of internet users who keep to spend extra time on social media, all the time spent on cellular gadgets and social media in the US on laptop, which increases 37 percent to 121 billion minutes within the US. 2012 in evaluation to 88 billion mins in 2011. On the other hand, when groups use social media internet web sites to discover and communicate with clients, businesses may be shown the benefits of social media Asghar, M. Z., Ahmad, S., Qasim, M., Zahra, S. R., & Kundi, F. M in 2016. Since social networks can be made public so without problems, the dissemination of personal statistics in a social context may be dangerous. E-exchange or e-commerce refers to the buying and promoting of services or products online through social networks which include Twitter, that is handy because of its 24/7 availability, ease of customer service, and world wide attain. In addition, D. Osimo, and F. Mureddu in 2010 also affirms that the lexical sentence is in detecting the sentence of the voice within the body after which in predicting the sentence expressed in the textual content. Vocabulary gear have proven that they have a simple model, namely:

(i) Preprocess every tweet, submit, cast off punctuation
(ii) Initialise the world wide vertex rating(s) to zero -> s = 0
(iii) Check if the signal is gift inside the dictionary, then if the signal is advantageous, s can be wonderful (+). If the keyis poor,might be bad.

4. Existing System

The existing machine, makes use of information primarily based method to categorise the tweets into both positive, negative or neutral, however using this method consequences in much less accuracy of the type. Sentiment analysis structures skilled on information are regularly a lot less correct when applied to statistics from other domains issue- primarily based Sentiment assessment (ABSA) goals to expect the sentiment polarity of various aspects in a sentence or file, that's a high-quality-grained project of natural language processing, most of the existing work specializes in the correlation among component sentiment polarity and close by context. The essential deep correlations among worldwide context and issue sentiment polarity have now not acquired sufficient attention Deep Sentiment analysis: A Case study on Stemmed Turkish Twitter information Sentiment evaluation the use of stemmed Twitter information from numerous languages is an growing studies subject depend. in this paper,we deal with 3 facts augmentation strategies mainly Shift, Shuffle, and Hybrid to growth the scale of the schooling information; and then we use 3 key forms of deep learning fashions specifically recurrent neural community convolution neural community , and hierarchical interest community to classify the stemmed Turkish Twitter information for sentiment evaluation.

5. Proposed System

In the proposed system, we tried to analyze the mode of Twitter. NLP (Natural Language Processing) primarily based K-Media set of rules turned into used to perform the analysis using the extracted capabilities. Due to the lack of predefined datasets to determine the sensitivity of every term, the top of the set of rules changed into decreased to a technique that directly led to efficiency profits are expressed, which may be high quality, negative or neutral. We recommend a brand new hybrid method that consists of the corpus- and vocabulary-based totally, this methods to determine semantic motive of the words in tweets. It has these days come to be a massive part of the Internet language. The proposed analysis of sentiment in Twitter records is based on two fundamental elements, namely records extraction, pre-processing of the extracted statistics and class.

Retrieval of tweets : Since social networking is the most important part of social networking, it's been suggested by using diverse blogs from numerous topics round the sector. Instead of taking entire blogs, we're going to search for a selected web site and download all its pages, extracting them as files using a mining tool like Weka, which offers a classifier sense.

Pre-processing extracted data, Filtering, Tokenization, Removing Stopword.

Parallel processing: A sentence classifier that classifies sentences is constructed the usage of sine classifier polynomials or help vector machines (SVMs). The important purpose behind this step is the class statistics formation. Each database incorporates hidden information that may be used to make decisions.

Module 1: Data Collection This is the first actual step in definitely growing a system getting to know model, information collection. There are numerous methods to collect information from distinctive datasets.

Module 2: Data Pre-Processing While making ready the statistics, it's far vital to extract the crucial records and put off the vital records from the cards.

Module 3: Feature Extraction A very useful characteristics referred to as extracting words from a tweet. In the function extraction method,we extract debts from the preceding Twitter dataset.

Module 4: Feature Selection Sentiment evaluation makes use of proprietary function selection strategies that play a crucial role in identifying attributes and enhancing classification accuracy (device learning).Natural language processing. Statistics, Clustering is based.

6. Implementation

In our implementation, the following step is to discover the person's tweets and examine their emotions. In that we need to open the home page and need to select the desired feature and after that we need to enter the user name in which we need to detect their emotions which can be also positive,negative or neutral.

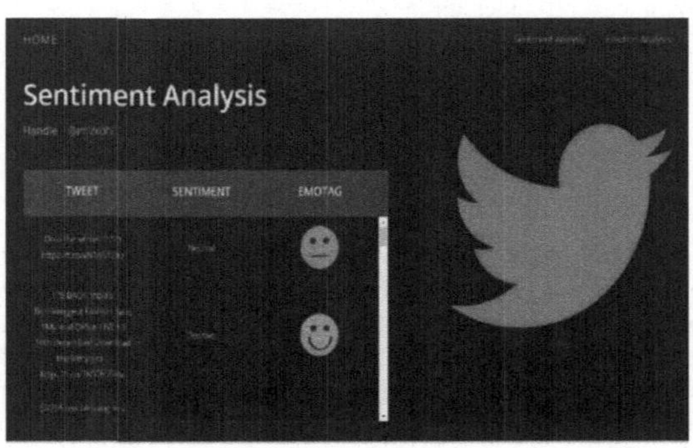

Fig. 3.1 Output image

7. Conclusion

Sentiment analysis is used to show human beings' ideals, attitudes and emotional states. People's opinions can be tremendous or terrible. Generally, elements of speech are used as a characteristic to extract textual content. The adjective plays a major function in figuring out the way of the parts of speech. Sometimes words which might be adjective and adverb are prepare, it's miles difficult to determine the temper and sentence. To perform sentiment evaluation on tweets, the system first extracts the target tweet messages from the user. The machine can calculate the frequency of each time period in a tweet. Using supervised system for a gaining knowledge of technique helps to reap effects. Need is a large supply of records,which makes it more appealing for sentiment analysis. We examine about 18 to 15 consistent with side to research the outcomes, apprehend the patterns, and observe people's opinions. We have seen that special components have exclusive outcomes depending on their development and paintings flow. We have additionally seen how any public event, speech, or come upon causes human beings' attitudes to differ. We also try to find out which plans get the maximum aid from people on each sides of the company. It has been shown that BJP is now the most a success political party based totally on people's critiques and K-means algorithm were used.

8. Future Enhancement

The application of sentiment analysis to extract opinion has end up an essential undertaking for plenty organizations and even people. Sentiment evaluation is a new discipline in decision making and is developing rapidly. The motive of our undertaking is to research the critiques on the subject matter extracted from it and to decide their nature(nice/negative/impartial) with appreciate to sure arguments. The development of document stage evaluation strategies is one of the most vital parts of this subject. Recently, human beings have began to explicit their evaluations at the Internet, which has elevated the need for numerous actual-international applications for the subjective evaluation of on line content material. There are many studies within the literature at the detection of meaning in textual content. However, there is massive room for improvement in sentiment analysis in existing models.

Reference

1. M. Rambocas, and J. Gama. (2013). Advertising Research the Role of Sentiment Analysis. The 5th SNA-KDD Workshop '11. University of Porto.
2. A. K. Jose, N. Bhatia, and S. Krishna. (2010) Twitter Sentiment Analysis. National Institute of Technology Calicut.
3. P. Lai. (2012). Extracting the Strong Sentiment Trend from Twitter using recent trends, Stanford University.
4. Saif H, Fern M, & He Y. (2013). Evaluation Datasets for Twitter Sentiment Analysis A survey and a new dataset the STS-Gold. Proc. 1st ESSEM Workshop. Turin, Italy.
5. D. Osimo, and F. Mureddu. (2010). Research Challenge on Opinion Mining and Sentiment Analysis, Proceeding of the 12th conference of Fruct association, United Kingdom.
6. A. Pak, and P. Paroubek. (2010). Twitter as a Corpus for Sentiment Analysis and Opinion Mining, Special Issue of International Journal of Computer Application, France: Universitede Paris-Sud.
7. Taboada, M., Brooke, J., Tofiloski, M., Voll, K., &Stede. (2011). Lexicon based methods for sentiment analysis. Computational linguistics.
8. Asghar, M. Z., Ahmad, S., Qasim, M., Zahra, S. R., & Kundi, F. M. (2016). Senti Health: Creating health-related sentiment lexicon using hybrid approach. SpringerPlus.
9. M. Comesaña, A. P.Soares, M. Perea, A. P. Piñeiro, I. Fraga, and A. Pinheiro. (2020). Author private replica computer systems in Human conduct ERP correlates of masked affective priming with emoticons, computer systems in Human conduct, 29, 588–595,IEEE.
10. P. Bhoir, S. Kolte. (2015). Sentiment Analysis of Movie Reviews using Lexicon approach, IEEE International Conference on Computational Intelligence and Computing Research, Madurai.

Note: The figure in this chapter was made by the author.

Disruptive Technologies for Sustainable Development – Dr. G. Nagappan et al. (eds)
© 2024 Taylor & Francis Group, London, ISBN 978-1-032-55000-8

Enhanced Graphical Password Authentication System Based on Puzzles

4

Poonkuzhali S[1]
Professor, Computer Science and Engineering,
Rajalakshmi Engineering College, Tamil Nadu, India
Abuthahir A.[2], **Dinesh S. P.**[3], **Daadreyaa D**[4]
Computer Science and Engineering, Rajalakshmi Engineering College, Tamil Nadu, India

Abstract: Technology advancements have led to a surge in cybercrime and data breaches, making password protection crucial for maintaining online and offline privacy. However, alphanumeric passwords can be difficult to remember and vulnerable to guessing attacks. Studies suggest that graphical passwords can overcome these limitations, as the human brain finds it easier to recall images than alphanumeric characters. As a result, graphical passwords offer a promising alternative to traditional password authentication. Thus, a RESTful web service that makes use of a puzzle-based graphical password authentication system is required. Users of this system solve a series of patterns and puzzles by employing recognition and recall techniques. Web developers can use this service to provide an extra degree of protection to their websites. They can also select which puzzles to employ and save the solutions to the puzzles that their users solve utilising an interactive dashboard service.

Keywords: Authentication system, Password , Web service, Puzzle-based

1. Introduction

Authentication is a critical aspect of computer security that ensures secure access to devices and data. Passwords are commonly used for authentication and fall under the category of knowledge-based authentication. However, remembering long and complicated alphanumeric passwords can be challenging for users, leading to weak passwords and compromised security. Graphical password authentication systems provide a solution to this problem by using images or puzzles as passwords. These systems are designed to capitalise on the human brain's ability to recall and recognize images easily. Graphic passwords are considered more secure than conventional alphanumeric passwords as they lower the risk of keyloggers, brute force attacks, and shoulder surfing. This paper focuses on a novel graphical authentication system, which operates as a web service and utilises an advanced puzzle engine. The system employs a hybrid approach of graphical authentication methods to enhance security.

2. Literature Review

Graphical password strategies can be categorised into the following categories:

Recognition based graphical password system

Graphical password schemes based on recognition such as Dhamija and Perrig's (2000) Deja vu technique and Brostoff and Sasse's (2000) Passfaces method are two popular recognition-based graphical password schemes that have their own strengths

[1]poonkuzhali.s@rajalakshmi.edu.in, [2]abuthahir.a.2019.cse@rajalakshmi.edu.in, [3]dinesh.sp.2019.cse@rajalakshmi.edu.in,
[4]daadreyaa.d.2019.cse@rajalakshmi.edu.in

DOI: 10.1201/9781003428473-4

and weaknesses. Por(2017). proposed a new method in which users click on the pass image provided in the set using the Ligature Replacement Rule and then return the pass image to improve security against shoulder-surfing and video camera attack. The disadvantages of this are the lengthy login process and memory.

Recall based graphical password system

Recall based graphical password systems have gained significant attention in recent years. The Draw-A-Secret (DAS) method proposed by Jermyn et al.(1999) allows users to create a password on a 5x5 grid using a stylus or mouse, while PassShapes introduced by Weiss and De Luca (2002) requires users to create simple geometric designs using eight distinct strokes. PassShapes decreases password size while increasing password memorability, making it a promising alternative to traditional text-based passwords. GEAT is a gesture-based authentication method proposed by Shahzad et al.(2013) Users produce sample motions on a mobile phone's touch screen to create a multi-touch gesture password. Password effectiveness depends on complexity and memorability, and users must safeguard their credentials from attacks when gathering training data.

Cued Recall based graphical password scheme

Cued recall methods have gained attention as a user-friendly alternative to traditional text-based passwords. PassMap, proposed by Sun et al. (2012), uses a map-based authentication system where the user selects two click places on a world map, while HapticPoint (2018), extends PassPoint by providing haptic feedback to the user. PassBYOP, proposed by Bianchi et al. (2015), uses image points selected by the user on a token's live video while entering a password to enhance security against shoulder surfing. However, its effectiveness can be limited by attack cost and other factors, and password schemes should be chosen based on individual needs and risks. Cued recall methods offer an alternative to text passwords, but their efficacy depends on the specific context of use.

Hybrid based graphical password scheme

Several hybrid techniques such as Pristine PixCaptcha (2018) and TCpC (2020). Other proposed hybrid techniques include dynamic graphs, Passhint, and text-based GPS systems. These techniques offer increased security and usability over traditional graphical passwords, but often require longer authentication times or more complex registration processes. In response, we propose a highly scalable and attack-resistant puzzle-based authentication system that enables users to easily recall their passwords through graphical representation and provides more security.

3. System Architecture

The system architecture is made up of three levels. The service API and its database form the first layer. The service operates in total transparency; it does not record user puzzle passwords, but rather stores the puzzle id mapped to a user id given by the developer in order to increase security. The developer integration layer makes up the second layer. When a developer uses our service, we will provide him with a dashboard from which he can control the puzzles he selects from the puzzle engine as well as the users of the developer's site. The third layer is for the user authentication layer, where the user authenticates by solving the puzzle picked during the signup process. The diagrammatic representation is shown in Fig. 4.1.

3.1 Web Service Layer (Puzzle Engine)

The service's puzzles are integrated into the graphical puzzle engine. Six unique puzzles are used to authenticate users. The problems are meant to be easy to recall but difficult to solve for an intruder or attacker. The puzzles are click-based, making it difficult for the attacker to even guess our pattern, and they are combined at random each time to prevent shoulder surfing. The web service does not store the developer's user login credentials. After each user authentication, an API request is made.

Padlock Puzzle

A pattern-based problem called the padlock puzzle requires users to join the dots to create any design they wish. Jumping can also be used to connect dots that don't have any middle points i.e., if the user wants to connect 1 and 9 without an intermediary . This puzzle is incapable of highlighting the click that successfully prevents shoulder surfing. This puzzle comes under np-hard. The complexity is unpredictable.

Polygonal Puzzle

Polygonal puzzle is a shape-based puzzle in which the user can select any shape and create a password using a colour palette. The number of lines in the shape will determine how the shape is divided into four quadrants. A user can make their own patterns by giving the same or different colours to one or more quadrants. The puzzle is C to the Power Q complex, where C is the number of colours and Q is the number of quadrants.

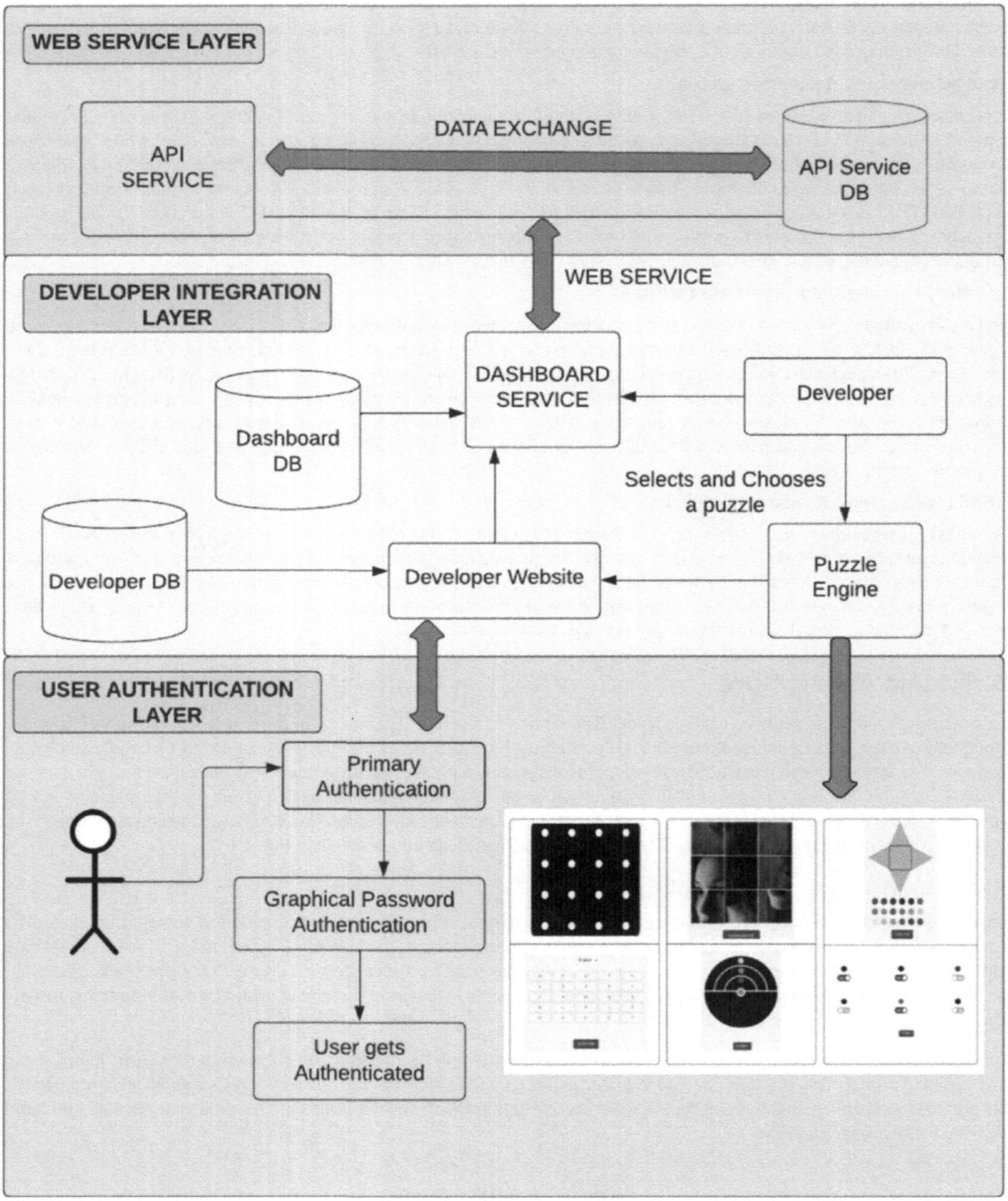

Fig. 4.1 System architecture

Click Puzzle

Click Puzzle is an image-based puzzle that allows users to create puzzles using either their own images or the default images provided by the service, select the grid size they desire, and then click to generate a password. As a password, the user can click the tiles an endless number of times. This puzzle comes under np-hard i.e., unpredictable.

Multi-lingual Puzzle

Multilingual puzzle is a text-based password that allows the user to create a password in a variety of languages. The key distinction between a normal text password and this puzzle is that for normal text password a complex password can be created in a specific format, such as by using special characters and condition-specific passwords, a multilingual password is what the user wants and should be created as such. This puzzle is considered np-hard because it can only be solved once in a lifetime.

Concentric-circle

Concentric-circle is a circle-based puzzle in which the circle is divided into two sections, the upper portion of which serves as the password and the lower portion of which is static. There are three layers, with a set of small circles on each layer. The colour of the circle determines which layer's circle is to serve as the password. The password has a complexity of n choose 1 to the power three, where n is the total number of colours.

3.2 Developer Integration Layer (Dashboard Service)

This layer offers the infrastructure necessary for developers to explore the puzzle and incorporate the service. Developers who have already signed up and are utilising the service can choose the puzzle during registration and receive dashboard access. Using the dashboard, a developer selects a collection of puzzles for his website, which they will administer. Once he authenticates himself, he will have access to it. The chosen puzzles, logs, and success rates may all be viewed by the developer. He can update the puzzle set on the dashboard if there are any changes. The dashboard shows a graphic representation of the user count in relation to the puzzle to demonstrate his analysis of the puzzles he had selected. This dashboard will show any suspicious login activity if a hacker attempts to guess a user's password by repeatedly entering it. The user will be notified via email. The developer will be alerted and notified as well if someone else adds a new puzzle or modifies an existing puzzle.

3.3 User Authentication Layer

The User Authentication is handled by this layer. Users register the developer's website by entering their primary authentication credentials, such as their email address and password, and then select a puzzle from a set of puzzles chosen by the developer and provide their own solution to the puzzle they selected. When logging in, the user must enter his primary authentication credentials. He will be presented with his chosen puzzle only after primary authentication has been confirmed. If the solution provided is correct, he is authenticated. If he forgets his password or answers incorrectly, he will be given five chances to retry. After two attempts, he will be notified that someone is attempting to access his account via email. His account will be locked once all of his attempts have been exhausted. After two attempts, he will receive an email informing him that someone is attempting to access his account. Once all of his attempts have been exhausted, his account will be locked.

3.4 Features

Scalability: The graphical password authentication system is served as a web service along with the easier integration using dashboards to the developers. If it attracts the developers to integrate, scalability is of ease.

Usability: The puzzles are created in such a way, the solutions are "easy to remember and hard to crack".

Resilient to attacks: The system is resistant to attacks like shoulder surfing, dictionary attacks, rainbow tables, keyloggers and malware attacks.

Higher Complexity: The puzzles contain an infinite number of solution combinations, making it difficult to guess and crack.

Web Services: This Graphical Based Authentication System is provided as RESTFUL Web Service, which lets our system communicate with any application.

Shoulder Surfing: Since the puzzles are getting shuffled every time , it becomes hard to shoulder surf the user.

4. Experimental Results

We conducted an experiment of around 215 users for analysis. The users are of diverse ages below 18 , 18 -25 , 25-40 , 40-60 and above 60 years old. The puzzles chosen by the users and the total login time taken to successfully authenticate are shown in Fig. 4.2 and 4.3, and their attempts, recognition and recall in Fig. 4.4 and 4.5 respectively.

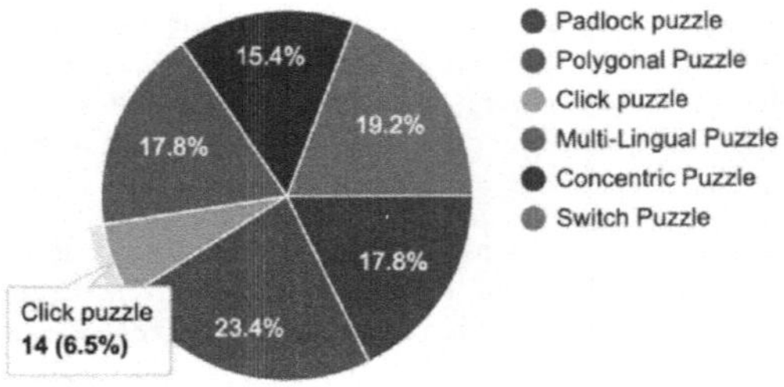

Fig. 4.2 Puzzles chosen

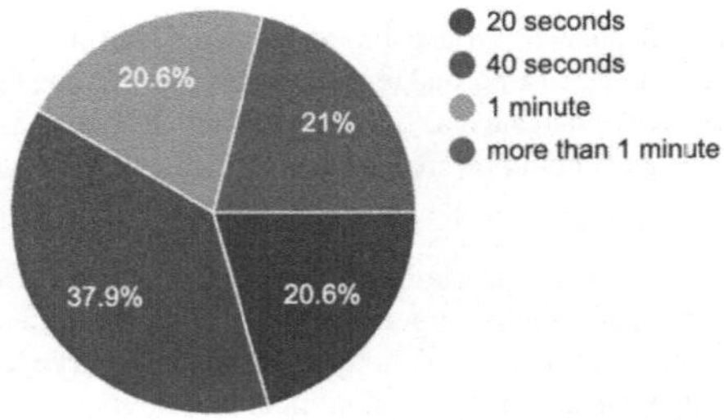

Fig. 4.3 Login time

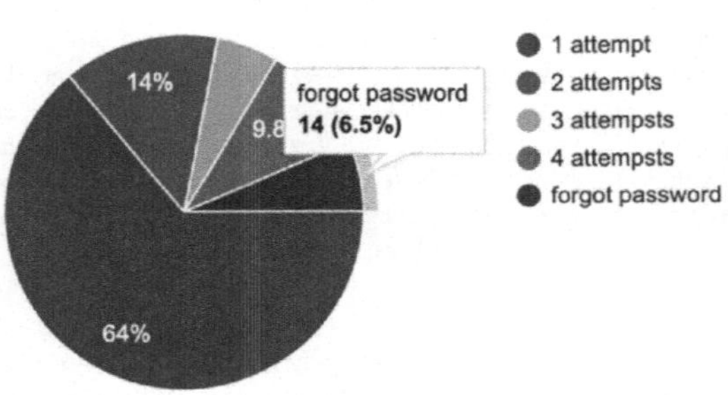

Fig. 4.4 Attempts taken to sign in

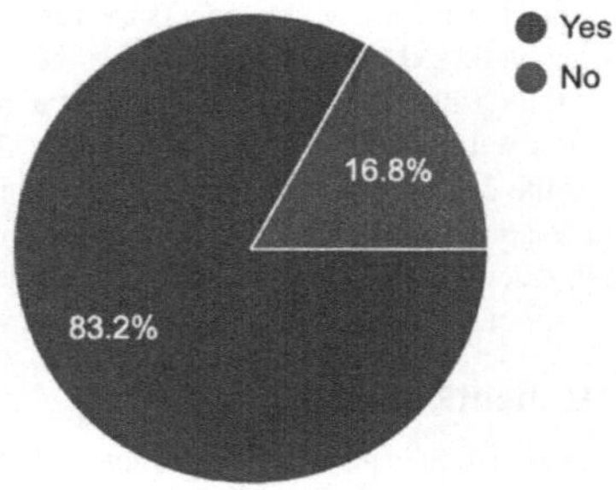

Fig. 4.5 Ability to recall and recognize puzzles

5. Conclusion

In summary, the system has been designed with interconnected databases to enable seamless communication between the service provider and developer. Its dependence on the puzzle ID and password ensures that only authorised users can access it, making it completely secure. Furthermore, the puzzles have been created and tested with scalability in mind to ensure that the system can handle high volumes of traffic without any issues. In conclusion, the system's design prioritises robustness and scalability, making it an ideal solution for organisations looking for a secure and efficient system.

References

1. Dhamija, R., & Perrig, A. (2000). Deja vu: a user study: using images for authentication. In Proceedings of the 9th USENIX Security Symposium, Denver, CO, pp. 45–58.
2. Brostoff, S., & Sasse, M. A. (2000). Are passfaces more usable than passwords? A field trial investigation. In People and Computers XIV-Usability or Else! London, pp. 405–424.
3. Por, L. Y., Ku, C. S., Islam, A., & Ang, T. F. (2017). Graphical password: prevent shoulder-surfing attack using digraph substitution rules. Frontiers of Computer Science, 11(6), 1098–1108.
4. Jermyn, I. H., Mayer, A., Monrose, F., Reiter, M. K., & Rubin, A. D. (1999). The design and analysis of graphical passwords. In Proceedings of the 8th USENIX Security Symposium, Washington, DC.
5. Goldberg, J., Hagman, J., & Sazawal, V. (2002). Doodling our way to better authentication. In Proceedings of the Human Factors in Computing Systems (Extended Abstracts), Minneapolis, MN, pp. 868–869.

6. Shahzad, M., Liu, A. X., & Samuel, A. (2013). Secure unlocking of mobile touch screen devices by simple gestures: you can see it but you can not do it. In Proceedings of the 19th Annual International Conference on Mobile Computing & Networking, Miami, FL, pp. 39–50.
7. Sun, H. M., Chen, Y. H., Fang, C. C., & Chang, S. Y. (2012). PassMap: a map based graphical-password authentication system. In Proceedings of ACM Symposium on Information, Computer and Communications Security, Seoul, Korea, pp. 99–100.
8. Ratchasan, T., & Wiangsripanawan, R. (2018). HapticPoints: the extended passpoints graphical password. In Information Security Applications. Cham, Switzerland: Springer, pp. 16–28.
9. Bianchi, A., Oakley, I., & Kim, H. (2015). PassBYOP: bring your own picture for securing graphical passwords. IEEE Transactions on Human-Machine Systems, 46(3), 380–389.
10. Chithra, P., & Sathya, K. (2018). Pristine PixCaptcha as graphical password for secure e-Banking using Gaussian elimination and cleaves algorithm. Proceedings of 2018 International Conference on Computer, Communication, and Signal Processing (ICCCSP), Chennai, India, pp. 1–6.
11. Matta, P., & Pant, B. (2020). TCpC: a graphical password scheme ensuring authentication for IoT resources. International Journal of Information Technology, 12, 699–709.

Note: All the figures in this chapter were made by the author.

Disruptive Technologies for Sustainable Development – Dr. G. Nagappan et al. (eds)
© 2024 Taylor & Francis Group, London, ISBN 978-1-032-55000-8

Audio-Driven Talking Face Generator

5

Pooja V.[1], Sivakami Raja[2]
Department of Computer Science and Engineering,
Rajalakshmi Engineering College, Chennai

Abstract: The creation of talking faces powered by audio has made tremendous strides in recent years. These models seek to produce expressive computer-generated characters, realistic face animations that match speech input, and lip-syncing in dubbed movies and TV shows that are more believable. These models also attempt to provide more immersive virtual reality experiences. Recurrent neural networks and convolutional neural networks are common deep learning models used by audio-driven talking face generators to understand the correlation between audio and facial movements. For these models to understand the intricate mappings between the two modalities, a lot of training data is needed. One popular technique is to employ motion capture equipment to capture a speaker's face while they talk. The model is then trained using the obtained data in order to produce the desired face motions for a specific audio input. This method is costly and time-consuming, and the models it produces are frequently restricted to a small number of speakers and speech patterns. Recent research has concentrated on creating talking face generators that are independent of the speaker and adaptable with speech styles to solve these constraints. In order to provide precise and expressive face animations for every speaker or speech style, these models seek to understand the link between audio and facial motions in a more generalised fashion.

Keywords: Deep learning, 3d animations, Audio-driven

1. Introduction

Modern machine media, such as computer games, heavily rely on the utilisation of emotional face movements. Perception motion capture, which includes animating a face with observed human actor actions, is already a common component of production workflows. The cost of producing high-quality facial animation is still significant even if capture system quality is continually improving. Initially, complex setups, as well as labour-intensive cleaning and other processing stages, are required for computer vision systems. An additional, less obvious issue is that actors must be present whenever new pictures are taken, and they should ideally maintain their appearance. This could be challenging if, for instance, a different job requires you to develop a beard.

1.1 Conditional Variational Autoencoders For Mesh Creation

According to our two-step methodology, the first stage includes converting certain methods of contextual words S t to the mesh posture at epoch t, i.e. at = f. (S t). The last speech window in a series of k + 1 in our investigation is S t = Stk:t. However, if the posture were simply regressed on the speech signal, it would ignore the multimodality of the face and tongue postures related to speech and would instead default to the average position. In order to address this, we offer a probabilistic method based on conditional variable automatic encoders (cVAEs) [SLY15]. It is assumed by cVAEs that a latent variable zt N (0,I) exists that encodes speech-independent differences in facial postures, such as distinct speaking trajectories.

[1]Poojavijayaraj510@gmail.com, [2]sivakami.r@rajalakshmi.edu.in

DOI: 10.1201/9781003428473-5

1.2 cVAE inference

After training, animated recordings of fresh speech are produced using the decoder D. With enough breathing room, the speech signal coincided with the beginning of the voice recording, a0 (S 0 = Sk:0). If the cVAE was trained using several quality datasets, qt has been adjusted to the maximum level of accuracy that was utilised during training. In order to ensure the correct face expression, our latent optimisation technique modifies the latent vector zt to reflect a certain speech style while keeping the latent vector zt constant during inference (i.e. zt = z). Since we don't think fine-tuning tongue poses is necessary, we configure z equivalent null for the cVAE related to tongue movements.

1.3 PYTORCH Usage

Users may create natural-sounding speech from unprocessed transcripts using a text-to-speech system that combines the Tacotron 2 and WaveGlow models without the need for extra prosody data. The Tacotron 2 model uses an encoder-decoder architecture to generate Mel spectrograms from input text. Mel spectrograms are consumed by a flow-based model called WaveGlow (also accessible at torch.com) to create speech.

2. Existing System

A generic three-dimensional humanoid facial model is constructed and linked with an operational framework of expressive text-to-speech (TTS) composition employing a set of relevant generators. Earlier face animations had only lip movement based on audio synchronization. Backgrounds were not focused on the existing system. The drawbacks of the existing systems are stated below

- Lag in facial movement After training.
- Given the audio characteristics, it can only anticipate the visual parameters.

3. Proposed System

The proposed system aims to take into account the various strengths and weaknesses of facial mesh. Using 3D Facial mesh and Rig, two different meshes are used for the tongue and face. Using facial animation algorithm, individual face features are identified. Conditional variational autoencoder generates mesh animation. This technique is used for lip synchronisation. Some of the advantages are listed below

- Takes factors apart from academics like personality and natural intelligence into account
- Entertains a wider range audience i.e.; seminar preparation and youtube video

4. Literature Review

4.1 Realistic 3D Facial Animations with Subtle Texture Changes

Realistic facial animation must include the inclusion of lighting variation in the material. Conventional face animation methods mostly focus on changing the geometry of facial features while ignoring the intricacies of texture changes caused by lighting. This research introduces a revolutionary method that incorporates tiny lighting changes to produce realistic 3D face motion. The Partial Expression Ratio Image (PERI) idea and its acquisition procedure are introduced by the proposed technique. Within the MPEG-4-based facial animation framework, the PERI is parameterized in accordance with Facial Animation Parameters (FAPs). The multi-view PERI technique's advancement makes it easier for 3D facial models to realistically express emotions from a variety of angles. The results of the experiment support the usefulness of the suggested method in boosting facial expressions' expressiveness and plausibility.

4.2 Real-time Individual 3D Facial Animation by Combining Parameterized Model and Muscular Model

In order to provide precise and realistic results, the real-time system for individual 3D facial animation presented in this research incorporates a number of methods. To localise face areas and characteristics on a frontal facial picture, the system first combines AdaBoost, CamShift, and the Active Appearance Model. An appearance mesh model is created by first adapting a general 3D facial mesh model to each individual facial feature. This appearance mesh model is then combined with each internal articulatory mesh to produce a full 3D facial mesh model. The suggested method utilises a facial action area-splitting approach that combines the parameterized model and muscle model to generate face animation that strikes a compromise between performance and realism.

5. Module Description

5.1 Data Collection by Capturing Video

Developing smart applications, backed by Computer Vision, NLP, and Deep Learning technologies can be tricky. While textual, acoustic, and graphic datasets play a role, training models for proactively identifying video-specific elements require stringent monitoring and the availability of top-rated insights. Nothing against Image data collection but video datasets lend an additional sense of continuity to the machine learning models, making them way more perceptive and accurate in time. We utilize a variety of dynamical, realistic, and impromptu facial video datasets, including demographics, locations, size, posture and motion of the spectator within the frame, and the illumination of the facial region across multiple ethnicities, life stages, sexual identity, and so forth.

5.2 Converting User-provided Text to Speech

In NLP, speech recognition is a crucial task. The sole method for making computers comprehend human spoken language is voice recognition. Using different feature extraction techniques, computers can quickly interpret written language by turning it into features (such as numerical characteristics). Here, the user inputs the text that has to be spoken. The training video and this speech are then mixed.

5.3 Training the Video

By each actor, the instruction package is divided into two parts pangrams and in-character data. This collection aims to depict the variety of facial movements that may occur during normal speaking in a given chosen language, in this case, English. In order to adequately cover the range of expression, the actor delivers one to three pangrams, or utterances designed to contain as many diverse phonemes as feasible.

5.4 Using Pytorch, Combine Audio and Visual Data

In this section, the text will be turned into an audio file with the help of a trained voice, and the lips will move in time with the spoken words. All other facial expressions would be monitored using the training video. The first stage is to reconstruct a chaotic mesh containing texture and optical flow data from the 10 images obtained for each frame. The optical flow, which was constructed prior to the capture process using a unique photogrammetry approach, is then linked with the unstructured mesh, which is then projected using a settled-topology template mesh. When instructed by text or speech, the virtual head should provide speech-synchronized facial movement. The animation frame for a phoneme is known as a viseme. The foundation of facial animation is the viseme.

6. Methodology

In order to reduce the effects of lighting and person dependence, we first introduce a web demeanour three-dimensional cranial motion detection automated system. This system uses multiple appearance-based measurements that are fused, particulate sieving, local enhancement, and flexible interpolation to improve gesture filtration, and it also uses the three-dimensional facial model and the web appearance model to deal with thrombus. Additionally, an ordinary 3-D face model is distorted from a single-head image of a human for facial features, adaptation, and the textures of the face and hair are continuously changed. To speed up the creation of 3-D facial animation, the parameterized model and the muscular model are finally combined. Ultimately, monologue retinal co-articulation uses the tri-phone method.

6.1 Facial Animation from Video

Facial animation is created utilising the facial motion technology based on the face visual medium b gathered from the frames, and sound backed by visual is transmitted simultaneously. With just one image of the anterior face, it can create a three-dimensional facial model for a specific person using the feature extraction results, and the texture map is continuously updated.

This outcome closely resembles the subject of the film sequence, setting the framework for subsequent face animation, and it is appropriate for the creation of the synthetic head.

7. Conclusion

To put into practice a deep learning model that generates tongue and face motions from text input. In our approach, a model that transforms mesh animations into rig controller space and a cVAE that creates mesh animation are seamlessly combined. To show that quality conditioning may be used to train on datasets that contain animations of different quality levels without causing a drop in animation quality. Additionally, a method to optimise the latent variable of the model in order to change the overall speech style of our animations. Additionally, to offer a fresh method for assessing lip closure in order to rate the effectiveness of lip sync. This model generates animations without regard to the user's identity, in contrast to past studies.

References

1. D L Jiang, WGao, Z Q Wang, Y Q Chen. (2004). Realistic 3D Facial Animations with Subtle Texture Changes. IEEE.
2. Jun Yu, Lingyan Li, Jie Zou. (2020). Real-time Individual 3D Facial Animation by Combining Parameterized Model and Muscular Model.IEEE.
3. Dongmei Jian, Ilse Ravyse, PeizhenLiu, Hichem Sahli Werner Verhelst. (2006). Realistic mouth animation based on a articulatory DBN model with Constrained asynchrony. IEEE.
4. Fu Jie Huang and Tsuhan Chen. (2008). Real Time Face sync animation driven by human voice. IEEE.
5. Pengyu Hong, Zhen Wen and Thomas S ,Huang. (2002). Real-Time Speech-Driven Face Animation With Expressions Using Neural Networks 32-674.
6. Fu Jie Huang and Tsuhan Chen. Real Time Face sync animation driven by human voice. IEEE.
7. R. Gutierrez-Osuna, Member, IEEE P. K Kakumanu,Student Member, IEEE. (2016). Speech-Driven Facial Animation With Realistic Dynamics
8. Mancini, Roberto Bresin, and Catherine Pelachaud. A Virtual Head Driven by Music Expressivity.
9. Xin Geng, Zhi-Hua Zhou, Senior Member, IEEE, and KateSmith-Miles,Senior Member,IEEE.(2018). Individual Stable Space: An Approach to Face Recognition Under Uncontrolled Conditions.
10. Thomas Gerig, Andreas Morel-Forster, Clemens Blumer Bernhard Egger. (2005). Morphable Face Models - An Open Framework. IEEE.
11. Hao Tang, Student Member, IEEE, Yun Fu, Student Member, IEEE, Jilin Tu, Student Member, IEEE, Hasegawa Johnson, Senior Member, IEEE, and Thomas S. Huang. Humanoid. (2004). Audio–Visual Avatar With Emotive Text-to-Speech Synthesis.
12. Yun Fu and Nanning Zheng, IEEE. (2008). M-Face: An Appearance Based Photorealistic Model for Multiple Facial Attributes Rendering.
13. Hynek Hermansky S't, Nelson Morgan t, and Hans Gunter Hirsch IEEE. (2017). Recognition of speech in additive and convolutional noise based on RASTA spectral Processing.
14. Zhao Guliang IEEE, Wang Shoujue IEEE. Hypersausage. (2019). Neural Networks and its Application in Face Recognition.
15. Jun Yu and Zeng-Fu Wang. (2009). A Video, Text, and Speech-Driven Realistic 3-D Virtual Head for Human–Machine Interface. IEEE.

Disruptive Technologies for Sustainable Development – Dr. G. Nagappan et al. (eds)
 ISBN 978-1-032-55000-8

Medical Waste Classification Using Deep Learning for Recycling

6

Revathy P.[1]
Professor and head, Computer Science and Design,
Rajalakshmi Engineering College, Chennai, India

A. Alagu Gomathi Brindha[2], **B. Ancy Tera**[3], **J. Avinesh**[4]
UG Students, Computer Science and Engineering, Rajalakshmi Engineering College, Chennai, India

Abstract: The mass production of medical waste ultimately outpaces the burden as the need for healthcare rises, presenting a threat to the ecosystem. The pandemic produced an enormous amount of medical waste, which pollutes the environment. Medical wastes after the sterilization process can be recycled into useful products. Recycling is an efficient means of preventing medical waste from being dumped into the environment. As a result, our method suggests a Deep Learning (Ana Rebelo, 2019) based categorization for garbage sorting facilities. This system uses Convolutional Neural Network (CNN) for classification, after which, the application is designed such that the user can get the knowledge of methods and places where wastes can be recycled. The Medical Waste Classification (Xiangyu Yu, 2020) for Recycling (MWCR) is a medical waste classifier that provides suggestions for recycling medical waste. MWCR's ambitious goal is to improve medical waste recycling and establish processing units with a revenue model.

Keywords: Classification, Medical waste, Neural networks, Recommendation, Recycling

1. Introduction

Medical Waste Classification deals with the process of separating waste into different categories and identifying whether it is toxic. India produces around 54 thousand tonnes of the 300 million tonnes of medical waste created globally. Given that the global market for medical waste is expected to be worth 19.94 billion dollars and is constantly expanding, raising awareness of medical waste recycling and the threats it poses to humanity if ignored has become critical (T. Mythili, 2021). Machine learning techniques have been employed in recent years to improve the classification and recycling of medical waste. One such approach that has proven effective in image categorization applications is the LeNet convolutional neural network. In this context, LeNet is used to categorize medical refuse into various groups based on composition, allowing for more efficient recycling and reducing the environmental impact of medical waste.

2. Literature Review

2.1 Classification of Medical waste Using Yolov4

Yupeng Mou et al. (2022) address in this paper, the Medical Waste Classification Method using Yolov4, which can be used to increase the speed of target detection. This research concentrates on the improvement of the Yolov4 classification algorithm

[1]revathy.p@rajalakshmi.edu.in, [2]alagugomathibrindha.a.2019.cse@rajalakshmi.edu.in, [3]ancytera.b.2019.cse@rajalakshmi.edu.in, [4]avinesh.j.2019.cse@rajalakshmi.edu.in

DOI: 10.1201/9781003428473-6

by adding a Convolutional Block Attention Module(CBAM) below the backbone layer of the network. Instead of the mesh activation function, the hard mesh activation function is employed. This approach has an accuracy of 85.71%.

2.2 Recycling and Reusing Biomedical Materials

SinaKheirabadil et al. (2022) address in this paper, the critical requirement for medical waste management involving the identification of existing treatment methods which includes the type of solvent used and devices that can be utilized. It also focuses on reusing and reprocessing of medical materials and devices which are the most essential step in waste management. This explains the sterilization techniques and the issues related to various sterilization methods for a particular type of material. This study also concentrates on various recycling techniques that are sustainable for the environment and cost-effective. Thus, this study technically identifies the different kinds of processes for any particular type of medical waste.

2.3 AI-based Solution for Medical Waste Sorting

Nallapaneni ManojKumar et al. (2021) address in this paper, the problem of general municipal wastes contaminated with COVID-related wastes thus addressing the need for COVID waste management with an artificial intelligence-based automation solution. The dataset images are collected from Flickr. The waste images are given as input to support vector machine (SVM) classifiers which classify covid-related waste images from other wastes. It uses image texture-dependent features. It achieved an overall accuracy of 96.5% which is verified by a simulated experiment in MATLAB.

2.4 Medical Waste Classification with Enhanced Segmentation Network

T. Mythili et al. (2021) address in this paper a method using "Enhanced Segmentation Network with Deep Neural Classification of Network Garbage" for classifying medical wastes with maximized accuracy. Using an Encoder-Decoder Network (EDN), for image segmentation, a layer for pixel-level categorization is used. The proposed EnSegNet-DNN-TC architecture, considering 100 distinct classified biological pictures from the garbage image dataset, obtains 88% accuracy over the DNN-TC.

2.5 Medical Waste Separator Using Computer Vision

Vrushali Sule et al. (2021) address in this paper, medical waste has been classified into gloves, syringes, cotton, and masks and it has been detected using a medical waste separator. It is an automatic computer vision-based. A model can be trained using AlexNet's deep learning network. As a result, it identifies and detects waste. The hardware is designed to detect the opening and closing of the input bin lid and then capture an image of the waste item dropped into the bin. The waste object is dropped into the correct container using servo motors that after the model has classified it. This model came up with an accuracy of 86.17%.

3. System Architecture

As shown in Fig. 6.1, there is a database which consists of different types of medical wastes that have been taken from ImageNet and the remaining web scraped from websites. Data is then analysed by separating it into six types of medical wastes namely, metals, electrical instruments, plastics, pills, fabrics, and glass. The dataset preprocessing operations are performed, namely resizing and reshaping the image, removing the background and conversion to an array (Vasumathi Ayyallu Madangopal et al. 2016). Then the dataset is separated into test and train. The CNN networks were trained with images of individual medical waste items. Three different CNN architectures have been used namely manual, alexnet and LeNet. CNN architecture with better accuracy is taken into account as a mode to identify medical waste. The website is developed using the Django framework. Then the prediction of the medical waste is identified and its respective recycling method would be visible on the website.

4. Methodology

4.1 Data Collection and Analysis

As none of the information sets relating to medical waste was publicly accessible (Shuaung liang, 2021), the data collection procedure was carried out manually. The researchers scraped content from ImageNet, Flickr, and other databases. The collected data is labelled manually for training purposes.

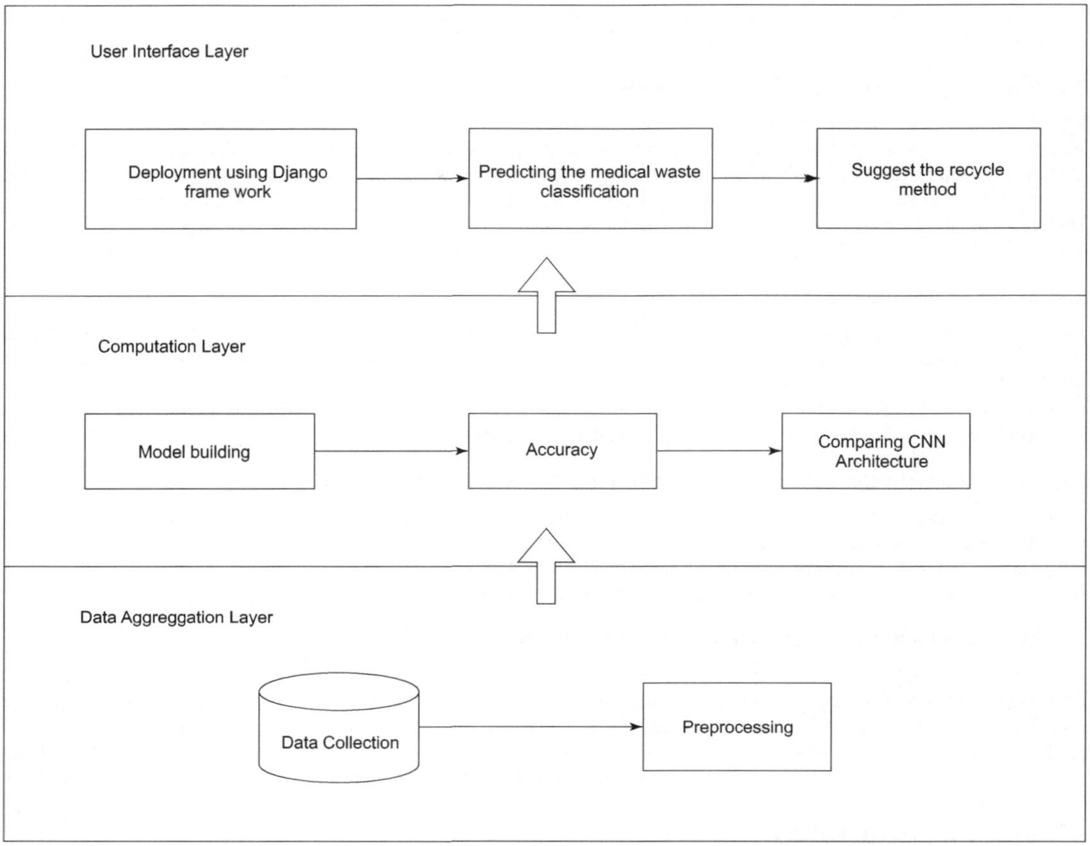

Fig. 6.1 Architecture diagram

4.2 Preprocessing

The set of images is preprocessed, which includes picture scaling, image reshaping, and array form translation which can produce improved effect on training(Liyan Zhang, 2022). The same processing is performed on the test dataset. Every picture from a collection of approximately six different types of medical detritus can be used as a test image for the software.

4.3 Model Training

Conv2d: The incredibly basic 2D convolution method begins by using the kernel. It is a very small matrix. It condenses the findings into a single output pixel after travelling across the 2D input data and performing element-wise multiplication with the area of the input it is currently on. This procedure is repeated for each area the kernel crosses, resulting in the generation of a second 2 dimension matrix. The outcome data are the weighted averages of the i/p features and the characteristics of the i/p are essentially positioned on the input layer in an approximately similar location as the pixel of output. The number of direct input characteristics received is determined by the kernel size

MaxPooling2D layer: By selecting the maximum value for each channel of the input over a window, the input has been expanded along its physical dimensions, namely height and breadth. The window measure is then shifted one step at a time.

With the appropriate padding option, the output's spatial shape (number of rows or columns) can be calculated as follows: output shape = math.floor ((input shape - pool size)/steps) + 1 (provided that input shape is greater than or equal to pool size).

Flatten layer: It is used to flatten the image's dimensions after it has been convolved. Dense: This is the hidden layer that is utilized to build this fully linked model. Dropout: This technique is used to prevent the dataset from being overfit, and it's sparse since the output layer only has one neuron that determines which category a picture belongs in. Using flatten, the data is compressed.

Dense layer: The Dense function in Keras uses a specific formula to compute its output. The formula consists of an activation function, which is specified by the user and applied element-wise, a weight matrix known as the kernel, and a bias vector. Once a layer has been created, its properties cannot be changed. If an input shape is provided as a common keyword argument, it will automatically form an i/p layer and place it above the instant layer, which is equivalent to explicitly specifying an InputLayer.

Dropout layer: The Dropout layer randomly adjusts the units of input with 0 at a certain amount of rate at each learning stage to avoid overfitting. Inputs that are not negative are adjusted increasing by 1/ to maintain the total of all inputs(1 - rate).

4.4 Comparing CNN Architecture

The different types of Convolution Neural Networks considered for the process were LeNet, Alexnet (M. Shanthalakshmi,2022) and a manually built CNN. The manual CNN architecture you used may not have been optimized for the specific task of Medical Waste classification (Christian Cabrera, 2020), leading to lower accuracy. Similarly, AlexNet was designed for general image classification(Vrushali sule, 2021) and did not suit our data samples. LeNet is used and is appropriate for the medical waste classification process because of its architecture and effectiveness in image classification tasks. LeNet was particularly developed for image classification and is made up of numerous convolutional and pooling layers, and also completely connected layers.

4.5 Model Building

The architecture includes multiple convolutional and pooling layers, as well as dense layers for classification. The input image is passed through the layers and a final softmax activation layer provides the predicted class probabilities. Backpropagation is used to modify the parameters of the network after it has been trained using a labelled dataset to minimize the loss function. The trained model can then be used for the classification (Le Hoang Son. 2019) of new images.

4.6 Development using Django Framework

The user interface is an essential part of the medical waste classification(Chen Huang, 2018) system as it provides a means for users to interact with the application and access the results of the classification process. The development process of the user interface involves designing an intuitive and user-friendly interface that allows users to input images for classification and view the recycling suggestion(Jemy James et al. 2021) involved with the classified waste category. The Django framework provides an excellent foundation for developing such an interface, and its built-in features make it easy to create and deploy the interface.

5. Experimental Results

The experimental results over the LeNet show 98% accuracy in correctly classifying the test instances of the Medical Waste dataset which can be represented graphically in Fig. 6.2.

6. Conclusion

To sum up, classifying medical waste is an essential first step towards effective waste management and recycling. Medical waste can be categorized into several classes with the aid of machine learning (Varadharajan et al. (2022)) techniques, such as the LeNet convolutional neural network, enabling more effective recycling and lowering the environmental impact of waste. LeNet categorization can be included within a Django framework to ease the mapping of the recycling process associated with the different classes of Medical Wastes, leading to more effective and affordable trash(Mahesh Jangid ,2021) management. Overall, this strategy presents a viable answer to the problems associated with managing medical waste, which have significant repercussions for the sustainability of the environment and public health.

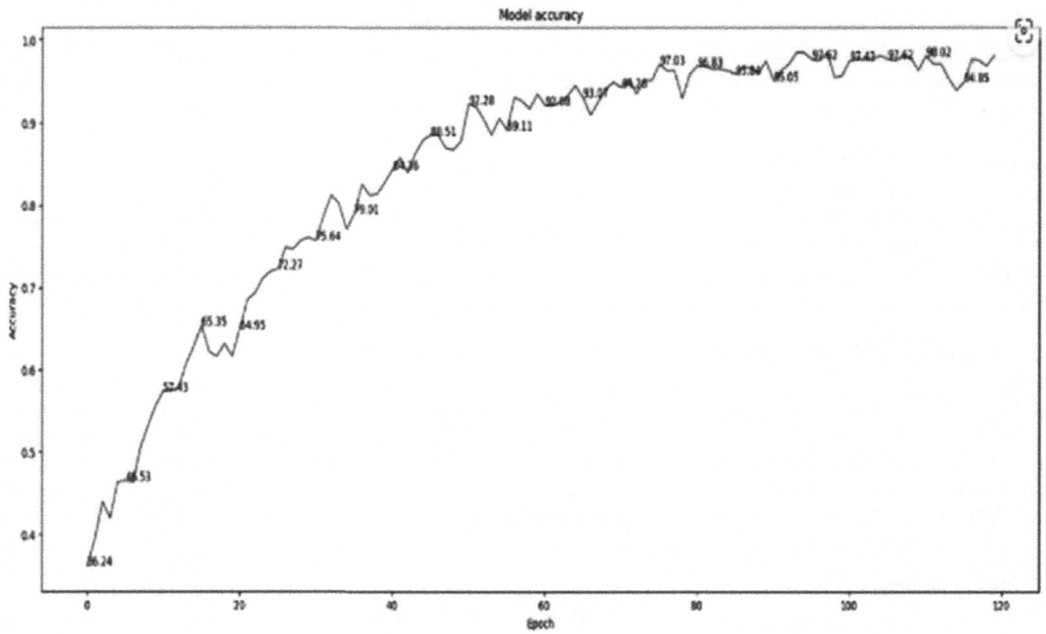

Fig. 6.2 Graph results of LeNet architecture

REFERENCES

1. Liyan Zhang et al. (2022) A Medical Waste Classification Method Based on Yolov4, Journal of 201822 (ITME).
2. T. Mythili et al. (2021) Enhanced Segmentation network with deep learning for Biomedical waste classification, IJST, Volume 14.
3. Vrushali sule et al. (2021) Computer Vision based medical waste separator, K. J. S College of Engineering, Vidyavihar, Mumbai, Fourth B-412.
4. Varadharajan et al. (2022), Segregation of Medical Wastes using Feedforward neural Networks and Image Processing for a New classification, (2021) FETBE.
5. Shuaung liang, Yu Gu (2021), A Deep Convolutional neural Network simultaneously localise and recognize waste types in images, Science direct, Volume 126.
6. Jemy James et al. Advanced Industrial and Engineering Polymer Research, (2021) Kingfa Scientific and Technological Co. Ltd. Elsevier B.V.
7. Xiangyu Yu et al. (2020) A deep learning approach for medical waste classification
8. Christian Cabrera et al. (2020), WasteNet: Waste Classification at the Edge for Smart Bins
9. Le Hoang Son et al. (2019), A Novel Framework for Trash Classification Using Deep Transfer Learning,*3* Duy Tan University, Da Nang 550000, Vietnam
10. Mahesh Jangid et al. (2021), Scrap An Efficient Approach to Trash Classification, Volume 9, 2021 Creative Commons Attribution 4.0 License
11. Ana Rebelo et al. (2019), Automation of Waste Sorting with Deep Learning, IEEE
12. Chen Huang et al. (2018), Multilayer Hybrid Deep-Learning Method for Waste classification and Recycling, Volume 2018, Computational Intelligence and Neuroscience.
13. Vasumathi Ayyallu Madangopal et al. (2016), A Bio Medical Waste Identification and Classification Algorithm Using Mltrp and Rvm, Iran J Public Health. 20016
14. M. Shanthalakshmi et al. (2022) Identification of Casting Product Surface Quality Using Alex net and Le-net CNN Models, Journal of Physics, Publisher: IOP

Note: All the figures in this chapter were made by the author.

Disruptive Technologies for Sustainable Development – Dr. G. Nagappan et al. (eds)

Air Quality Monitoring System Using ESP8266

7

K. Valarmathi[1], M. Hema[2]
Information Technology, Easwari Engineering College Chennai, India
Sivaramkumar V.[3], Sooriya Kumar N.[4], Vishwanathan J. R.[5]
Information Technology, Eeswari Engineering College Chennai, India

Abstract: The air pollution is a one of the most dangerous and unstoppable pollution that has been haunting us in our day to day lives there are always some protective measures to not get affected by pollutions other than air pollution but this pollution is the only type of pollution that cannot be stopped with some preventive measures. There must be drastic changes to be made and done to completely stop this pollution. So, the only measure we can take are that individuals should protect themselves by taking preventive measures using current technology. We think air pollution is mainly caused in the outdoors because of the emission in factories and vehicles but the harmful gases mixed with air can travel further places due to the air current once it reaches near the area of people living there is a high possibility that it can enter indoors which is just a personal and a small confined area for the people it is also the space where we spend most of our time. So, the preventive measures should start from indoor, we should protect ourselves from harmful gases that get along with air to our home. This air quality monitoring system using the ESP8266 is the most effective device to detect harmful gases that enter our homes this device can detect all the harmful gases such as Carbon Dioxide, Nitrogen, Alcohol, Benzene and more the gases are calculated according to their levels of intake and lively updated in cloud system and alerts the user in the mobile or pc/laptop.

Keywords: Multi-Point, Internet of things, Data representation, Thing speak, Air quality index, Indoor air quality monitoring

1. Introduction

According to recent studies and research we got to know that we spend most our time in our desired personal space that is our home. Because of this information we get to know that the most vulnerable space to start with the preventive measure is indoors. With these harmful gases that the air current brings indoors, the exposure for a long time and while sleeping or any other activities might lead to rapid or dangerous diseases to occur. There are some short-term problems that can occur those are consistent eye and throat irritation the people cannot be sure to say where does he/she get these problems from we also can get respiratory disease, even heart disease. These gases might slowly take our life span away if left unnoticed, these are only short-term problems that occur then think about what happens if these changes into long term problem and illness they can produce so the people should take an effective preventive measure to protect themselves.

2. Literature Survey

Ajithesh Kumar (2020) implemented Design and Analysis of IoT based Air Quality Monitoring System. The System is built in the consideration of particulate matters is very much important particulate matters within the size of 2.5 PM to 10 PM is also

[1]valarmathi.k@eec.srmrmp.edu.in, [2]avenz700@gmail.com, [3]avenz700@gmail.com, [4]nsksooriya1207@gmail.com, [5]jrvishwanathan@gmail.com

DOI: 10.1201/9781003428473-7

thoroughly monitored in the atmosphere using the device. The device lively updates the reading levels in atmosphere that it has been placed on to the Thing-Speak IOT analytic platform which helps the user to display the currently updated reading in the graph format in the user's PC/laptop using the Hyper Text Transfer Protocol. The Thing-speak plays a major role in getting the data from the system and updating it with 13-15 seconds time interval and the graphs are updated simultaneously. The device is placed in a densely populated area where there are a lot of Industries and factories to get to know the results of the built system. Sensors used they are MQ2 sensor the gas sensor which is used for getting the level of all the gases and particulate matters that is present in the atmosphere it can also distinguish individual gas compounds for example carbon dioxide, carbon monoxide etc. some of the gases are not at the point harmful for the people these gases get harmful when they get mixed with other gas compounds in the air so this sensor can get the readings of the harmful level and update them.

3. Proposed System Architecture

The System is built using the Node MCU ESP8266 Wi-Fi module as the center brain of the system it is purely based on Internet of Things technology. It connects all the sensors together and process their data. There are totally two sensors that plays a vital role in completing the system. The gas sensor i.e., the MQ135 sensor is used for detecting the distinguished gases present in the atmosphere. The DHT-11 sensor is the second sensor used for detecting the atmosphere's temperature and humidity. The reading values are passed by both sensors to the ESP8266 then the module gets programmed to establish TCP connection to pass on the data through internet to other platforms for data representation. The network connection is established and the data from the sensors are live updated in graph representation using Thing-Speak platform and in gauge interface in Blynk.

4. Methodology and Model Specifications

Initially and the main connection that is to be established is the Node MCU ESP8266 connection it intakes a power that is 3.3V this type of Wi-Fi module is an upgraded version of other Node MCU board versions it can directly be programmed to connect to the Wi-Fi connection that is need to be connected with just the connection's SSID and Password coded into the board it has a total of 30 pins 15 on each side. This module is the vital part of the entire system because it reduces the requirements for building the system using the Arduino UNO, the system totally has 2 sensor modules those are MQ135 gas sensor and DHT11 Temperature and Humidity sensor. The gas sensor i.e., the MQ135 sensor is the upgraded sensor module in the MQ sensor series it can catch the reading of multiple gases. It is sensitive to many distinguished of gases including carbon monoxide, alcohol, Benzene, smoke, $CO2$ and so on. Some of these gases are not harmful until they reach their saturation amount that one person can intake. The temperature and humidity sensor i.e., A thermistor for measuring temperature and a capacitive humidity detecting element make up the DHT11 sensor. This sensor is mainly made up of semiconductor so that it can get larger resistance value for the smallest change in power and temperature. This sensor has a temperature detection range of with a maximum 2-degree accuracy, 0 to 50 degrees Celsius. This sensor has a humidity range of 20 to 80% with a 5% accuracy. This sensor has a sampling rate of 1Hz, which means that it provides one reading each second. The DHT11 is a little device with an operational voltage range of 3 to 5 volts. The most current that may be utilized when measuring is 2.5 mA.

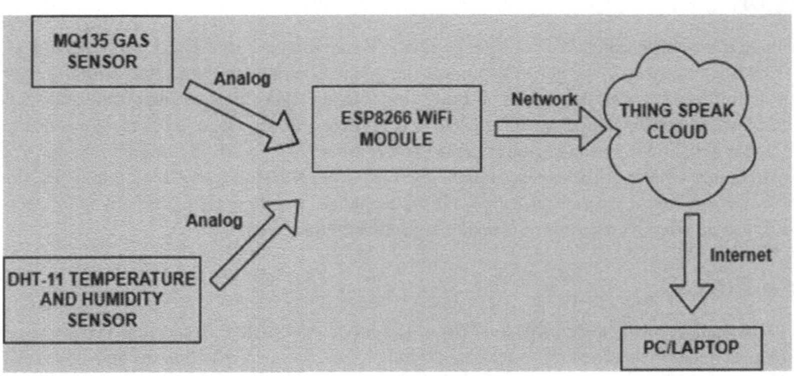

Fig. 7.1 Bar diagram of the system

Model Requirements

- Breadboard
- Connecting Wires
- DHT11 Temperature and Humidity sensor
- MQ135 gas sensor
- Node MCU ESP8266 Wi-Fi Module

Connection Establishment

For MQ135 connection establishment we need 3 of its pins.

- VIN: It should be connected for circulation of power.
- GND: Ground pin to be connected.
- A0: Operating mode/Standby mode
- For DHT11 sensor connection, it has 3 pins.
- 3V: It should be connected for circulation of power.
- GND: Ground pin to be connected.
- D5: Operating mode for the sensor.

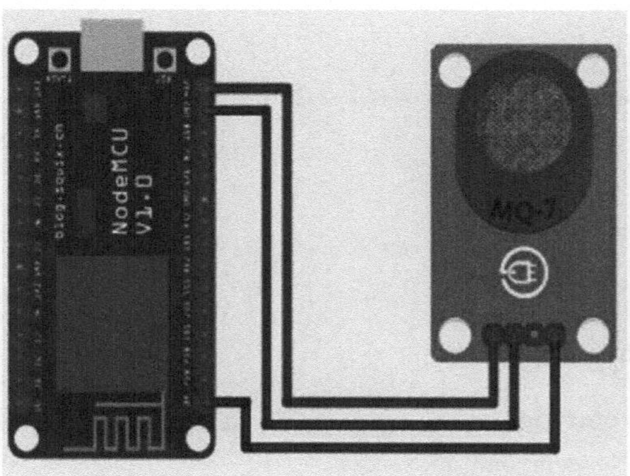

Fig. 7.2 MQ135 sensor connection to ESP8266

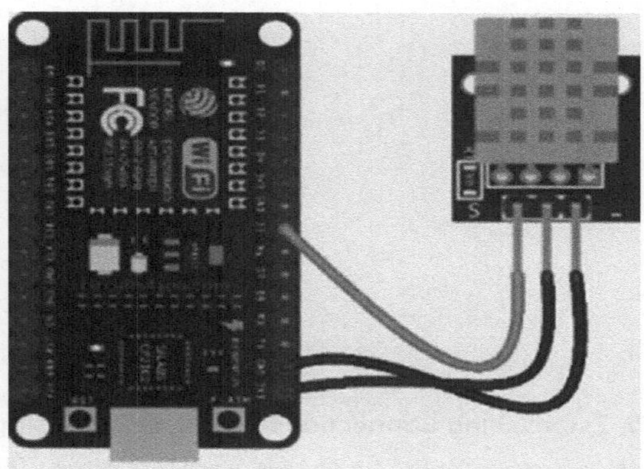

Fig. 7.3 DHT11 sensor connection to ESP8266

For DHT11 sensor connection, it has 3 pins.

- 3V: It should be connected for circulation of power.
- GND: Ground pin to be connected.
- D5: Operating mode for the sensor.

5. Prototype Results

The Device is Tested indoor in our college lab by establishing the device using the Blynk application and the output can be seen in the Blynk applications results.

5.1 Establishing Connection to Thing-Speak Using Prototype

The Network module ESP2866 which gives access to the functionality of uploading live data to the cloud system i.e., Thing-Speak system. The cloud system is accessed using the API key for the particular user that created the cloud system the API key gives us the functionality of connecting the cloud to our built system to upload live data. The Graphs are created while we create a thing-speak account for the cloud system, the graphs are created by creating new project and assigning graph fields.

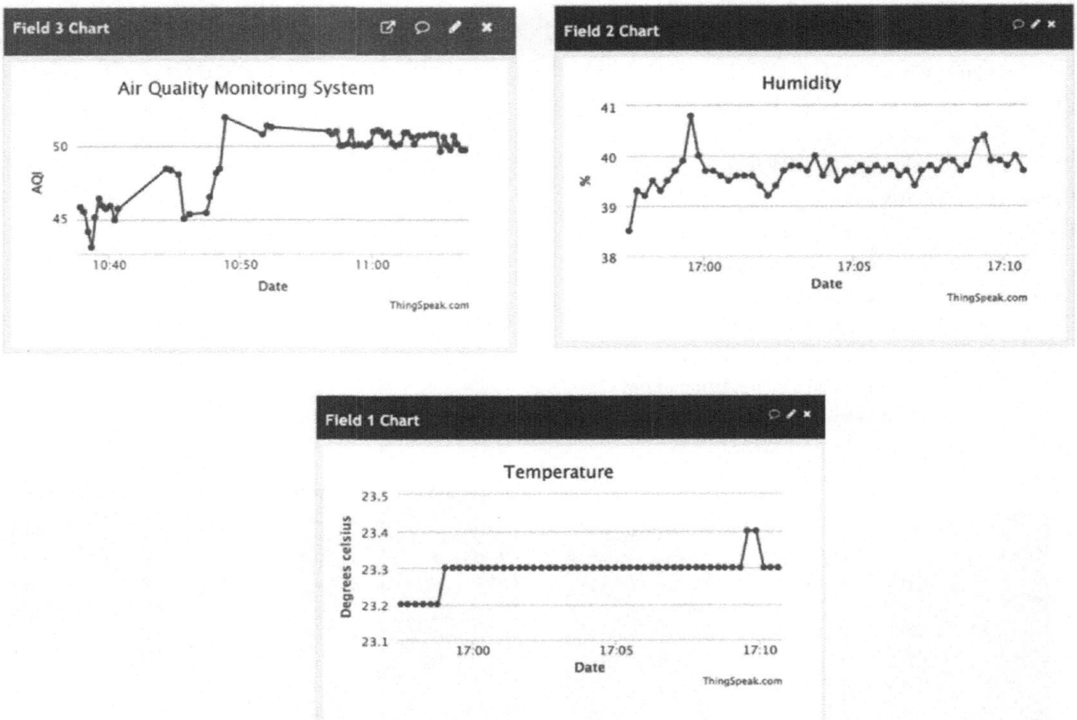

Fig. 7.4 Thing speak graphs

5.2 Establishing Connection to Thing-Speak Using Prototype

Blynk is also an IOT platform which is somewhat similar to Thing-Speak the only difference is Blynk can be used via mobile phones, it is compatible with both IOS and Android OS mobile phones. It is mainly used for creating Graphical User Interface by establishing connections by the device or widgets address or through network. We have done the Thing-Speak model for the user's information display in the PC/Laptop. And for the mobile phone Blynk can be easily used for creating UI for the IOT based air quality monitoring system, connection between the app and the prototype it is used in the code and run in the Arduino IDE.

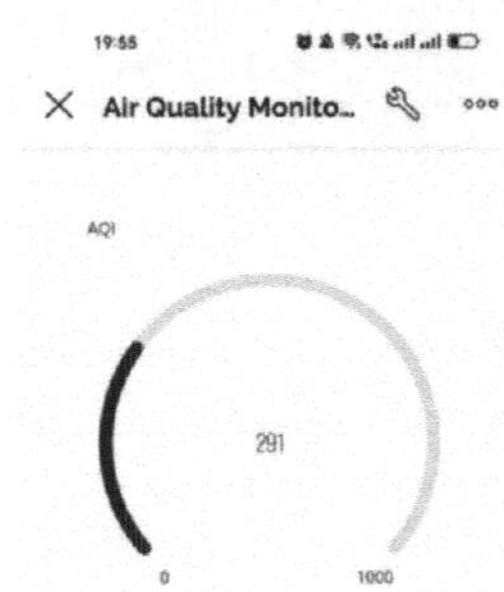

Fig. 7.5 Blynk applications results

6. Conclusion

Simple system which can be honed even more perfectly with upcoming technological advancements but for now the monitoring system we made using ESP8266 is purely based on IOT. With the use of appropriate sensor, we built the system to not only detect harmful gas but for other environmental data too.

REFERENCES

1. Virendra Barot, Viral Kapadia (2020) shilong India "Air Quality Monitoring System Using IOT " in IEEE 20(2) 123–139.
2. Iwan Kustiawan on (2018) India "IoT based Air Quality Monitoring" published by ISMEE 384(1).

3. Anand Jaya kumar, Arumugham, Praviss Yesyand, Venkatesh Prashanth (2021) India “IoT Based Air Pollution Monitoring System” published by IRJET 8(3) 2458–2461.
4. Yomi Debo Saiye, Babatunde M.Ajose-Ismail (2020) “IoT based Air Quality Detection and Monitoring System” published by IJRIAS 5(12) 66-68.
5. Esaki Rajavel (2018) “IOT Based Environmental Monitoring System ” published by AJAST 2(2) 461–467.
6. FN Setiawan1 and IK ustiawan1(2018) “IoT based Air Quality Monitoring ” published under the license by IOP publishing Ltd 384(1) 1–3.
7. Feresu ZTT (2022) “DHT11 Based Temperature and Humidity Measuring System” published by SciTechnol 11(5) 3–3.

Note: All the figures in this chapter were made by the author.

Disruptive Technologies for Sustainable Development – Dr. G. Nagappan et al. (eds)
© 2024 Taylor & Francis Group, London, ISBN 978-1-032-55000-8

Drowsiness Detection Using Deep Learning and Raspberry Pi

8

T. Saraswathi[1]
Information Technology, Easwari Engineering College, Chennai, India
Vijay kumar S.[2], **Srinivasan D.**[3], **Sushanth V.**[4]
Information Technology, Easwari Engineering College, Chennai, India

Abstract: Drowsiness and fatigue are constituting to many road accidents. Such accidents can be avoided with early detection and warning by the driver. In this project, we implemented a drowsiness tracking system using MobileNet framework. The behaviour of the human brain is replicated by these neural networks. The system uses a raspberry pi and a camera to record the driver's expressions on the face, collecting frames from the camera at specified time intervals. These frames are then processed on cloud server to extract the facial features required for analysis, such as eyes, mouth and other important features, and process them using the MobileNet model. As a result, if the result shows drowsiness, an audible alarm is activated in parallel.

Keywords: Deep learning, Transfer learning, MobileNet, Image processing

1. Introduction

Drowsy driving is a major contributor to accidents on the roads and can have serious consequences for both the driver and other road users. To address this problem, we have developed a drowsy detection project that utilizes advanced machine learning technology to detect when a driver is exhibiting signs of drowsiness. By providing an alert to the driver, our system aims to help prevent accidents caused by drowsy driving. Our project involves the application of machine learning methods to analyze a variety of factors, including the automobilist eye motion, expressions on the face, and driving behaviour, to determine when drowsy driving is occurring. We think that this novel strategy has the potential to considerably lower the frequency of crashes brought on by fatigued driving and raise road safety.

2. Literature Review

Ayman Altameem (2021) implemented Sleepiness detection and image segmenting in real time by utilising machine learning methods. In the suggested work, facial expressions were employed to build a support vector machine-based emotion identification technique (SVM). Joao Ruivo Paulo (2021) covers the pre-event timeframe and evaluates two techniques of detecting tiredness using EEG signals test centred on cross-subject zero-calibration. Without representation, the balanced accuracy is 69.99%. Chen (2020) using neuromorphic vision sensors, this study describes the initial event-based driver sleepiness detection (EDDD) system. Since the event stream lacks colour information, it might be challenging to identify and follow the eyes and lips. Wanghua Deng(2019) with the help of 68 key points, have created a new facial region identification technique. The authors suggest the MC-KCF algorithm. According to the findings of the experiments, DriCare had an accuracy of about 92%. Linlin Zhang (2022) there is still a chance that private information could be intercepted when federated learning is used to identify

[1]saraswathi.t@eec.srmrmp.edu.in, [2]vijaysaran0707@gmail.com, [3]srinivasan5655@gmail.com, [4]sushanthkumar252001@gmail.com

DOI: 10.1201/9781003428473-8

driver drowsiness. When compared to the traditional sleepiness tracking FL systems the findings from the driver drowsiness experiment dataset demonstrate that PFTL-DDD approach performs superbly. Feng You(2019) offer a new algorithm that takes into account the unique characteristics of the drivers in order to detect driving fatigue. DCCNN is used in this study. According to experimental findings, face detection accuracy can reach 98.8%. Soares(2019) propose a drowsiness detection system by calculating the proportion of eyelid closure without contact with devices. The application ran more smoothly and with 93.09% more frames per second after reducing the interest region and the search window. Mika Sunagawa(2020) has developed a sleepiness detection model that can recognize all levels of fatigue, from light to severe. Drowsiness was classified with 94% accuracy using machine learning techniques like neural networks. Jongmin(2019) implemented a driver sleepiness detection system on the basis of 3D CNN by employing feature combination technique on the basis of tensor product method they gathered spatio-temporal depiction and put it together with the vectors that portray the scene findings. Dasgupta (2019) introducing a system for smartphones that recognizes drowsy driving. There are three degrees of fatigue detection in the suggested framework. Using the framework, 93.33% of sleepy states can be classified.

3. Proposed Methodology

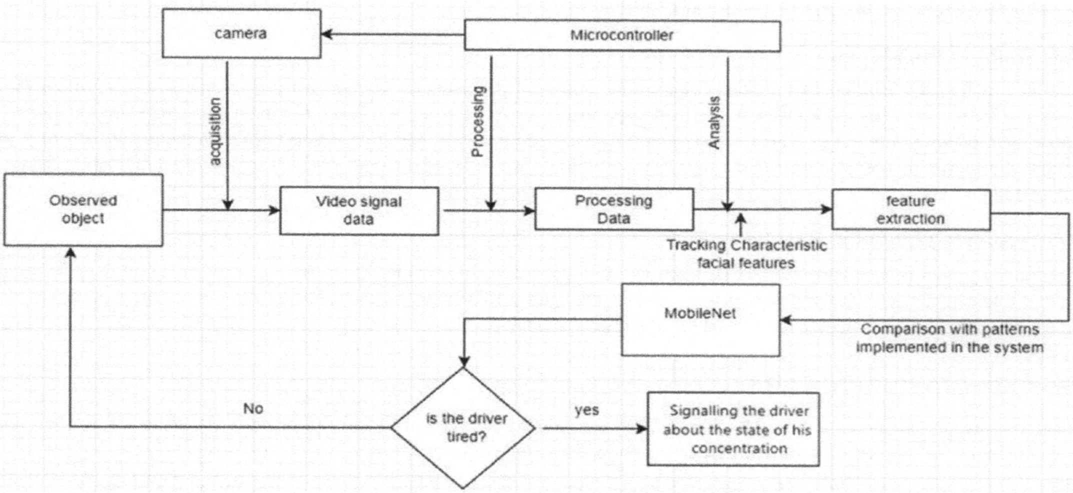

Fig. 8.1 Proposed architecture

Video can be used to identify driver fatigue by extracting key features such as eye position, whether the driver is open or closed, and whether he is yawning. From the camera the video stream is gathered and the frames are extracted from the video for further analysis.

3.1 Data Collection

The first step is to collect data. The collection includes eyes closed, eyes open, mouth closed, mouth open, and head moving, and the cleaned data are Divided into two sets. Raw pixel data was used for initial image classification. Therefore, the image is decomposed into its pixel components by the algorithm. The training data are data collected for training the model have image that are taken on various environments that involves variation in lighting, with glass, without glasses and other similar scenarios

3.2 Pre-Processing

The Pre-processing includes image scaling, grayscale conversion, and image augmentation of intensity images using Open CV. The video is recorded in 360p in order to reduce the memory and processing requirements. Frames are taken at a particular time intervals for processing the image.

3.3 Feature Extraction

The ratios of the mouth, eyes, and number of yawns are extracted using dlib library. The dlib is a facial landmark detector that forecasts the locations of the 68 facial points on a person's face are represented by a set of coordinates (x, y). using pre-trained models. For face detection, Dlib employs the Classifier using Linear SVM and the Histogram of Oriented Gradients (HOG). The aspect ratio of eye is determined by utilizing the formulae mentioned below

$$\text{Ear} = \frac{\|p2 - p6\| + \|p3 - p5\|}{2\|p1 - p4\|} \tag{1}$$

3.4 Building Model

For this purpose, we will employ a transfer learning model, MobileNet, a lightweight convolutional neural network architecture. There are two layers in MobileNetV1. The first layer is referred to as a depth wise convolution, and it performs minimal filtering with a single convolutional filter to each channel of entry. The task of developing additional features by calculating linear combinations of the input falls to the second layer, which is a 1x1 convolution, also referred to as pointwise convolution. There are 28 layers, excluding the first 2 layers, in a MobileNet.

3.5 Training the Model

The idea of transfer learning was applied because the generated datasets lacked enough training photos to train the object detection framework. The model was first trained with open-source datasets published by reputed universities and platforms. Later, the model was fine-tuned on our custom-made dataset specifically to detect drowsiness. To train the model, we used 30 epochs. Furthermore, we used early stopping to halt training once the model's performance on a holdout validation dataset stopped improving.

3.6 Deploying Model in Raspberry Pi

The trained model is then converted to TensorFlow lite model. TensorFlow lite framework enables to deploy deep learning models into edge devices like raspberry pi and mobile applications. If the system detects closed eye and yawning more than the threshold count, It alerts the user using an alarm to prevent accidents.

4. Experimental Results

We have trained, validated and tested the TensorFlow model on desktop and raspberry pi respectively using NTHU data set, which includes 36 subjects of various ethnicities captured in a range of circumstances.

The accuracy and loss were measured using the standard metrics. The formula for Accuracy is shown in figure below.

$$\text{Accuracy} = \frac{\text{TrueNegatives} + \text{TruePositive}}{\text{TruePositive} + \text{FlasePositive} + \text{TrueNegative} + \text{FlaseNegative}} \tag{2}$$

We also used another metric, called Precision, to gauge the prediction. The formula for Recall is shown

$$\text{Recall} = \frac{\text{TruePositive}}{\text{TruePositive} + \text{FalseNegative}} \tag{3}$$

The recall score that we attained was around 0.86. If an algorithm has a high recall, it will typically return the majority of relevant results.

The Accuracy and Loss were calculated for each epoch and visualised using matplotlib. These visualisations aid in determining whether a model is underfitting or overfitting and in adjusting hyperparameters.

The result of the Drowsiness Detection System is presented below in Fig. 8.4

5. Future Improvements

In future along with the analysing the image of the driver for drowsiness the current state of the automobile like speed of the vehicle, steering position, Rpm and other such parameters can be included in the decision making with the help of onboard diagnostic's tool to ensure maximum safety to the driver.

Confusion Matrix
[[2527 17]
[666 1662]]
Classification Report

	precision	recall	f1-score	support
Close Eyes	0.79	0.99	0.88	2544
Open Eyes	0.99	0.71	0.83	2328
accuracy			0.86	4872
macro avg	0.89	0.85	0.86	4872
weighted avg	0.89	0.86	0.86	4872

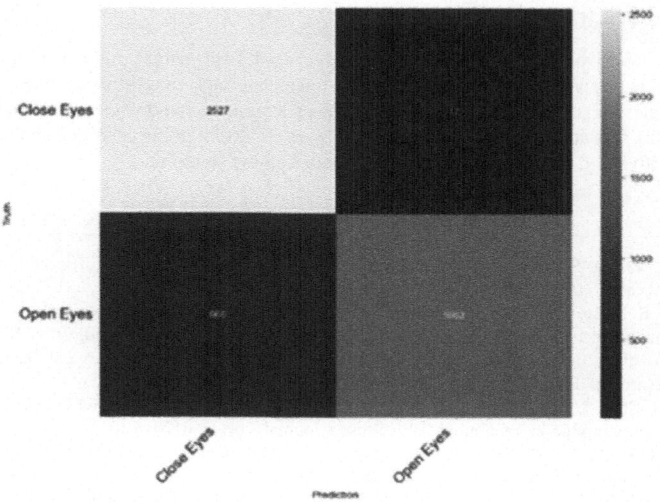

Fig. 8.2 Confusion matrix and classification report of the trained model

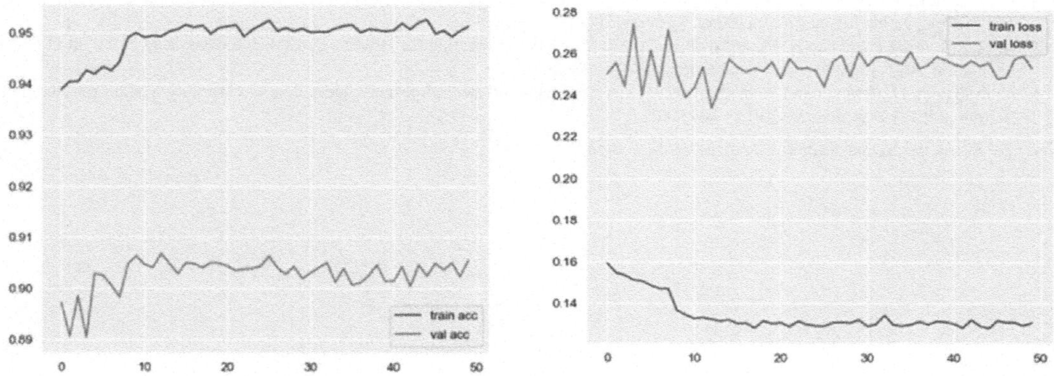

Fig. 8.3 Accuracy and loss graph of training and validation data

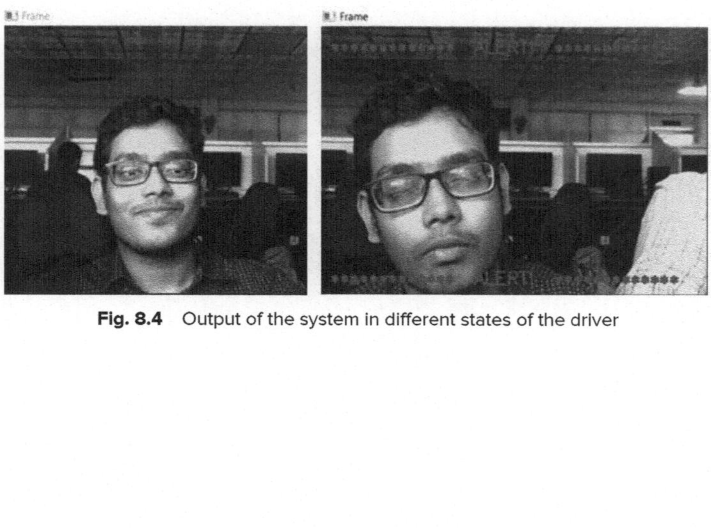

Fig. 8.4 Output of the system in different states of the driver

6. Conclusion

In order to drive safely, drowsiness detection is crucial. Using a transfer learning-based methodology, we developed an enhanced system in this work for reducing drowsiness-related mishaps. In order to increase the accuracy of detecting driver fatigue using real-time data, we also research and analyse deep learning methodologies and facial recognition technologies. A total accuracy of 86.0% was attained by the proposed system under ideal lighting condition and circumstances. Accuracy can be increased by using advanced models and training under poor lighting conditions.

REFERENCES

1. L. Zhang, H. Saito, L. Yang et al. (2022), Privacy-Preserving Federated Transfer Learning for Driver Drowsiness Detection, IEEE Access, 1(10): 80565–80574.
2. J. R. Paulo, G. Pires and U. J. Nunes, (2021), Cross-Subject Zero Calibration Driver's Drowsiness Detection: Exploring Spatiotemporal Image Encoding of EEG Signals for Convolutional Neural Network Classification, IEEE Transactions, 29, 905–915
3. A. Altameem, A. Kumar, R. C. Poonia, et al. (2021), Early Identification and Detection of Driver Drowsiness by Hybrid Machine Learning, IEEE Access, 9, 162805–162819.
4. G. Chen, L. Hong, J. Dong et al. (2020), EDDD: Event-Based Drowsiness Driving Detection Through Facial Motion Analysis With Neuromorphic Vision Sensor, IEEE Sensors Journal, 20(11): 6170–6181.
5. M. Sunagawa, W. Nakai, M. Mochizuki et al, (2020), Comprehensive Drowsiness Level Detection Model Combining Multimodal Information, IEEE Sensors Journal, 20(7): 3709–3717.
6. F. You, X. Li, Y. Gong et al., (2019), A Real-time Driving Drowsiness Detection Algorithm With Individual Differences Consideration, IEEE Access, 20(11): 4206–4218.
7. W. Deng and R. Wu ,(2019), Real-Time Driver-Drowsiness Detection System Using Facial Features, IEEE Access, 7: 118727–118738.
8. J. Yu, S. Park, S. Lee et al. (2019), Driver Drowsiness Detection Using Condition-Adaptive Representation Learning Framework, IEEE Transactions on Intelligent Transportation Systems, 20(11): 4206–4218.
9. G. Soares, D. de Lima and A. Miranda Neto, (2019), A Mobile Application for Driver's Drowsiness Monitoring based on PERCLOS Estimation, IEEE Latin America Transactions, 17(2): 193–202.
10. A. Dasgupta, D. Rahman and A. Routray, (2019), A Smartphone-Based Drowsiness Detection and Warning System for Automotive Drivers in IEEE Transactions, 20(11): 4045–4054.

Note: All the figures in this chapter were made by the author.

Disruptive Technologies for Sustainable Development – Dr. G. Nagappan et al. (eds)
© 2024 Taylor & Francis Group, London, ISBN 978-1-032-55000-8

Security-Apprehensive Dynamic Scheduling for Real-Time Optimization

9

G. Nagappan[1]
Head of Department, Department of Computer Science and Engineering,
Saveetha Engineering College, Chennai

Karpagakamatchi S.[2]
Department of Computer Science and Engineering,
Saveetha Engineering College, Chennai

Sujatha G.[3]
Department of Computer Science and Engineering,
Saveetha Engineering College, Chennai

Abstract: Having a secure file system has become a huge challenge in current century. Anonymous resources/ people are hacking and leaking the confidential files using various technologies. The various hacking techniques are Phishing, Sniffing, Social Engineering, Foot printing, SQL injection, Enumeration, Ettercap, Netsparker. To have a secure way of handling files can be done using cryptography and dynamic scheduling. This paper proposes a technique through which confidential files can be downloaded and viewed in a secure way only by authorized resources/people. Using AES algorithm, the cryptic code can be generated and shared only with authorized set of resources/people. Dynamic Scheduling determines the code while runtime and shared in the file system. This process can be carried out in cloud-based file system too for secure way of viewing the files.

Keywords: Cryptography, Dynamic scheduling

1. Introduction

Cloud-based administrations are applications or programming that are accessible from a distance and facilitated on the seller's server for the benefit of the client. One simple method for seeing this is to in any case consider the product being on-premises, yet it happens to likewise be accessible from a distance. It likewise fills in as a stage that supports early arrangement and planning applications which can utilize the gathered information to foster creation schedules. In these ventures we have proposed to plan for Modern web application. A pall garcon is a pooled, centralized garcon resource that's hosted and delivered over a network generally the Internet and penetrated on demand by multiple druggies. Pall waiters can perform all the same functions of a traditional physical garcon, delivering processing power, storehouse, and operations.

2. Literature Review

Jiachengxie and Xuwenwang (2019). Proposed An Integrated Cloud CAE Simulation System for Industrial Service Applications. The pall-based computer-backed engineering (CAE) technology described in this work widens the operational range of CAE, resolves the issue of uneven distribution, and enhances the efficacy of the simulation coffers. The integrated

[2]nagappan.cse@saveetha.ac.in, [2]anusethuraman1998@gmail.com, [3]sujathagopinath1899@gmail.com

DOI: 10.1201/9781003428473-9

Pall CAE simulation system for artificial service operations is established in this study and is based on the ASP.NET, VB.NET, and ANSYS Parametric Design Language. Stoners can now successfully bypass the ANSYS software thanks to the system's network connection of the dispatching director and clustering of the waitstaff before dispatch based on weight and operation. The parameters of the parametric model can be entered using the runner, or a three-dimensional computer-aided design (CAD) model that complies with the format requirements can be uploaded.

K.L. Keungand and C.K.M. Lee, (2020). Proposed Pall- Grounded Cyber-Physical Robotic Mobile Fulfillment Systems. In order to assure virtually real-time adaptation, the Internet of Things (IoT) and Cyber-Physical Systems (CPS) have accepted a virtual identity during their rapid development and performance in the engineering and manufacturing sectors. In order to provide a thorough knowledge of conflict avoidance technique in thematic-layers multi-deeps storage arrangement, the paper presents a pall-predicated CPS architecture. This essay examines six RMFS conflict groups and offers a case study of the real-world setting. A brand-new conflict type that is emerging in multidisc RMFS is wharf grid conflict. In order to illustrate the collision finding and result, a script analysis using actual customer orders is used.

Saad Mubeen and Alma Didic (2017). Proposed Constructs Delay Mitigation in Offloaded Cloud Controllers in Industrial IoT. The interaction of cloud computing, fog computing, and the Internet of Things (Internet of Things) in control operations aimed at robotization assiduity is examined in this work. To investigate the application of IoT bias that communicate with a pall-grounded regulator—that is, a regulator that is unloaded to pall or fog—a prototype has been created. The effects of having a pall garcon between the end device and the regulator are investigated through a number of trials.

3. Existing System

When resources allocated to the concerned users separately, require more resources. Though a particular number of resources are available, want to schedule the accessibility. Sometimes, resources are allocated there is a change to access by unknown users. The Technique used the existing system is PSO Algorithm Flyspeck mass optimization (PSO) is bio-inspired algorithms. Flyspeck mass optimization algorithm is an easy way to find an optimal solution in the result space. PSO is different from other optimization algorithms in such a way that only the objective function is demanded, and it isn't dependent on the grade or any discrimination form of the ideal. It also has veritably many hyperactive parameters.

4. Disadvantage

1. It's easy to fall into original optimum in high- dimensional space.
2. It has a low confluence rate in the iterative.

5. Proposed System

Whenever coffers are demanded, diurnal request wants to give to a sanctioned person. After finishing their services, the authority will remove the access control to the members. They give access control again to them when entering the request from the druggies or concerned persons. To view the report of request a key is generated and only using the key the train can be opened. This paper proposes encryption and decryption grounded on Advanced Encryption Standard (AES) which is a trusted algorithm. AES is vastly considered impregnable to all attacks Anyhow; numerous internet security experts believe AES will ultimately be regarded as the go- to standard for cracking data in the private sector. AES has erected- in harshness of vital length, which allows a degree of 'unborn proofing' against progress in the capability to perform total vital quest.

6. Methodology and Result

The client login with the registered username and password. The username and password are stored in the database and retrieved from the DB whenever needed. The client then places a package delivery order; the details of these orders are stored in picker hub. The picker hub contains all the order request details. The delivery hub team has access to picker hub and assigns each order to concerned person.

The concerned person will deliver the order to sender location and notifies the same. The manager in the picker and delivery hub enters all the details in a file and generates a report. These reports are encrypted and uploaded in the database. Only authorized persons can decrypt, a key will be generated using the key the file or the report can be viewed. The modules are listed below,

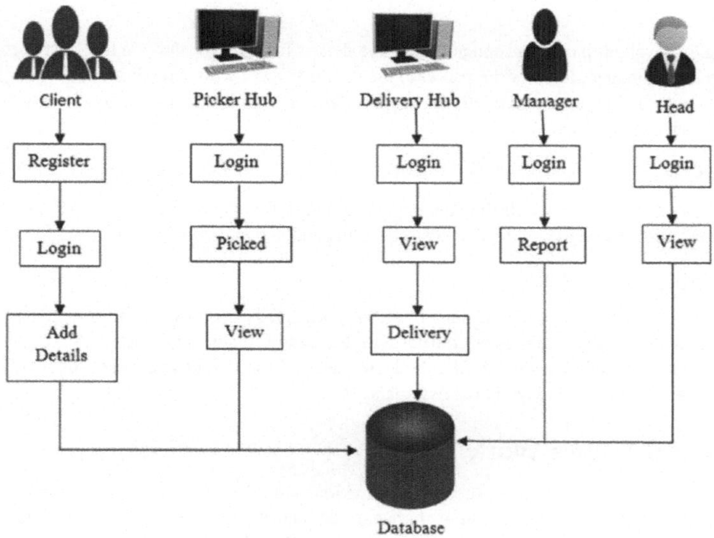

Fig. 9.1 Architecture diagram

- Registration
- Add Details
- Picker Hub
- Delivery Hub
- Report Upload
- Report View and Download

6.1 Registration

The user needs to fill all the mandatory details to register. The user can be client, manager, team member or anyone. All as to register based on their roles with the all the details required. The user will have username and password through which he/she can access the application at any time. This module gives the way to enter main page after login with valid input such as username or email id and password. The entered username and password are stored in database and the entered and stored details are checked and if they are match. If they are a match the user is followed to enter the application. If the user is not a valid user, then the entry is banned. If the user doesn't remember his/her password, they can reset it using forgot password button. After resetting the password. The user should login again with the new password set.

6.2 Add Details

In this module, gives the way to client enter the details in form. Fill the required details in the form. The data collected in this page are username, nickname, valid email id, phone number, address, security questions, age and gender. The entered details are stored in the database and used for later use. The data can be fetched when the valid user is logged. In the form all the details are updated and stored in database. The user can update if needed in the user details page.

6.3 Picker Hub

In this module, the pickup details like the username, user mail id, delivery location, the size of the package and the pickup placed date with the time. All the details here are stored in database. The parcel details are then shared to delivery hub in the city mentioned in the address. This module picker up collects the client parcels.

6.4 Deliver Hub

This module uses to delivery client parcel in send address. The details from picker hub are transferred to delivery hub through database. The delivery hub manager allocates the parcel to the delivery agent who is responsible for that location. The delivery agents pick the parcel from the hub and delivers to the sender address. The data of parcel received in the delivery location is updated by the delivery agent through app.

6.5 Report Upload

The branch manager uploads the report of all the delivery happened in the day at the branch. All the reports are shared with heads of the office. If any file needed to be added to the report it can added and downloaded using the key.

6.6 Report View and Download

This module is to help the head officer to view the daily branch report. Head officers view the branch manager's uploaded report and check the details. If there are any changes or conflict with the details. They are informed to branch managers. The result is when a file is downloaded/opened; to view the file, the generated code should be entered in the text box and submits. If the entered code is correct the file can be viewed by the externals.

7. Conclusion and Future Work

We have proposed for constant provisioning of modern applications stored in cloud. It pointed toward getting a streamlined execution cost progressively with the thought of client characterized security arrangement constraints in this cycle picker center receive the client parcel and send conveyance address everything is checking in lastly branch supervisor send the day to day picking parcel and conveyance in branch center point each detail to send the administrative center. The branch supervisor report will be sent secure the administrative center. Report will be encoded first utilizing AES Algorithm.

In future work planned to implement a genuine data set framework with random private keys. To improve the productivity of conventions, as far as number of messages traded and concerning their sizes, also. To implement using two are more algorithms like Blowfish algorithm and DES algorithm.

References

1. S. Zhang, X. Li, Z. Tan,T. Peng, and G. Wang (2019). A hiding and spatial K- obscurity driven sequestration improvement scheme in nonstop location based services, Future Genre, Computer System.
2. Q. S. Hua et al. (2020), faster resemblant core conservation algorithms in dynamic graphs, IEEE Trans. resemblant Distrib.Syst.
3. A. Jolfaei, X.-W. Wu, and V. Muthukkumarasamy (2016). On the security of permutation-only image encryption schemes, IEEE Trans. Inf. Forensics Secure.
4. W. Gong,L. Qi, andY. Xu (2018). Sequestration- apprehensive multidimensional mobile service quality vaticination and recommendation in distributed fog terrain Wireless Common Mobile Computer.
5. S. Zhang, G. Wang, M. Z. A. Bhuiyan, and Q. Liu (2018), Binary sequestration conserving scheme in nonstop position-grounded services, IEEE Internet effects.
6. X. Xu, X. Zhang, H. Gao,Y. Xue, L. Qi, and W. Dou (2019), Come Block chain- enabled calculation unpacking for IoT in mobile edge computing. IEEE Trans. Ind. Informat., vol. 16, no. 6, pp. 4187–4195.
7. S. Wang, L. Hu,Y. Wang, Q. Z. Sheng, M. Orgun, and L. Cao (2019). Modeling multi-purpose sessions for coming- item recommendations via mixture channel purpose routing networks.
8. B. Zhao,Y. Wang,Y. Li,Y. Gao, and X. Tong (2019). Task allocation model grounded on worker friend relationship for mobile crowdsourcing, Detectors.

Note: The figure in this chapter was made by the author.

Disruptive Technologies for Sustainable Development – Dr. G. Nagappan et al. (eds)

Healthy Food Recipe Recommendation System Using Deep Inverse Cooking Algorithm

10

V. Uma Rani[1]
Associate Professor, Department of Computer Science and Engineering,
Saveetha Engineering College, Chennai

K. Gunasundari[2]
M.E Scholar, Department of Computer Science and Engineering,
Saveetha Engineering College, Chennai

P. Vijay Karthick[3]
Professor, R. L. Jallapa Institute of Technology, Bengaluru

Abstract: Because of advances in technology, online food purchasing and preparation is becoming increasingly popular. People enjoy trying different foods from different nations. Eating is crucial to living a healthy existence. Users ordering food without understanding what it contains leads to an unhealthy lifestyle. Healthy food recipe recommendations are required to provide consumers with recommendations for choosing healthy diet foods. In this work, a personalised mobile wallet recipe recommendation system is built to help consumers obtain dish recommendations from food images and choose foods based on the calories in the recipe contents. The suggested Personalized recipe recommendation system employs a Deep inverse cooking algorithm to find food ingredients based on their images. It rank the healthy food according to user preferences.

Keywords: Recipe recommendation, Deep inverse cooking algorithm

1. Introduction

Now a days, People like to cook and eat various kinds of food within a place through internet technology. They are eager to taste and to know healthy ingredients at the same time. Sometimes, people order or cook the foods without knowing health ingredients of it. It provides health issues to people. The food recommendation system is a kind of retrieval system which selects recipe details for corresponding food images from heterogeneous food dataset or webpages. Nowadays, a day, Food recommendation system is developed to assist people to choose healthy food to balance their diets. In this paper, This study develops an algorithm for recommending foods based on user profiles and inverse cooking techniques. This system provides the user with a recommended list of nutritious foods based on their past food consumption and the number of calories in those foods.

The main goal of this research is to

1. To find the recipe of food based on images
2. It assists people to know the calories of food based on its ingredients
3. It provides the mobile wallet to purchase the food using smart card

[1]umaranibharathy@gmail.com, umaraniv@saveetha.ac.in; [2]sundariguna37@gmail.com; [3]principal@rljit.in

DOI: 10.1201/9781003428473-10

2. Literature Review

Several research works were carried out to develop a food recommendation system. Early work focuses on finding the food recipe based on food image or food name. They use several content-based information retrieval systems to extract ingredients from food images. Due to the popularity of machine learning techniques, little bit of research takes place to retrieve food recipes based on food images using those techniques. Some of them are discussed below

Morol M.K (2022) have developed a food ingredients recognition system using Convolution Neural Network. They have used the datasets food101, fruit 360, and UECFOOD256 from Kaggle to train the system. The pretrained ResNet50 model classify images into 1000 categories. After food ingredients identification, system rate the food based on kinds of food. Dataset contain 9856 images and their analysis shows that CNN based Resnet50 provide 92.6% accuracy for ingredient identification.

Mehrdad Rostami (2022) create a hybrid meal recommendation system. The model has been examined against LDA, HAFR, and FGCN methods. The food details were obtained from "Allrecipes.com". The developed system produced the best results in terms of basic metrics.

Naoki Shino (2015) suggests a strategy for suggesting substitute components based on co-occurrence relationships in recipe databases. They produce "exchange-input." The role of the substitute component and the exchange ingredient in a dish was the subject of the other algorithm. Both approaches leverage co-occurrence relations on the recipe database. According to the evaluation studies, over 90% of the required elements for both algorithms were found to be acceptable.

Diya Li (2021) pre-trained recipe embeddings are used to generate nutrition-guided recipes. The largest recipe dataset, Recipe1M, which has higher nutritional data, was used to generate meal suggestions for consumers utilising relevant embeddings and nutrition facts. Then contrast the traits of recipes that were located using recipe embeddings with those that were located using a typical keyword search. Results showed that 86% of respondents valued nutritional diversity and wellness.

Amaia Salvador (2019) uses inverse cooking algorithm, they created a system for creating recipes from food photographs.

In this research, we takes a food photograph from camera and turns it into a recipe with a heading, ingredients, and a list of steps for cooking and calorie information forit. Results revealed that our system outperformed state-of-the-art image-to-recipe retrieval techniques in 80% of the tasks, proving its superiority.

3. Proposed Methodology

Our proposed methodology is based on a machine learning technology that helps individuals have an easier time buying. This methos is used for people to know what are the ingredients can be bought based on their budget set amount. Nowadays many people don't know the amount of items needed for day to day lives and amount of ingredients to be used for day to day food which we make. They just buy everything with the money they have and still most of the food becomes stored and it becomes unhealthy for people to consume. So to make people understand their needs a app can be made for user which records the amount attached in the app and based on the amount in the M-Wallet a items list are recommended to users to buy for their food for that day and if those items are accepted by user it means user has these items then the user is given with necessary food recipe based on the ingredients selected .so it makes it user for calculate the amount to be bought for per day food which reduces food storage and wastage and it also gives ease of experience in food making easily for their day to day life.

- Data collection
- Filtering
- Recipe Recommendation

3.1 Data Collection

The process of acquiring information on a subject is called data collection. Mostly, it gathers the datasets and requirements for the project. It is the process of collecting quantitative and qualitative information about the project's requirements with the goal of evaluating results. To ensure that the data gathered is accurate, trustworthy, and consistent, good data gathering necessitates a clear strategy.

3.2 Filtering

In this research uses collaborative filtering which provides numerous methods for identifying comparable users or things and numerous methods for calculating information based on dataset matching. The similarity between users is calculated using user-based collaborative filtering, examining the diet-frequency matrix of the user. Then it locates a group of users with comparable dietary histories and makes suggestions for items that the target users have not tried but could enjoy. The K-means algorithm is used to dividing a dataset into K distinct subgroups (clusters) for quick retrieval.

3.3 Recipe Recommendation

The inverse cooking algorithm is used to suggests a pipeline with an intermediate phase where the set of ingredients is first gathered and it is processed by layers rather than directly getting a recipe from a picture. This enables the development of the instructions to consider both the image and the components

4. Results and Analysis

The majority of the analysis and outcome focuses on the system's performance. The data in the database is filtered using a cluster filtering technique in the current system, which uses machine learning filtering algorithms. This produces effective results, but it can only do so from a database that also only contains a small number of databases. We have combined a number of methods to make the system more sophisticated and efficient in order to address this issue.

We incorporate algorithms like inverse cooking, cluster analysis, and collaborative filtering into our system. In this, data from the database is filtered using collaborative filtering and cluster analysis filtering for display based on user requirements. Recommendations for cooking dish titles and related preparation instructions are made using the inverse cooking algorithm. It advises the dish based on the chosen ingredients. It can also quickly process large amounts of data. This is so user-friendly that even users may easily add ingredients to the database. The performance and efficiency of the current system have improved because to the use of these methods. Moreover, the loss during running decreased. The running time is used to calculate the performance of the. Together, the filtering and recommender algorithms boost system performance, enabling total calories to be calculated and results to be displayed in a matter of seconds. Fine training has improved the system's accuracy and efficiency over the current system by producing accurate results without delay or loss.

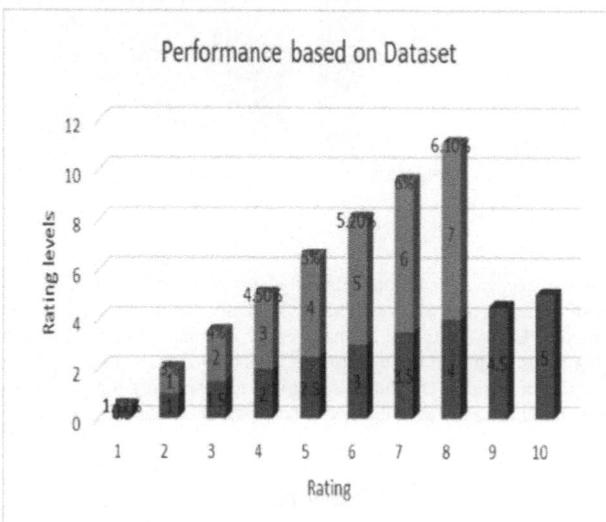

Fig. 10.1 Performance based on dataset

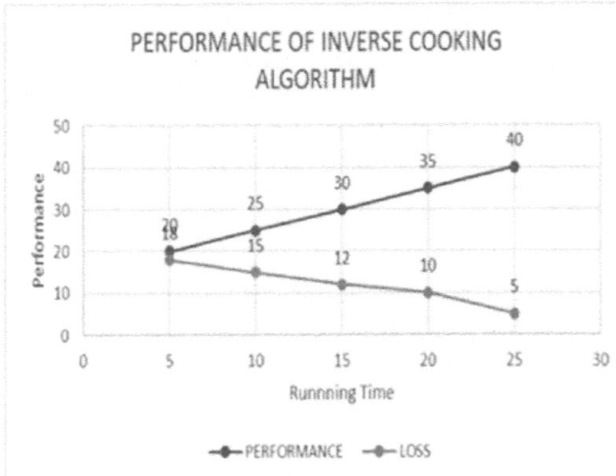

Fig. 10.2 Performance of inverse cooking

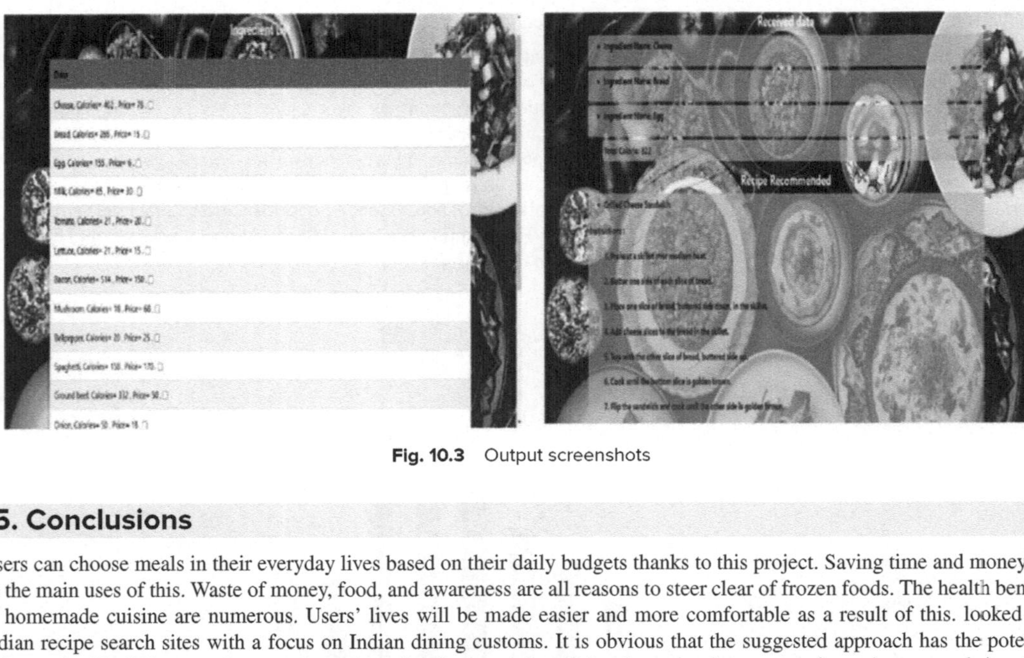

Fig. 10.3 Output screenshots

5. Conclusions

Users can choose meals in their everyday lives based on their daily budgets thanks to this project. Saving time and money can be the main uses of this. Waste of money, food, and awareness are all reasons to steer clear of frozen foods. The health benefits of homemade cuisine are numerous. Users' lives will be made easier and more comfortable as a result of this. looked into Indian recipe search sites with a focus on Indian dining customs. It is obvious that the suggested approach has the potential to be used with different recipe search engines. Our future study will involve using it to search for recipes on websites from different nations, revealing the cultural diversity of culinary customs from the perspective of active co-occurrence patterns for ingredients.

REFERENCES

1. Salvador, Amaia, Michal Drozdzal, Xavier Giró-i-Nieto, and Adriana Romero (2019). Inverse cooking: Recipe generation from food images. In Proceedings of the IEEE/CVF Conference on Computer Vision and Pattern Recognition,.
2. Desai, Ms Soundarya, Ms Pooja Patil, Mr Pratik Shinde, Mr Azhar Sayyed, and Rohini Bhosale (2019). Recipe recommendation based on ingredients using machine learning.

3. Shah, Lipi, Hetal Gaudani and Prem Balani (2016). Personalized recipe recommendation system using hybrid approach. International Journal of Advanced Research in Computer and Communication Engineering.
4. Trattner, Christoph, and David Elsweiler (2017). Food recommender systems: important contributions, challenges and future research directions. arXiv preprint arXiv.
5. Toledo, Raciel Yera, Ahmad A. Alzahrani, and Luis Martinez (2019). A food recommender system considering nutritional information and user preferences. IEEE Access.
6. Kul, S., & Sayar, A. (2022). A Smart Recipe Recommendation System Based on Image Processing and Deep Learning. In Innovations in Smart Cities Applications The Proceedings of the 6th International Conference on Smart City Applications.
7. D. Li, M. J. Zaki and C. -H. Chen, *(2021).* Nutrition Guided Recipe Search via Pre-trained Recipe Embeddings, IEEE 37th International Conference on Data Engineering Workshops (ICDEW),
8. Morol, M.K., Rokon, M.S.J., Hasan, I.B., Saif, A.M., Khan, R.H. and Das, S.S, (2022). Food recipe recommendation based on ingredients detection using deep learning. In Proceedings of the 2nd International Conference on Computing Advancements.

Note: All the figures in this chapter were made by the author.

Disruptive Technologies for Sustainable Development – Dr. G. Nagappan et al. (eds)
© 2024 Taylor & Francis Group, London, ISBN 978-1-032-55000-8

Cloud-Based Multimedia Content Protection System

11

Rekha K. S.
Assistant Professor, Department of Computer Science and Engineering,
Saveetha Engineering College

Kanimozhi S.[2], Srimathi N.[3]
Department of Computer Science and Engineering,
Saveetha Engineering College, Chennai

Abstract: In today's digital age, the protection of multimedia content has become increasingly important as the internet has made it easier than ever to share and distribute such content. With the rapid growth of cloud computing technology, the protection of multimedia content has taken on a new level of complexity. A cloud-based multimedia content protection system is a solution that provides the necessary security measures to protect multimedia content stored and distributed through the cloud. Such systems utilize encryption and digital rights management (DRM) technologies to safeguard against unauthorized access, copying, or sharing of multimedia content. This system has become particularly relevant for content creators, media companies, and other organizations that produce and distribute valuable multimedia content. In this paper, we will explore the challenges facing multimedia content protection and the role of cloud-based systems in mitigating these challenges. The exploratory examination and security confirmation both demonstrate that our proposed plot is safer and more effective than the current plans.

Keywords: Cloud computing, Encryption, Unauthorized access

1. Introduction

Because of the daily increase in the number of users & online transactions, data security is a top issue. Many different types of information are processed and stored using cloud platforms. For the purpose of offering stable and scalable carriers for information storage and communication, the cloud computing prototype has received a lot of attention. Customers are increasingly adopting cell phones to access cloud apps and communicate verbally, which increases the security risk associated with data loss. Our system structure uses cloud infrastructures to support various workloads with cost effectiveness, rapid deployment, portability, and mobility. This system is used to cozy distinct multimedia statistics kinds, inclusive of 2d and three-D films, jpg/jpeg files, mp3 and mp4 documents. The system can be set up on private and/ or public clouds.

Any private information that has to be safeguarded in order to prevent identity theft and fraud, such as phishing scams. Any sort of delicate information should be protected from unapproved access to shield the security or wellbeing of the association.

Effective data protection and fact protection procedures should limit access to sensitive documents in order to prevent data breaches and leaks. To keep away from information breaches, make certain the facts are encrypted before they're being sent. In our proposed net software the enterprise's touchy information which has to be handiest to their employees. Consequently,

[1]rekhaks@saveetha.ac.in, [2]sivalingamkani@gmail.com, [3]srimathieswari@gmail.com

DOI: 10.1201/9781003428473-11

for security of statistics that to be shared, public key cryptography and the adoption of RSA cryptography, the primary sensible public key cryptography, solved the key sharing issue in secret key cryptography.

2. Literature Review

Nileshkumar Kakade, Utpal Kumar Patel (2020) assembles the implementation of Secret sharing of data grounded on Homomorphic Encryption. This paper made-up of three phases. Paillier encryption is used to translate the central authority secret. This secret will be encrypted and divided into n figures of secret shares. The Paillier cryptography set of rules, also referred to as an asymmetric probabilistic non-deterministic method that maintains accretive homomorphic assets, allows one to cipher the encryption of m1 & m2 without having to first decrypt it. In the suggested approach, the system becomes more tolerant of assault when we expand a little bit in length. It may take more time to compute as bit period increases, which creates obstacles for the use of big bit lengths like2048 or 4096 bits.

The study of Dongjie Liu, Jong-Hyouk Lee (2020) construct Malicious Website Detection by Invalidating Multiple Web Spam's based on CNN. They begin by illustrating three distinct web spam strategies that vicious websites employ, including content-hiding spam, redirection spam, and hidden I Frame spam, in the proposed scheme. Additionally presents a new detection technology that mimics user perspective and screenshots vicious websites to disprove Web spam.

S. Sundari and M. Anandhi (2015) assemble secure multi-party computation in differentially private data with Data Integrity Protection. A multi-party protocol for the exponential technique has been proposed to do this. The majority of software program and data servers need constant access to the data they use. Corrupted data might result from errors. Despite the regularity of disc breaches, which cause further data loss. To ensure that any subset of data is accurate and free from severe or widespread corruptions using distributed data integrity (DDI) protection. In another study Paul J. Taylor et. al (2020) construct A systematic literature review of block chain cyber security. The outcomes of this observation demonstrate how the Internet of Things (IoT) is ideally suited for cutting-edge block chain applications, network and device virtualization, public key cryptography, web applications, document systems, and the safe archiving Personally Identifiable Information (PII).

Keiic Keiichi Iwamura and Ahmad Akmal Aminuddin Mohamed Kamalhi (2021) construct Secure Computation by Secret Sharing using Input Encrypted with Random Number. on this paper, The authors suggested that with $k \leq n < 2k - 1$, It is possible to use information-theoretic security for safe computing against a semi-honest attacker. Similarly, we filter the advantage of presenting a safe computation with mystery information that has been encrypted using an arbitrary integer. The dashing up of computation time is one of the advantages. As a result, in assessment to standard techniques, we acquire a hastily on line length for computations which includes internal product operations.

3. Proposed System

In the existing system authors used bilinear pairing and cryptographic algorithms to generate a public and private key, which takes a long time to method a feature. Hence to overcome these issues, In the plot we propose, Using the RSA algorithm, the cloud server can easily change the mark without using more sanitizer when sharing sensitive information. It's an asymmetric cryptography algorithm, which makes use of a private key (i.e one-of-a-kind, mathematically linked keys).As their names recommend, a public key is shared openly, while a confidential key is a mystery key and should not be imparted to anybody. In the example proven above, the RSA approach makes sure that the keys are as safe as viable, and the datasets manage the usage of SQL procedures.

3.1 Implementation

The system architecture establishes the basic structure of the system. We use the Hash Code Solomon algorithm, and in order to safeguard privacy, we are able to store a small amount of data on the local device and cloud server. The algorithm calculates the proportion of the distribution that is stored in the local device and the cloud, respectively based on the computational intelligence. The viability of a plan that is a truly potent addition to the cloud storage schemes that are currently in use has been demonstrated through theoretical security analysis and experimental evaluations.

Only the particular employees of the organizations are allowed to login into the web application by entering their login credentials and new users are able to register their credentials by their respective team leaders, and these team leaders are registered by the management or admin. The team leader is responsible for uploading the files into the cloud server. The RSA cryptographic algorithm is used togenerate the public and private keys necessary to safeguard those files. Following encryption, a public key is generated and made accessible alongside uploaded files. Employees can request to download the files by contacting the team leader once management grants approval. The uploaded files are decrypted using the private key; the private key is obtained by scanning the QR code generated.

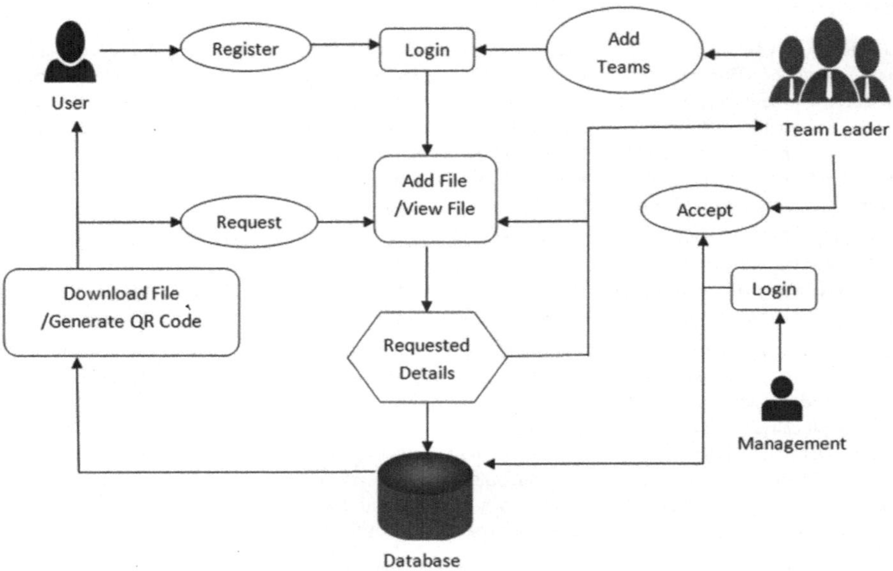

Fig. 11.1 Architecture diagram for content protection system

3.2 Modules

- Registration
- Login
- Files upload and download
- File view

3.3 Registration

In the registration process the team leader and the team members were registered into the database by the management and the team members itself respectively. This module gives a calculated structure to enter information in both the team members and the team leaders in a way that: ease us accuracy and ease of data entry by matching the data entered to the database. The process of registering team members creates individual staff records that link registers to staff data and gathers data elements to facilitate improved monitoring. Similar to this, team leader registration generates individual records that link registers to group pioneer information and collect information components for improved group program monitoring.

3.4 Login

The three primary users of our web application will be users (i.e., team members), Team leaders, and the management of the company. Upon successful validation the users were allowed to access the web application's features. The team members can download the file by sending the request to the team leader. The team leaders are responsible to upload the file to their respective teams. And the management can add the team leaders and accept the team leader responses. They can also view and manage the team leaders list.

3.5 Secret Key Generation

The team leader can then upload a file to the server by selecting it from their desktop. The Java file uploaded into the server will then keep a copy of that file. It will be put away in the data set. Once the files get uploaded, the team members can download the files by sending the request to the team leader. On receiving the request from the team members, the team leader is to accept the request and even the management may also accept the team leader's response so that management can generate public and private keys for security purposes.

3.6 Files Download

Once the team members' requests are accepted, they can download the files by decrypting the QR code generated after the file encryption process. The decryption process is done by entering the public key and private key which is generated while scanning the QR code.

4. Conclusion

Data sensitivity refers to information that, because of its sensitivity, has to be protected from unauthorized access or disclosure. For some, the team leader or staff details records may be that. Delicate information is classified data that must be safeguarded and kept out of the hands of any pariahs except if they have consent to see it. By sufficient information security and data security methods planned to get it, admittance to classified data ought to be confined.

We intended to implement a genuine anonymous database system in our upcoming work. Increasing the number of switched dispatches and their sizes, as well as the efficiency of protocols.

References

1. Paul J. Taylor, TooskaDargahi, Ali Dehghantanha, Reza M. Parizi, Kim-Kwang Raymond Choo, (2020). A systematic literature review of block chain cyber security,vol. 6, no. 2, pp. 147–156.
2. Dongjie Liu, Jong-Hyouk Lee, (2020).CNN - based Malicious Website Detection by Invalidating Multiple Web Spam's. IEEE Access, vol. 8, no. 1, pp. 97258–97266.
3. Martin Wester Kamp, Fried helm Victor, Axel Kupper, (2019). Tracing manufacturing processes using block chain-based token compositions. 6(2), 167–176.
4. Puthal Deepak, NishaMalik, S. P. Mohanty, E. Kougianos, GautamDas,(2018). Everything you wanted to know about the Block chain: Its Promise, Components, Processes, and Problems. IEEEConsumerElectronics Magazine,7(4), 6–14.
5. LiPeng, WeiFeng, and ZhengYan. (2020). Privacy preservation in the permission less block chain: A survey. Available: https://doi.org/10.1016/j.dcan.2020.05.008.
6. NileshkumarKakade, UtpalkumarPatel, (2020). Secure Secret Sharing Using Homomorphic Encryption,in Proc. 2020 11th International Conference on Computing, Communication and Networking Technologies 1–7.
7. Sundari S, Ananthi M, (2015). Secure multi-party computation in differential private data with Data Integrity Protection, in Proc. 2015 International Conference on Computing and Communications Technologies, 180–184.
8. Shuming Jiao, Ting Lei, Yang Gao, ZhenweiXie and Xie Yuan,(2019). Known-Plaintext Attack and Cipher text-Only Attack for Encrypted Single-Pixel Imaging.IEEE Access, 7(2), pp. 119557–119565.
9. S. Kaushik, and S. Puri, (2012). Online transaction processing using enhanced sensitive data transfer security model in Proc. 2012 Students Conference on Engineering and Systems, 1–4.
10. Weilin Zheng, ZiBin Zheng, Xiang Chen, Kemian Dai, PeiShan Li and RenFei Chen, (2019). NutBaaS: A Block chain-as-a-Service Platform.IEEE Access, 7, 134422–134433.
11. Fran Casino and ConstantinosPatsakis, (2020). An Efficient Block chain-Based Privacy-Preserving Collaborative Filtering Architecture. IEEE Transactions on Engineering Management, 67(4), 1501–1513.
12. Dmitri Chkliaev, JozefHooman and Peter van der Stok, (2000). Mechanical verification of transaction processing systems inProc. ICFEM 2000. Third IEEE International Conference on Formal Engineering Methods, 89–97.
13. Shijie Zhang, Jong-HyoukLee, (2020). Mitigations on Sybil-based Double-spend Attacks in Bit coin. IEEE Consumer Electronics Magazine, 7(2), 1–1.
14. Xin Wang, Qing Feng, JianpingChai, (2018). The Research of Consortium Block chain Dynamic Consensus Based on Data Transaction Evaluationin Proc. 2018 11th International Symposium on Computational Intelligence and Design, 214–217.
15. Shijie Zhang, Jong-HyoukLee, (2019). A group signature and authentication scheme for block chain-based mobile-edge computing. IEEE Internet of Things Journal, 7(5), 4557–4565.
16. Iwamura K & Mohamed Kamal, A. A. A.(2021). Secure computation by secret sharing using input encrypted with a random number in Proceedings of the 18th International Conference on Security and Cryptography, SECRYPT 2021, 540-547). SciTePress. https://doi.org/10.5220/0010552305400547

Note: The figure in this chapter was made by the author.

Disruptive Technologies for Sustainable Development – Dr. G. Nagappan et al. (eds)
© 2024 Taylor & Francis Group, London, ISBN 978-1-032-55000-8

Classification of Gastrointestinal Tract Abnormalities from Endoscopic Image

12

M. Vijay Anand[1]
Associate Professor, Department of Computer Science and Engineering,
Saveetha Engineering College, Chennai, Tamilnadu, India

Raja Sahaya Renuka S.[2]
PG Scholar, Department of Computer Science and Engineering,
Saveetha Engineering College, Chennai, Tamilnadu, India

Abstract: The gastrointestinal tract (GI tract) is a passageway in the body that allows nourishment to pass and solid waste to exit. The GI tract is comprised of several parts, including the throat, abdomen, small and large intestines, the anus, and the oesophagus. There are many different illnesses that can impact the gastrointestinal tract. These irregularities, if not identified early, have the chance of evolving into gastric cancer, a frequent condition of malignancy across over one million confirmed cases world wide each year. Endoscopy is an established technique for identifying and treating GI tract problems. Endoscopy is a microsurgical procedure used to examine the digestive tract. Gastroenterologists may miss some abnormalities during the examination for a variety of reasons, including irregular morphologies, a large number of frames, and exhaustion. Thus, the use of automated methods to categorize abnormalities found in endoscopic images is increasingly important in helping with medical diagnoses and cutting down on medical expenses and time. In this project. We intend to use the Kvasir medical datasets that are freely available to the public. It is made up of visuals from eight different gastrointestinal endoscopic imaging classes. Using Convolutional Neural Networks, we have simplified the recognition of GI tract markers and pathogens in Kvasir classes (CNNs). Because they can capture local features and are more highly scalable than fully connected networks, CNN models are broadly used in image perception.. We suggest using a dense model for the dataset that is based on the Densenet architecture. We strive to achieve a higher rate than that achieved in the reference papers.

Keyword: Gastrointestinal tract, Xception, CNN, Deep learning

1. Introduction

As time has passed, our approach to treating patients and the medicines we use have evolved. Initially, we relied on interpreting signs and symptoms in basic clinical situations, but imaging technologies have since enabled us to diagnose patients more quickly and accurately. The integration of endoscopy into everyday medical practice has further improved patient outcomes, particularly in cases like colorectal cancer (CRC) where screening programs have been implemented. Endoscopy has evolved into a crucial diagnostic and therapeutic tool for the gastrointestinal tract. Continuous investigation of available technologies has resulted in a groundbreaking promising foundation for exploring new options for the patient's condition. This has created a new opportunity for patients and medical practitioners to utilize evidence-based medicine in their daily practices. Improving the diagnosis of gastrointestinal tumors has been a major focus in endoscopy, and it has become a crucial tool for detecting these types of tumors in their early stages. Additionally, endoscopy is continuously exploring new therapeutic methods, which are

[1]vijayanand@saveetha.ac.in, [2]renuka.mariyans@gmail.com

DOI: 10.1201/9781003428473-12

certified for treating cancerous gastrointestinal lesions. Screening for pre-cancerous lesions using endoscopy and treating them has become an increasingly popular trend, with exponential growth in many specific types of lesions. This can be attributed to the integration of new technologies or the adaptation of current surgical procedures. Endoscopy within the gastrointestinal tract has become ingrained as a welcome advancement for both diagnosis and therapeutic paths.

2. Related Work

A recent study describes a novel deep learning model that uses a Faster R-CNN to detect esophageal abnormalities in endoscopic images. The proposed detection system combines handcrafted Gabor features and CNN features extracted using a densely connected convolutional networks architecture, which improves feature propagation and addresses the vanishing gradient problem. To improve texture details during the detection stage, the authors concatenate the extracted Gabor features with CNN features to detect abnormal complex regions more effectively.

The model achieves outstanding results on the KVASIR dataset, with a recall of 90.2% and a precision of 92.1%.

In a separate paper, the authors propose using an ensemble of deep features to recognise eight types of digestive tract diseases. As feature extractors, the ensemble consists of pre-trained DenseNet-201, ResNet-18, and VGG-16 CNN models, followed by a GAP layer. The authors create image textures with regional binary patterns and Haralick features, then train models for each feature with regression models and kernel multiple regression with spectral regression. Predictions are then subjected to the ensemble technique. The outcomes demonstrate an encouraging accuracy of more than 96%, surpassing state-of-the-art methodologies.

3. Proposed Methodology

The implementation of the project can be described in order as modules in the following block diagram.

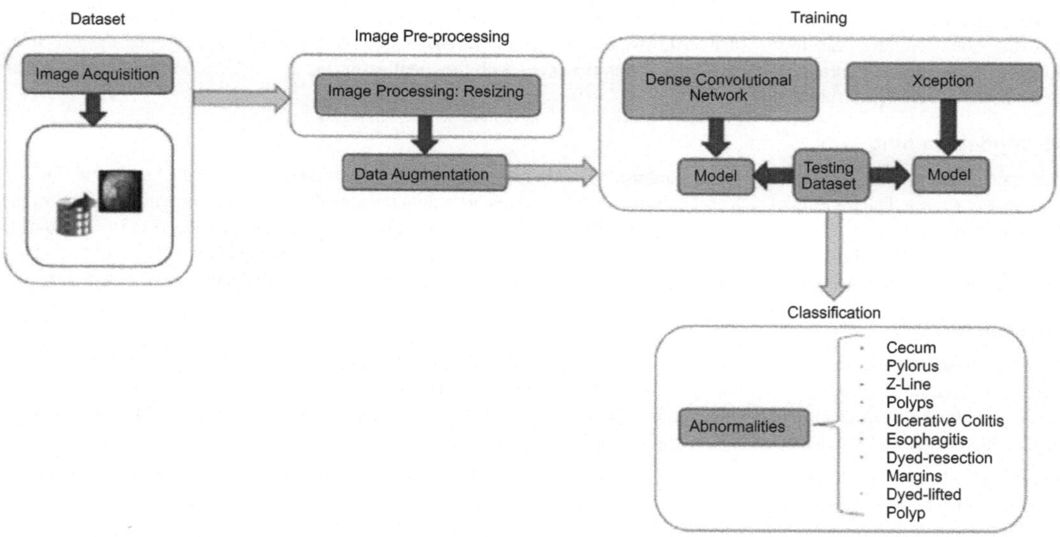

Fig. 12.1 Block diagram

This project's objective is to produce a model that can classify images of the GI tract captured during endoscopy into one of eight categories, which consist of three anatomical landmarks and five pathological findings. We constructed two models using different architectures, DenseNet and Xception, and evaluated their performance to determine which one yields better results. The models take the endoscopy images as input and apply their respective architectures to classify them.

4. Module Description

4.1 Dataset Acquisition

The detection of diseases using computerised methods is an important but under-researched field. In this case, we're using the Kvasir dataset, which contains images of the gastrointestinal (GI) tract. The dataset contains approximately 8000 images that have been classified into eight categories, including three anatomic structures, three important clinical findings, and two image classifications relating to endoscopic polyp removal. Healthcare professionals have thoroughly verified and annotated these images. The visuals range in resolution certain pixels, and some include a green image that shows the endoscope's position and configuration within the bowel.

4.2 Image Pre-processing

The goal of pre-processing is to enhance the quality of image data by either removing undesired distortions or amplifying significant image features that play a crucial role in the succeeding stages of processing.

4.3 Image Resize

To ensure consistency in the images being processed by our AI algorithm, it's necessary to establish a standardized size for all images, as the size of images captured by a camera can vary. This can be achieved through image resizing, which involves scaling the images. Resizing has various benefits, such as reducing the pixel in an image, which can improve the training of a neural network. To accommodate images of different sizes in our dataset, we resize them to 256*256 for DenseNet and 299*299 for Xception. The resized images are then divided into training and validation datasets in an 8:2 ratio, with data augmentation only applied to the training dataset, which comprises around 6400 images.

4.4 Image Augmentation

Image augmentation entails artificially generating training images using a variety of techniques or a mixture of them, such as spontaneous spinning, transitions, shear, and twists, among others. These methods are used to change the original image in various ways, making certain that our model is not presented with the identical image twice. This prevents overfitting and improves the model's generalizability.

4.5 Model Training

After data augmentation, the resulting images are used to train two separate models utilizing the DenseNet and Xception deep learning algorithms. The training of a Deep Learning model involves providing training data to a Deep Learning algorithm (i.e., the learning algorithm), with the correct answer, also referred to as the target or target attribute, included in the test dataset. The classification algorithm seeks out training data in order to identify trends that correspond to character traits of the training dataset to the target (the answer to be estimated) and produces a model using Deep Learning that can seize these structures.

4.6 DenseNet

According to recent research, convolutional neural network (cnn) with shorter connections between layers near the input and output can be trained to be deeper, more precise, and more efficient. The Dense Convolutional Network (DenseNet) is a pelleting network that links all layers to each other. DenseNets utilise the algorithm's possibility via function reprocess rather than relying on extensive underground or expansive frameworks for representational power.

DenseNets have several obvious advantages, including addressing the resulting in quicker, trying to improve functionality progression, trying to encourage feature reuse, and substantially decreasing the total amount of parameters.

4.7 Xception

We introduced our model with Modified Depthwise Separable Convolution, which involves performing two separate convolutions, namely Pointwise Convolution and Depthwise Convolution. The model comprises three sections or flows, namely the entry section, the middle section (repeated eight times), and the exit section. Additionally, it has residual or skip connections, which enable gradients to pass directly through the network without traversing Non-Linear Activation functions. This simplifies the architecture definition and modification process. The input entry flow of our model is 299 x 299 x 3, and it extracts a total of 2048 features. We incorporate non-linearity using Pooling and Activation layers, with the Rectifier Function (ReLU) used to increase the non-linearity of the image.

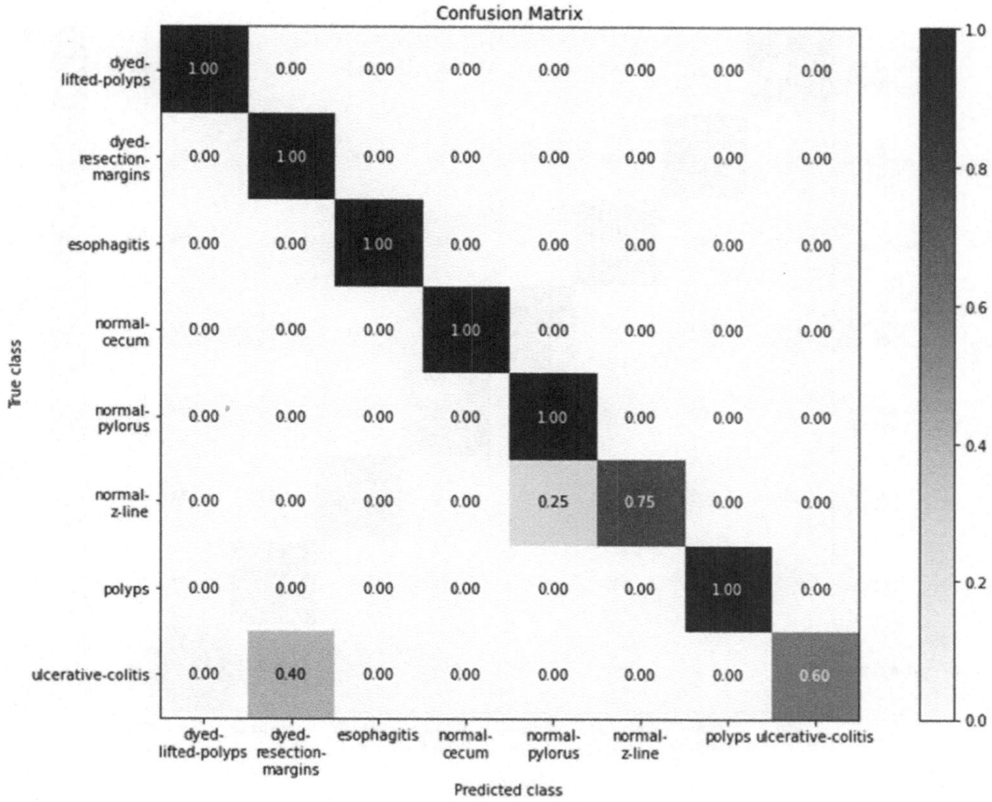

Fig. 12.2 Confusion matrix of DenseNet architecture

5. Module Description

We evaluated the chosen approach by randomly splitting the augmented dataset into a training set of 80% and a test set of 20%. We compared the performance of our proposed CNN networks in terms of accuracy, precision, recall, and F1-score. The results are presented in a table below for comparison.

Top of Form

Bottom of Form

6. Conclusion

Our project involved developing and testing two models, DenseNet and Xception, for classifying GI tract abnormalities. The success of these models was evaluated based on accuracy, precision, recall, and F-1 scores, with DenseNet achieving scores of 92.5%, 95%, 94%, and 94%, respectively, and Xception achieving scores of 92%, 94%, 95%, and 94%. During the training process, we observed that Xception had a faster training speed, completing an epoch in 107s, while DenseNet took 340s. Additionally, Xception required fewer epochs to achieve maximum accuracy compared to DenseNet, which required 140 epochs. Overall, we found that Xception outperformed DenseNet, although DenseNet had slightly higher accuracy. In the future, this approach could be extended to detect diseases in the GI tract using Wireless Capsule Endoscopy (WCE) and for video analysis of GI diseases.

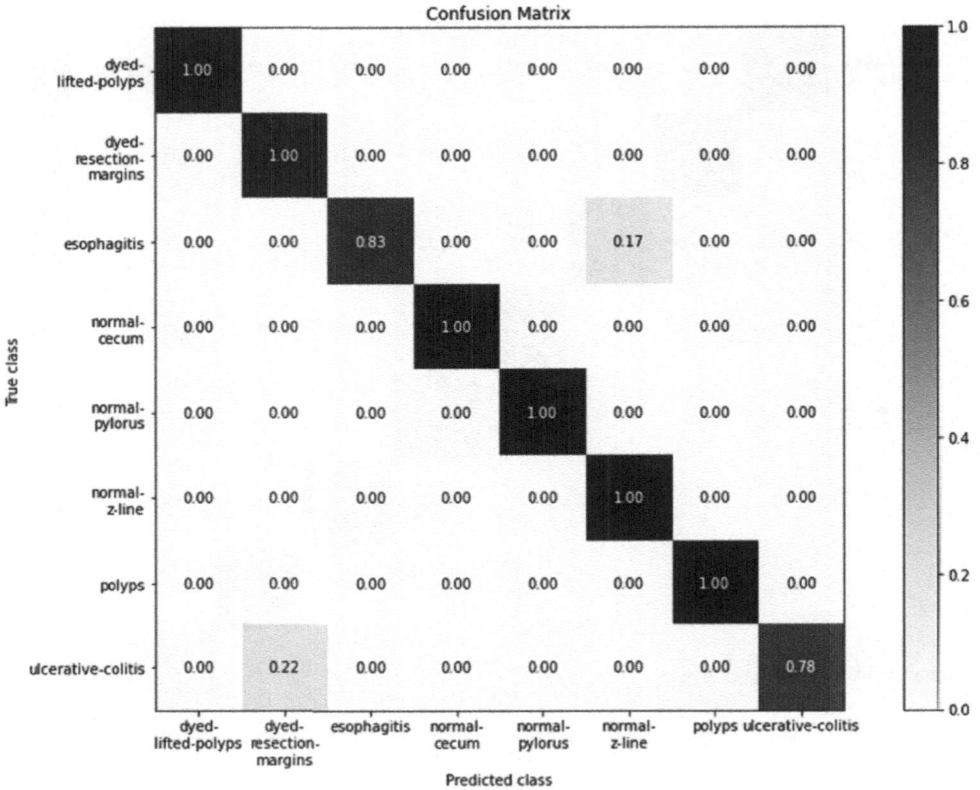

Fig. 12.3 Confusion matrix of Xception architecture

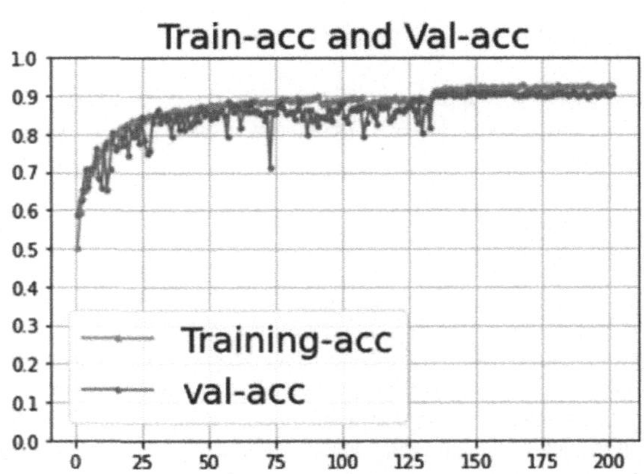

Fig. 12.4 Accuracy curve of our DenseNet model

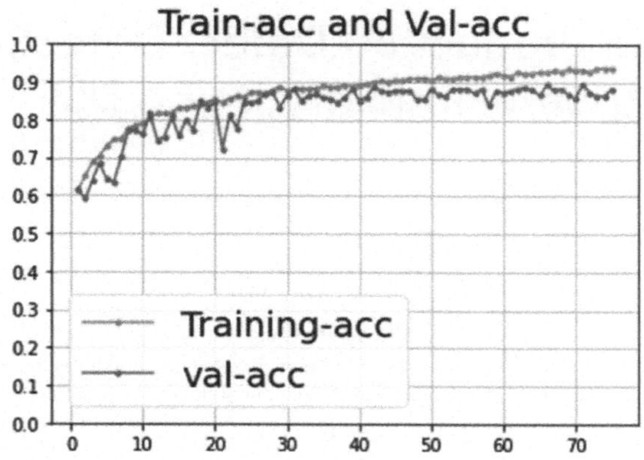

Fig. 12.5 Accuracy curve of our Xception model

REFERENCES

1. N. Ghatwary, X. Ye and M. Zolgharni, "Esophageal Abnormality Detection Using DenseNet Based Faster R-CNN With Gabor Features," IEEE Access, vol. 7, 2019 1(99): 1–1.
2. Konstantin Pogorelov, Kristin Ranheim Randel, Carsten Griwodz, Sigrun Losada Eskeland, Thomas de Lange, Dag Johansen, Concetto Spampinato, Duc-Tien Dang-Nguyen, Mathias Lux, Peter Thelin Schmidt, Michael Riegler, Pål Halvorsen, "Kvasir: A Multi-Class Image Dataset for Computer Aided Gastrointestinal Disease Detection", Multimedia Systems Conference (MMSYS), 2017 DOI: 10.1145/3083187.3083212.
3. S. Lafraxo and M. El Ansari, "GastroNet: Abnormalities Recognition in Gastrointestinal Tract through Endoscopic Imagery using Deep Learning Techniques," *2020 8th International Conference on Wireless Networks and Mobile Communications (WINCOM)*, Reims, France, 2020, pp. 1–5, doi: 10.1109/WINCOM50532.2020.9272456.
4. F. Chollet, "Xception: Deep Learning with Depthwise Separable Convolutions, "2017 IEEE Conference on Computer Vision and Pattern Recognition (CPVR), Honolulu, HI, USA, 2017, pp. 1800–1807, doi: 10.1109/CVPR.2017.195
4. C. Gamage, I. Wijesinghe, C. Chitraranjan, and I. Perera, "Gi-net: anomalies classification in gastrointestinal tract through endoscopic imagery with deep learning," in 2019 Moratuwa Engineering Research Conference (MERCon). IEEE, 2019, pp. 66–71.
5. Asperti, Andrea & Mastronardo, Claudio. (2018). The Effectiveness of Data Augmentation for Detection of Gastrointestinal Diseases from Endoscopical Images. 199–205. 10.5220/0006730901990205.
6. Zhang, Xu & hu, Weiling & Chen, Fei & Liu, Jiquan & Yang, Yuanhang & Wang, Liangjing & Duan, Huilong & Si, Jianmin. (2017). Gastric precancerous diseases classification using CNN with a concise model. PLOS ONE. 12. e0185508. 10.1371/journal.pone.0185508.

Note: All the figures in this chapter were made by the author.

Disruptive Technologies for Sustainable Development – Dr. G. Nagappan et al. (eds)
© 2024 Taylor & Francis Group, London, ISBN 978-1-032-55000-8

A Fabric and Production Defect Detection in the Apparel Industry Using DLT

13

V. Loganathan[2]
Assistant Professor, Department of Computer Science and Engineering,
Saveetha Engineering college, Chennai, Tamil Nadu, India

Suganya K.[2]
M.E Scholar,Department of Computer Science and Engineering,
Saveetha Engineering College, Chennai,Tamil Nadu, India

Abstract: Considering production, investment and employment, India's textile industry is large and old and therefore plays an important role. The increasing demand for quality items that are reliable, durable, and free of defects is one of the criteria used to identify suitable clothing manufacturers. All industries today must find ways to improve the quality of their products in order to survive in the face of increased competition from globalization. The textile industry urgently needs a system that can identify and classify defects in textiles. Using a visual inspection system is time consuming and error prone. If fabric problems are not discovered before garment production begins, they can lead to significant losses. This paper suggests a method employing GLCM and HF to choose the best features for a specific defect type by applying four directions averaging to the various texture features contained in the image increase. Each GLCM can produce 14 more statistics with HF to aid in the identification and categorization of errors. The models that were developed were assessed using a variety of machine learning approaches. According to the study's findings, 85% of the time, participants accurately identified key linked to faults in materials.

Keywords: Flaws detection, Fabric, Convolutional neural network

1. Introduction

It is impossible to overstate how crucial the textile sector is to India's economic independence. Several factors determine whether a clothing manufacturer is appropriate for customers. It has been determined that flaws cause textile costs to drop by 60% to 80%. We will discuss the defect if the buyer does not accept the material's flaw. The key challenges are the high time and financial costs and the subjectivity of people's perceptions.

2. Literature Review

High-contrast tissue defects are diagnosed using this method. Muhammad Adnan Shafry Untoro (2018) Due to the fact that strong signal contrast leads to dispersion, this approach compares grayscale data to a reference threshold. Simple thresholds, two-level thresholds, and multi-level thresholds are frequently used. Simple gray-level thresholding is an easy method to find faults in high contrast images. Antonio J Plaza; Nasser Kehtarnavaz; Demetri Terzopoulo (2021) This kind of image's intensity histogram typically contains a single maximum that designates a crucial place. The area is deemed faulty if the grey value is higher than the specified threshold. If not, it is not. Using this method, it is simple to find tissue flaws like large holes and dark

[1]loganathan@saveetha.ac.in, [2]suganya20006@gmail.com

DOI: 10.1201/9781003428473-13

spots, and tissue inspection systems have successfully used it to do so. The simplicity with which this technique can be put into practice is equally compelling. M. Patil, S. Verma and J. Wakode (2017) Error detection cannot be done with it, though. Many things, including dust particles, camera distortion, and lighting, can result in false alarms. In the study, both the Otsu or auto-threshold approach and the straightforward threshold method were employed. A quick adaptive threshold limit has been found to be useful in the study of low-contrast flaws in galvanised metal strips.

3. Methodology and Model Specifications

The steps taken to create the sectorization model are described in this section. An explanation of how the sectorization model works.

Table 13.1 A Comparison of manual and automated defect inspection

Inspection Type	Manual Inspection Versus Automated Inspection	
	Visual	Automated
Fabric Types	100%	70%
Defect Detection	70%	80%+
Reproducibility	50%	90%+
Objective Defect Judgment	50%	100%
Statistics Ability	0%	95%+
Inspection Speed	30 m/min	120 m/min
Response Type	50%	80%
Information Content	50%	90%+
Information Exchange	20%	90%+

3.1 Property Insulation

A discrete fourier transform (DFT) image is made up of complex coefficient values. Data on both magnitude and phase are shown in Fig. 13.1. The actions to make a feature vector database are as follows:

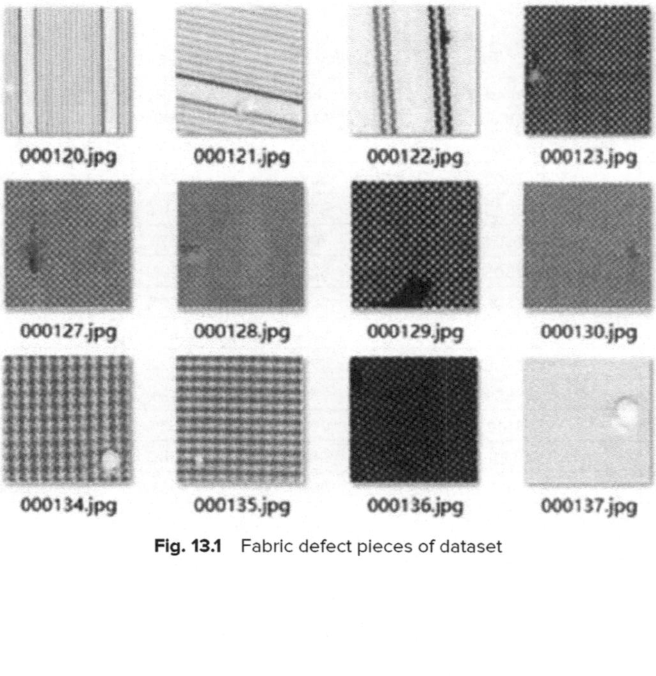

Fig. 13.1 Fabric defect pieces of dataset

3.2 Review and Considerations

The analysis of our test results is provided below. From the database's fabric database, 500 defect classification photos were used in this investigation. Each form of fault is represented here 200 times. Several error image classification techniques, such as the CNN algorithm, decision tables, support vector machines, full training set, and cross-validation to validate an algorithmic model, there are seven main sorts of errors.

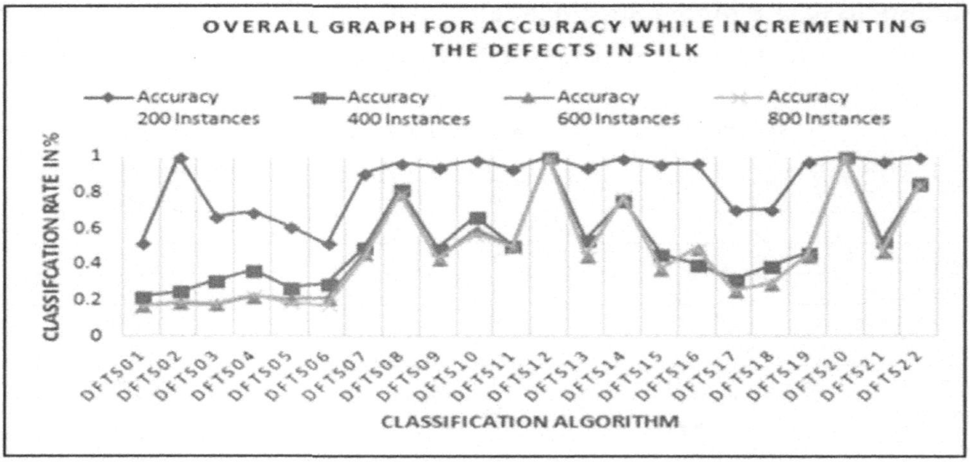

Fig. 13.2 Parent diagram showing spectral sectoring

Different classification strategies were assessed using impact, oil stains, condensation on threads, foreign matter, wrinkles, lighting, camera distortion, classification accuracy, and confusion matrix. We used four separate fabric data sets with 200 photographs in each category to adjust for the rise in poor images. 200 more photographs were added to the training dataset, which was then trained using the whole training dataset and assessed using several classification methods. The results were used to determine whether the material was good or bad.

Table 13.2 Accuracy errors in database

Classification Algorithm	Models Generated (Notations Algorithms-Graph)	Accuracy 200 Instances	Accuracy 400 Instances	Accuracy 600 Instances	Accuracy 800 Instances
SMO-RBF-10FCV	DFTS01	52%	23%	18%	17%
SMO-RBF-FTS	DFTS02	100%	25%	19%	19%
SMO-POLY-10FCV	DFTS03	67%	31%	18%	18%
SMO-POLY-FTS	DFTS04	70%	37%	23%	23%
Naïve Bayes-10FCV	DFTS05	62%	28%	21%	19%
Naïve Bayes-FTS	DFTS06	52%	30%	22%	18%
Bagging-10FCV	DFTS07	92%	50%	47%	45%
Bagging-FTS	DFTS08	97%	82%	80%	80%
Boosting-10FCV	DFTS09	95%	49%	44%	44%
Boosting-FTS	DFTS10	99%	67%	60%	57%
Random Forest-10FCV	DFTS11	94%	51%	51%	50%
Random Forest-FTS	DFTS12	100%	100%	100%	100%

Classification Algorithm	Models Generated (Notations Algorithms-Graph)	Accuracy 200 Instances	Accuracy 400 Instances	Accuracy 600 Instances	Accuracy 800 Instances
J48-10FCV	DFTS13	94%	54%	45%	45%
J48-FTS	DFTS14	99%	76%	78%	78%
Decision Table-10FCV	DFTS15	96%	46%	38%	38%
Decision Table-FTS	DFTS16	97%	41%	49%	49%
Logistic Regression-10FC	DFTS17	71%	32%	26%	26%
Logistic regression-FTS	DFTS18	71%	39%	30%	30%
LibSVM-10FCV	DFTS19	98%	47%	45%	45%
LiBSVM-FTS	DFTS20	100%	100%	100%	100%
Grid Search-10FCV	DFTS21	98%	53%	47%	47%
Grid Search-FTS	DFTS22	100%	86%	85%	85%

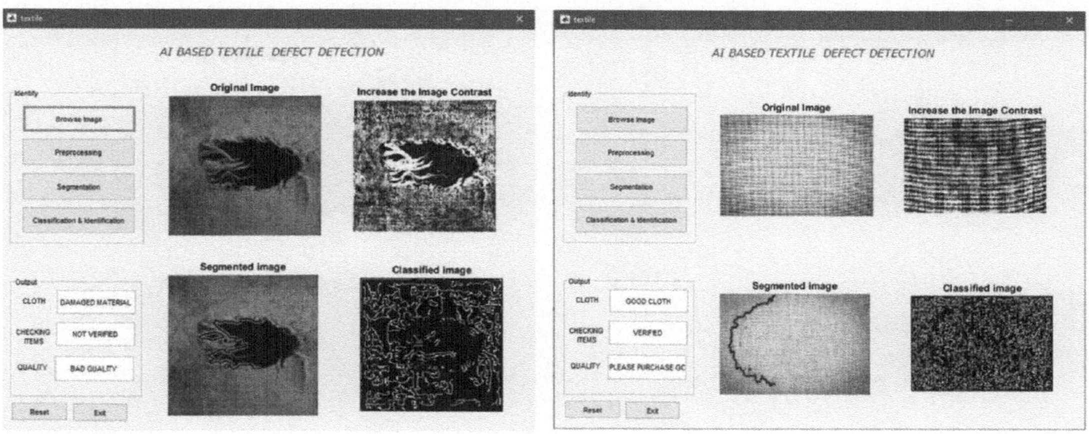

Fig. 13.3 Instances rendered with spectral sectorization

4. Conclusion

According to LibSVM tests on fabrics, classifying defects can be aided by a high cost value (50) and a low gamma weight (0.1), which give the model more leeway in selecting samples. I understand. However, a search of the grid shows that a cost of 10 and a gamma of 0.01 work best for our purposes. Having found that the random forest algorithm works well with the missing parameters in the datasets, we can conclude that it is the best option. Increasing the tree eliminates over fitting of the data and the tree structure gives more accurate results. It also works well with very large datasets.

In order to obtain more pertinent and reliable features for the upcoming research, we want to gather more data. Future research that automatically takes datasets in and detects quality and other features will use far less manual intervention.

A DFT sectorization process was used to extract features of TILDA textile images and classify defects. A Feature Vector Database (FVDB) is created using four different domains of DFT. Compared to the full training set, fabrics have been found to have lower classification accuracy using 10-fold cross-validation. Accuracy and kappa statistics for all characterization calculations are examined. Random forest has proven to be the most effective defect classification algorithm for fabrics due to its high classification rate.

REFERENCES

1. Fei-Fei Li, Justin Johnson and Serena Yueng (2017), "Lecture 9: CNN Architectures" MATLAB expo (2018), Deep Learning with MATLAB –
2. M. Patil, S. Verma and J. Wakode (2017), "A review on fabric defect detection techniques", Int. Res. J. Eng. Technol.
3. T.-Y. Lin, P. Goyal, R. Girshick, K. He, and P. Dollar (2020), "Focal loss for dense object detection," IEEE Trans. Pattern Anal. Mach. Intell., vol. 42, no. 2, pp. 318–327, Feb.
4. Durmuşoğlu and Y. Kahraman (2021), "Fabric Defect Detection Using Convolutional Networks," Elazig, Turkey: Intelligent Systems and Applications Conference (ASYU), pp. 1–5.
5. Alex Krizhevsky, Ilya Sutskever and Geoffrey E.Hinton (2015), "Image Net Classification with Deep Convolutional Neural Net-works"
6. Muhammed Fatih TALU, Kazım HANBAY, Mahdi HATAMİ VARJOVİ (2022), "CNN-based fabric defect detection system on loom fabric inspection".

Note: All the figures and tables in this chapter were made by the author.

Disruptive Technologies for Sustainable Development – Dr. G. Nagappan et al. (eds)
© 2024 Taylor & Francis Group, London, ISBN 978-1-032-55000-8

Fraud Detection in Credit Card Data Using Supervised Machine Learning Based Scheme

14

R. Kaladevi[1]
Associate Professor, Department of Computer Science and Engineering,
Saveetha Engineering College

N. Srimathi[2]
Master of Engineering, Department of Computer Science and Engineering,
Saveetha Engineering College

S. Kanimozhi[3]
Master of Engineering, Department of Computer Science and Engineering,
Saveetha Engineering College

Abstract: Credit cards are issued by banks or financial service firms and enable people who hold credit cards can utilize the option to obtain funds in order to cover the cost of purchases and expenses at credit card-accepting merchants. In today's world where everything is digitalized, the possibility of credit card misuse and financial loss to the account holder is a concerning issue. Therefore, it is crucial for credit card companies to identify and protect your customers from fraudulent charges by preventing fraudulent credit card transactions. The initial phase of the process consists of examining and preparing the data, followed by the utilization of diverse machine learning methodologies in practical scenarios for credit card records. The parameters of each algorithm will then be determined, and their performance metrics will be computed. There are several algorithms available for detecting fraudulent transactions using credit cards, each with its own set of benefits and accuracy. The algorithms are: Linear regression, Naive Bayes, Logistic Regression, Random Forest algorithm etc.

Keywords: Fraud detection, Accuracy, Random forest algorithm, Precision, Metrics

1. Introduction

Identification and prevention of credit card fraud through various detection methods are critical for ensuring the security of financial transactions and protecting the interests of both consumers and financial institutions. The financial industry continues to face the challenge of persistent credit card fraud, which leads to annual losses of billions of dollar. Financial institutions consider credit card fraud to be a significant concern and cardholders all over the world. Fraudulent transactions can result in significant losses for banks and inconvenience for customers. Therefore, it is important to develop effective fraud detection systems to identify and prevent fraudulent activities.

With the increasing availability of credit card data and the growing complexity of fraudulent techniques, traditional rule-based fraud detection methods are no longer sufficient to identify all fraudulent transactions accurately. Recently, there has been an increased focus on creating fraud detection techniques using machine learning that can analyze past data to recognize new fraudulent patterns. Our study introduces a supervised machine learning approach for detecting credit card fraud that aims

[1]kaladevi@saveetha.ac.in, [2]srimathieswari@gmail.com, [3]sivalingamkani@gmail.com

DOI: 10.1201/9781003428473-14

to enhance the precision and effectiveness of fraud detection while reducing the number of false positives. We evaluate the effectiveness of our suggested approach by comparing its results to that of other state-of-the-art methods using a publicly available dataset. Our findings show that our proposed method is effective at detecting credit card fraud and has real-world applications in the financial industry.

2. Literature Review

A Systematic Literature Review of Harish (2017) examine various ML algorithms used to detect such transactions. CCF is an issue in the financial sectors that continues to grow as more companies move towards online transactions. To prevent such activities, a fraud detection system is necessary. Companies offer customers the convenience of purchasing goods online, but this also exposes them to the risk of fraud but we mainly focus on the challenges with the credit card frauds related to the dataset which included samples and large number of transactions and a class imbalance. The research paper concentrates on the problem of increasing credit card fraud and the need for a fraud detection system due to the rise in online transactions. In this paper they are not mainly focus on the Trojan activities. So need to focus on the fraud detection techniques to detect the activities on phishing and trojan.

G P Rao, Rajkumar and Devika, et al (2019) construct this research, paper examine various techniques for detecting and preventing fraudulent activities related to credit card transactions. Different supervised machine learning algorithms which includes decision tree algorithm, logistic algorithm, Naive Bayes Classification are used in the creating the real time datasets and it used for detecting fraudulent credit card transactions. Even Though some of the ML algorithms which are implemented give good outcomes but some of them have imbalanced data. The insights provided in this research paper could be instrumental in improving the security of card transactions in the future, thereby providing a safer environment for credit card users. The future work mainly focuses on rectifying the imbalance data which is covered from the previous approaches and it should be handled.

In another study, Akanksha and Hitendra (2019) explore how online transactions continue to grow. It is crucial to consider security issues and the nature of credit card transactions. ML -based classifiers used to identify abnormal transactions, depending on the dataset and classifier used. Considering the security concerns and the occurrence of anomalies in credit card transactions, our proposed study presents a compilation of different methodologies employed to detect unusual transactions in credit card datasets. The outlook for the detection of credit card fraud involves exploring strategies in different associations and banks to ensure secure transactions. The proposed work in this paper is based on the ML based classifiers and depends on the datasets and the classifier used to find the efficiency of the problem and analyse the credit card frauds happening in the world and it will be helpful to the banker to analyse the fraud where it was happened. The data in this paper are imbalance and we mainly focused to get the balanced data to get the best results.

S. Abinayaa and H. Sangeetha et al (2020). They mainly aim to detect credit card fraud in real-world scenarios using the Random Forest algorithm for classification. The study involved the collection of credit card datasets and the execution of queries on user credit cards to assess the dataset's performance. The process involves collecting credit card datasets, running queries on user credit cards, and optimizing accuracy using an evaluated dataset. Attribute processing is used to identify factors that affect fraud detection through graphical representations. The effectiveness of the techniques is measured by their accuracy, flexibility, specificity, and precision. The primary objective of this study was to develop a system capable of detecting credit card fraud in real-world situations. To optimize the accuracy of the results, the researchers utilized the Random Forest algorithm for classification and an already assessed dataset. Due to the presence of imbalanced data in identifying fraud, the optimal level of accuracy could not be achieved and we mainly focus on the balanced data.

3. Dataset Parameters

The effective utilization of the dataset parameters is a crucial requirement for executing the classification method. The training and testing of a model can be affected by the size of the parameters. For our proposed characterization model, we utilized an enormous volume of default Visa extortion grouping information. The dataset contains a total of 284,500 credit card transactions, and in that 492 are labelled as fraudulent (0.17%). Each transaction is described by 30 numerical features, which were extracted from the original data using PCA transformation to preserve the confidentiality of the sensitive information. These features include time, amount, and V1-V28 (anonymous variables obtained after PCA transformation). There are two parts to the dataset: a test set (30%) and training set (70%) The test set is employed to assess the effectiveness of the supervised machine learning model after it has undergone training with the training set. The 30 features that were mentioned earlier were used to train this model in the training set. We optimized the hyper parameters of the model using a grid search method and evaluated its performance on the dataset.

4. Algorithms and Techniques

Supervised ML algorithms are a type of artificial intelligence (AI) that uses labelled training data to train a model to decide the predictions or classifications on new, unlabelled data.

4.1 Random Forest Algorithm

The Random Forest algorithm is a supervised machine learning approach utilized in the field of Machine Learning, which relies on decision trees. This method involves constructing multiple decision trees during the training phase, and the classification or regression outcome is based on the mode of the classes or mean prediction, respectively, of the individual trees. Additionally, it is an ensemble method. The algorithm is designed to mitigate the overfitting issue that may arise in a single decision tree model. To accomplish this, the algorithm creates multiple decision trees using random subsets of the initial dataset and random subsets of features. This results in a diverse set of decision trees, which are then combined to make predictions. This algorithm is commonly used in classification and regression tasks, where it has to be proven to be highly effective and robust.

The random forest algorithm is achieved in the following steps:

- From the dataset Select N records at random.
- Create a decision tree using these N records.

Repeat steps 1 and 2 for as many trees as you want in your algorithm.

In a regression problem, each dataset tree is used to predict a value for Y. The final value can be calculated by averaging all of the predicted values from each forest tree.

5. Proposed System

A proposed model aims to develop a fraud classification system that can distinguish credit card transactions that are fraudulent and those that are not. To accomplish this, a dataset of previous credit card cases is utilized to enable the machine to learn the problem. This paper uses reviews of products gathered from credit card transaction records as its source of data. This step focuses more on selecting a subset of all the data that can be used. Machine Learning problems begin with data, preferably a large amount of data for which you already have an answer.

Initially, the obtained data is formatted, cleaned and sampled using data pre-processing to get the correct dataset. Data pre-processing is done under three categories such as cleaning, formatting and sampling. The dataset could have null records which need to be removed. The selected data which is not in the suitable format needs to be formatted in a file format or likely in the relational database or vice versa. As the number of credit card fraud instances within the dataset is smaller than the overall number of transactions, the class distribution in credit card transactions is unbalanced. As a result, the sampling method is used to solve this problem. Once the data has been pre-processed, a classification model is constructed.

The dataset is divided into two segments - the first portion is used to train the model, while the second segment is reserved for testing purposes. The application of ML algorithms involves using the training data to train the model and to recognize patterns that can be applied to classify new data, such as test data, as fraudulent transactions or non-fraudulent transactions. To predict a value, the Random Forest algorithm also employs a linear equation with independent predictors. The predicted value ranges from negative infinity to positive infinity. The algorithm's output must be classified as variable data. By comparing the best accuracy, a Random Forest regression model predicts the result with greater accuracy. Data visualization is a method of representing data in a graphical and pictorial manner; data scientists depict a story using the results of data analysis and visualization.

6. Architecture Diagram

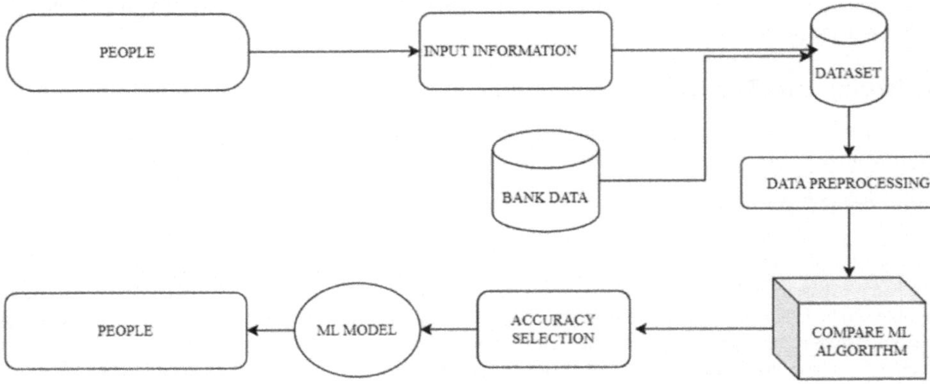

Fig. 14.1 Architecture diagram for credit card fraud detection

7. Results

This paper describes the most common fraud detection methods and algorithms findings in the credit card and describes how the various algorithms implemented in this credit card Detection mechanism to find the frauds happening around the world using credit cards. By Implementing the different ML Algorithms which includes Random Forest, Decision Trees classifier, Naive Bayes and Logistic Regression to find the best results .After Implementing all the algorithms, Random Forest Algorithm provide the best accuracy result based on the accuracy, performance and the specified metrics .The predicted value ranges from negative infinity to positive infinity. The prediction and accuracy serve as the foundation for the outcome. The accuracy, is correctly predicted based on the total number of observations.

8. Conclusion

Data processing and cleaning were the first steps in the analytical process. Subsequently, the analysis of missing values, exploration of data, and development and assessment of the model followed. A higher accuracy score is the best accuracy on a public test set. Based on the high accuracy the credit card fraud will be found out. Although we did not achieve the goal of 100 percent precision in detecting fraud, we did end up developing a plan with enough resources and time to come close. The design of this project makes it possible to include multiple algorithms as modules when combined, these approaches can enhance the accuracy of the ultimate outcome.

References

1. Paruchuri, H., (2017), Credit Card Fraud Detection using Machine Learning: A Systematic Literature Review. *ABC Journal of Advanced Research*, *6*(2), 113–120.
2. G P Rao, Rajkumar and Devika,et al, (2019). A Research on Credit Card Fraudulent Detection System," *Int J Recent Technol Eng (IJRTE)*, *8*, 5029–5032,
3. Akanksha and Hitendra, (2019). Efficient Techniques for Fraudulent detection in Credit Card Dataset: A Comprehensive study, 6(2), pp. 167–176
4. S. Abinayaa and H. Sangeetha et al, (2020) Credit Card Fraud Detection and Prevention using Machine Learning"IEEE Access, 9(4), 202–212.
5. D. Liu and J. Lee, (2019), A Review On Credit Card Fraud Detection Using Machine Learning, IEEE Access, 8(1), 265–274
6. L. Zheng, G. Liu, C. Yan, and C. Jiang, (2018)Transaction Fraud Detection Based on Total Order Relation and Behavior Diversity," IEEE Trans. Comput. Soc. Syst., 5(3), 796–806.

7. Sahin Yusuf and Ekrem Duman, et al, (2011), Detecting Credit Card Fraud by Decision Trees and Support Vector Machines, International Multi-Conference Of Engineers and Computer Scientists (IMECS 2011), 16–18,Hong Kong, 1, 1–6
8. Shiyang X and et al, (2018), Random Forest for credit card fraud detection, Int. conf. on Networking, Sensing and control, 2018, 1, 1–6.
9. Alejandro Correa Bahnsen and Djamila Aouada, et al, (2015). Example dependent cost-sensitive decision trees, Expert Syst. Appl., vol. 42, no. 19, pp. 6609–6619, 2015
10. Anuruddha T, Sasitha P and Shalitha M,(2019), Realtime Credit Card Fraud Detection Using Machine Learning, Int. Conf. on Cloud Computing, Data Science & Engineering.
11. Shuo Wang, Leandro L. Minku, and Xin. Yao, May (2015). Resampling-Based on ensemble methods for online class imbalance learning.
12. N. Mahmoudi and E. Duman, (2015) Detecting credit card fraud by modified fisher discriminant analysis, Expert Syst. Appl., 42(5), 2510–2516.

Note: The figure in this chapter was made by the author.

Disruptive Technologies for Sustainable Development – Dr. G. Nagappan et al. (eds)
© 2024 Taylor & Francis Group, London, ISBN 978-1-032-55000-8

Smart Door Design Based on Face Detection and Recognition Using CNN and Viola Jones Algorithm

15

Sheeba Joice C[1]
Professor and Deputy Head, Department of Electronics and Communication Engineering,
Saveetha Engineering College

Hitha Shanthini[2]
Research Scholar

Majety Dedeepya[3], **Murarishetty Masthan Roshini**[4], **N. S. Laya**[5]
Student, Department of Electronics and Communication Engineering,
Saveetha Engineering College

Abstract: It is important to safeguard our homes from unwanted intruders. Although technological advancement has resulted in many smart security systems, there is still a requirement for cost-effective and user-friendly systems. This paper proposes a face recognition and tracking based smart security system design using the Atmega328 Microcontroller and MATLAB software and alerts the user with a GSM module when an unrecognized person is found at door. The collection of data and training of images is very important for maintaining integrity. Viola Jones and Convolutional Neural Networks (CNN) algorithms that are fast and accurate are used for this purpose. The proposed design is implemented using a servo motor as a prototype, and its accuracy is tested. The result achieves low-cost implementation with efficient face recognition.

Keywords: CNN algorithm, Face recognition, GSM module, Microcontroller, MATLAB, Viola Jones algorithm

1. Introduction

Nowadays there is a great demand for security systems for homes because of the widespread proliferation of thefts throughout the nation. The National Crime Records Bureau (NCRB) reports that 2,44,119 occurrences of robbery occurred on residential property. So, it's time to reevaluate how safe our homes are. 70% of all thefts in India are home thefts, still 61% of people don't want to upgrade to high-tech safety for their homes because they can't afford it. In this paper, a home security system is described that detects and recognize unauthorized intruder and alerts the owner. At the same time, it meets the requirements of the proletariat.

Monitoring of authorized and unauthorized intruders is performed by face detection and face recognition techniques. Recognizing faces is an easy task for humans. The task becomes tedious when we try to detect it via machines. It requires complex steps to be undertaken. In order to detect and recognize the face, numerous varieties of algorithms have been proposed. The truly difficult problem is to develop a faster, more precise method for detecting and recognizing faces.

The proposed system uses the widely utilized Viola Jones Algorithm for face detection, which is known for its high face detection accuracy and quick processing time by Al-Tuwaijari and Shaker (2020). Convolution Neural Network (CNN), the best feature extractor with high computational efficiency and high accuracy, is utilized for face recognition by Hasan and Ahsan

[1]sheebajoice@saveetha.ac.in, [2]hithajas@gmail.com, [3]majetydedeepya0@gmail.com, [4]masthanroshini019@gmail.com, [5]layasuresh828@gmail.com

DOI: 10.1201/9781003428473-15

(2021). The MATLAB tool which is widely in image processing and embedded systems for its wide range of functions is used for both processes. GSM module helps in notify and alerts the owners whenever an unauthorized person tries to access the home security system. Servo motor is used as a prototype to visualize the door opening and locking system that takes place in real life. This paper is organized as follows. Similar works are described in section II. Hardware design is described in section III. The methodology is described in section IV. Results are discussed in section V and concluded in section VI.

2. Literature Review

2.1 Face Detection Using Model and Template

Clippingdale et al. (2012) at his work performs face recognition using person independent deformable template system in which, Every element in the same row must be less than the leading diagonal elements. Recognition error occurs when fails to meet his condition. Ribari et al. (2016) performed facet detection using the Deformable Part Model. Though it detects faces efficiently in at real-time environment.Still, it faces at lot of difficulties such as slowness or speed bottleneck and has problems in extending to newt objects or face categories.

2.2 Face Detection by Edge Detection

Devarajan et al. (2010) performed an edge-based face detection technique that uses the least amount of scanning possible. Though the system is cost-effective and robust in the case of noisy images it shows errors and does not examine edges on all scales.

2.3 Face Detection by Eye and Face Movement Detection

Kattner et al. (1995) performed at Gaze algorithm for facet detection using eye movement, where a multi-orientation Gaussian filter is used to create a crude depiction. Variations in the magnitude of the facet size, different orientations tilted faces, and modes illuminance fluctuations have not effect on accuracy. Yusuf et al. (2019) performed at line clustering algorithm in which at the beginning itself facet motion is detected. Detection will be unsuccessful if the primary axis is not perpendicular to the eye center connecting the line.

3. Hardware Design

3.1 Block Diagram

Block representation of the suggested design is shown in Fig. 15.1. The figure presents the hardware components and the connection between them and Fig. 15.2 shows the experimental setup.

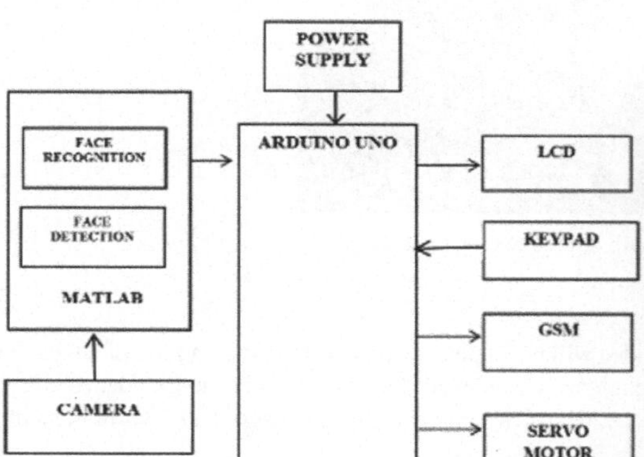

Fig. 15.1 Block diagram of the hardware

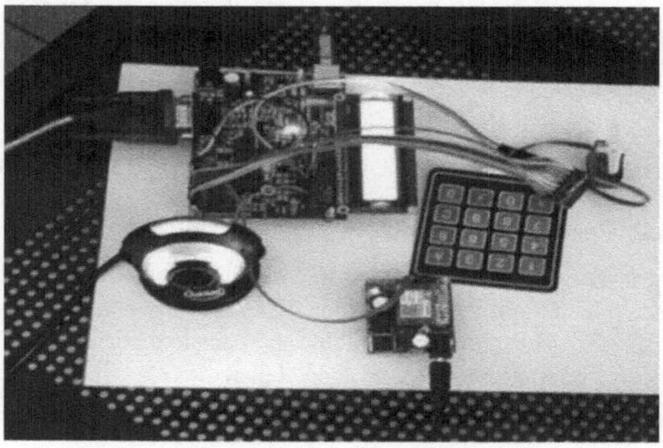

Fig. 15.2 Hardware setup

- **ARDUINO UNO:** It is an ATmega328 microcontroller board. The STK500 protocol is used for communicating.
- **Liquid Crystal Display:** When the face recognition is completed successfully, it displays a "verified successfully" message else it displays "authentication failed".
- **GSM:** It will be connected to ATmega328 microcontroller board through pin Rx/Tx. When an unauthorized intruder is recognized, alert messages will be sent to mobile devices using GSM Module.
- **Keypad:** A 4x4 matrix keypad is used. The ATmega328 microcontroller scans the keypad lines to determine whether the button is pressed. The design includes this module to unlock the door in the event of any atypical technological breakdown.
- **Servo Motor:** In this proposed design servo motor is used as a prototype instead of a door to demonstrate unlocking the door with face recognition is possible.

4. Methodology

4.1 Process of Tracking a Face

The main concept behind the proposed design is face detection. Viola Jones Algorithm is utilized for this purpose (Al-Tuwaijari and Shaker, 2020). Its strong nature and its detection speed of full upright front faces make it highly preferable in real-time situations. In a detection window, at a specific location, Adjacent rectangular sections are taken into account, As shown in Fig. 15.3. The average pixel intensity of a rectangle around the eyes is often lower than that of the adjacent rectangle. Two dark rectangles with a bright rectangle in the centre might stand in for two eyes and a nose because the bridge of the nose is typically lighter than the area around the eyes. A face is clearly present in a frame if there are numerous Haar-like characteristics present in the appropriate places. Integral images are quickly generated to determine the intensity difference between two blocks of pixels.

$$\text{Sum} = \sum(\text{pixels in a black area}) - \sum(\text{pixels in a white area}) \tag{1}$$

Extracted features are passed into a binary classifier, which outputs either positive or negative results depending on whether a face is present in the frame. Afterwards the window moves to the subsequent position, and the detecting procedure starts over.

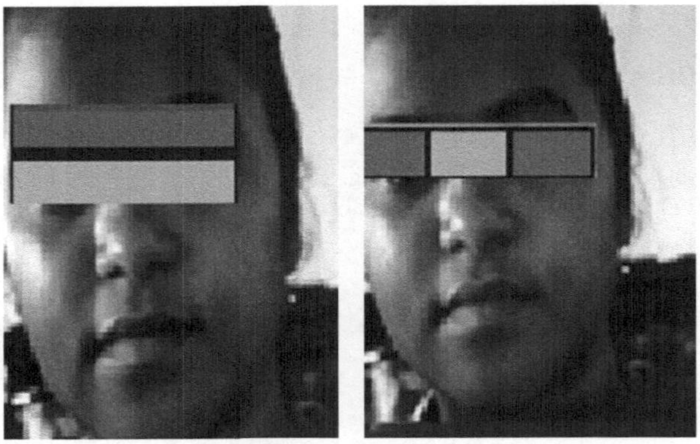

Fig. 15.3 The rectangular feature represents two eyes and a nose

4.2 Face Recognition

Face recognition plays a major role in this design. The proposed idea will be a failure without proper recognition of a face. For face recognition, the Convolutional Neural Networks (CNN) technique is employed. Deep learning is the foundation of this algorithm. This provides a scalable approach to recognition tasks and identifies patterns within an image. This algorithm mainly focuses on image, speech, or audio signals.

5. Simulation and Results

Fig. 15.4 Input image

Fig. 15.5 Detected face

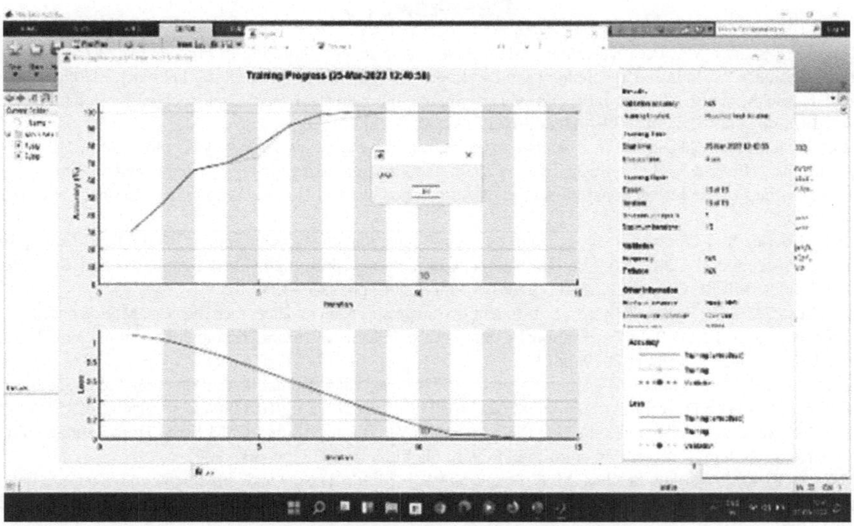

Fig. 15.6 Software output

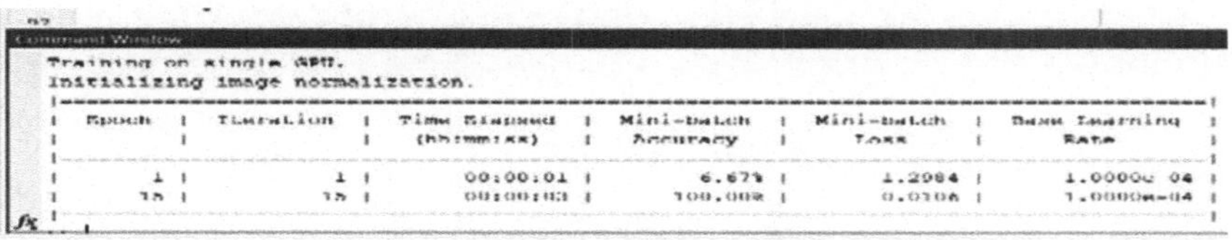

Command Window

Training on single GPU.
Initializing image normalization.

Epoch	Iteration	Time Elapsed (hh:mm:ss)	Mini-batch Accuracy	Mini-batch Loss	Base Learning Rate
1	1	00:00:01	6.67%	1.2984	1.0000e-04
15	15	00:00:03	100.00%	0.0106	1.0000e-04

Fig. 15.7 Terminal output

Face detection and Face recognition processing is done with the help of the MATLAB tool.

Input

Figure 15.4, Shows the input image. The web camera captures the image of the person who stands in front of it.

Software output

Once the image is captured, face detection is completed by Viola Jones Algorithm. Figure 15.5, shows the detected face image. Face recognition is done by identifying the right match from the given dataset. The dataset contains five images of different people. Using the CNN algorithm, it is trained to identify the accurate image. Figure 15.6, Shows the training and achieved accuracy. Figure 15.7, Shows that 100 percent achieved accuracy in the face recognition process. When the image is matched accurately, the servo motor rotated to 180 degrees is observed. When the image is not matched, the servo motor remained at 0 degrees is observed and an alert message is sent to the user using GSM Module.

6. Conclusion

In this paper, Unlocking the door with Face recognition is explained with the Hardware design. The design used a servo motor as a prototype, instead of the door. Simulation and results show successful face recognition. The user, wherever he might be, whether inside the home or outside the home, will be alerted with the message whenever an unauthorized person is found at the door. In the future, the proposed design can be implemented in the door.

References

1. Al-Tuwaijari, J.M.; Shaker, S.A. Face Detection System Based Viola-Jones Algorithm. In Proceedings of the 2020 6th International Engineering Conference "Sustainable Technology and Development" (IEC), Erbil, Iraq, 26–27 February 2020; pp. 211–215.
2. S. Clippingdale and M. Fujii, "Person-Independent Deformable Templates for Fast Face Recognition," *2012 IEEE International Symposium on Multimedia*, 2012, pp. 439–444, doi: 10.1109/ISM.2012.89.
3. Anila, S.; Devarajan, N. Simple and fast face detection system based on edges. Int. J. Univers. Comput. Sci. 2010, 1, 54–58
4. Herpers, R.; Kattner, H.; Rodax, H.; Sommer, G. GAZE: An attentive processing strategy to detect and analyze the prominent facial regions. In Proceedings of the International Workshop on Automatic Face and Gesture Recognition, Zurich, Switzerland, 26–28 June 1995; pp. 214–220.
5. Marčetić, D.; Ribarić, S. Deformable part-based robust face detection under occlusion by using face decomposition into face components. In Proceedings of the 2016 39th International Convention on Information and Communication Technology, Electronics and Microelectronics (MIPRO), Opatija, Croatia, 30 May–3 June 2016; pp. 1365–1370.
6. Hasan, M.M.; Yusuf, M.S.U.; Rohan, T.I.; Roy, S. Efficient two stage approach to detect face liveness: Motion based and Deep learning based. In Proceedings of the 2019 4th International Conference on Electrical Information and Communication Technology (EICT), Khulna, Bangladesh, 20–22 December 2019; pp. 1–6.
7. Hasan, M.K.; Ahsan, M.S.; Abdullah-Al-Mamun; Newaz, S.H.S.; Lee, G.M. Human Face Detection Techniques: A Comprehensive Review and Future Research Directions. *Electronics* **2021**, *10*, 2354. https://doi.org/10.3390/electronics10192354
8. R. Teymourzadeh, Salah Addin Ahmed, Kok Wai Chan and Mok Vee Hoong, "Smart GSM based Home Automation System," 2013 IEEE Conference on Systems, Process & Control (ICSPC), Kuala Lumpur, 2013, pp. 306–309.
9. G. M. S. Mahmud Rana, A. A. Mamun Khan, M. N. Hoque and A. F. Mitul, "Design and implementation of a GSM based remote home security and appliance control system," 2013 2nd International Conference on Advances in Electrical Engineering (ICAEE), Dhaka, 2013, pp. 291–295.

Note: All the figures in this chapter were made by the author.

Disruptive Technologies for Sustainable Development – Dr. G. Nagappan et al. (eds)
© 2024 Taylor & Francis Group, London, ISBN 978-1-032-55000-8

Comparison of Network Intrusion Detection System for Software-Defined Networks Using Tree-Based Classifiers

16

Deepika P.[1]
Assistant Professor, Department of Information Technology,
Easwari Engineering College, Chennai ,Tamil Nadu

Pritham Ganesh S.[2], **Sidharth Krishna**[3], **Rajasimman R.**[4]
Department of Information Technology,
Easwari Engineering College, Chennai, Tamil Nadu

Abstract: With the world moving towards automation, networks and their functions are no different. The advent of Network Function Virtualization (NFV) to provide basic networking functions and Software-Defined Networks (SDN) to orchestrate overall operations of the network has enabled more automated provisioning and policy-based management of network resources. Though these technologies increase efficiency and support automation, they still fall prey to online adversaries and attacks such as DDoS, XSS, brute force attacks, SQL injection attacks, etc. In this article, we provide a novel way to detect network intrusions by malicious threat actors that attempt to gain access or compromise the system using tree-based classifiers. The system is capable of detecting multiple types of attacks and also attacks that are similar. Using machine learning algorithms like random forest algorithm and XGBoost, which are based on trees, to make use of multiple features to detect malicious activity inside the network.

Keywords: Artificial intelligence, Intrusion detection, Random forest, XGBoost

1. Introduction

The number of people using the internet has grown exponentially over the past two decades. A total of 5.07 billion people across the world are connected to the internet which is identical to 63.5 percent of the world's total population (as of October 2022). This sheer number of users connected to the internet enables malicious attackers, hackers, and online adversaries to leverage the dependency. They leverage loopholes, vulnerabilities, and weaknesses of internet security to steal sensitive information of users, and disrupt services such as online banking, online shopping, websites, etc. According to Forbes, 17.02 billion records were lost in data breaches from 2004 to 2021, with 9.09 billion records from the web sector alone. In its 2022 mid-year cyber statistics, Forbes has stated that more than 90 percent of external threat actors can break into the perimeter of the organization and can access local resources of the network. This was from the findings of a new study from Positive Technologies. It also states that 82 percent of the CIOs (Chief Information Officer) believe that their own organization's supply chains are vulnerable to cyber attacks. Figure 1 shows the average cyber-attacks faced by organizations in 2021. Such findings indicate that a security mechanism to detect, prevent, and mitigate online adversaries is necessary. A NIDS (Network Intrusion Detection System) is mainly used to detect attacks and enhance the security systems that already exist. As stated by Pfleeger, a 'trusted system' can be developed through layered defenses and an intrusion detection mechanism is one of the layers.

[1]p.deepika@eec.srmrmp.in, [2]prithamganesh@gmail.com, [3]sidharthkrish7@gmail.com, [4]davidtarral78@gmail.com

DOI: 10.1201/9781003428473-16

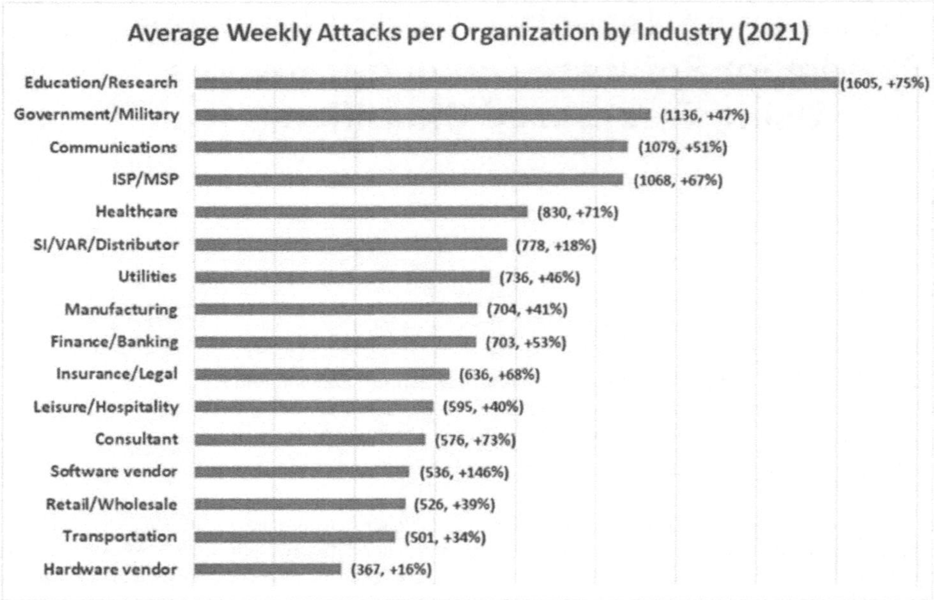

Fig. 16.1 Average weekly attacks

2. SDN (Software Defined Network)

SDN is a technique to physically separate the control plane and forwarding plane, where the former controls many devices. This type of architecture is optimal for high-bandwidth networks and the dynamic nature of modern world applications. Here the network control becomes directly programmable, and agile as administrators can alter traffic flow as per demand, and can be managed centrally using SDN controllers that have a global view and simplifies the network design and operations. SDN, therefore, is going to be the norm in the future of networking.

3. Intrusion Detection System

An IDS is a system that detects suspicious or malicious behaviour by anomalous users. IDS is classified into two categories by their method of detection of suspicious activities- one based on signatures of attacks and the other utilizes anomalies in the system. A signature-based IDS has records of signatures of past attacks. Using this, the IDS tries to match current computer activity to the past record to determine whether an event is malicious or not. An anomaly-based IDS first learns about anomalies and benign activities. Using the benchmarks, it classifies the activities into two categories. IDSs are usually deployed at three levels-at the application level, at the host level and at the network level. Each level has its own functions, advantages and disadvantages. Table 16.1 illustrates the summary of the features of the types of IDSs.

Table 16.1 Types of IDS

Network Based	Host Based	Application Based
Analyze data packet in real-time	Analyze operating system event log to look for aberrant patterns in the system	Analyze the application event log to look for aberrant pattern
Operate at network node	Operate at Host node	

A NIDS is usually the primary defence mechanism of the network. The system is placed behind the firewall. It inspects the incoming packets tagging malicious packets as threats. The system can be configured to trigger an intrusion prevention system (IPS) if available.

4. Machine Learning

ML (Machine Learning) enables the computer system to learn without being programmed. It enables the system to achieve the output when presented with a set of inputs based on the results from its learning. This is broadly split into supervised and unsupervised algorithms. Both are suitable for different scenarios and research work. Machine Learning gives the system the ability to differentiate between benign traffic and malicious or suspicious traffic by training using multi-classifiers. This makes the system capable of detecting modern and unknown attacks that traditional signature-based systems are unable to detect.

4.1 Tree-based Classifiers

These are supervised machine-learning algorithms. They use a series of conditional statements to split the data into subsets. Each subset added increases the complexity of the model, which is then used to make the prediction.

A classification model's performance is usually illustrated using a special kind of matrix known as a confusion matrix. This matrix contains information about the accuracy of the model which it illustrates through the number of correct predictions and the number of wrong predictions. The matrix has 4 components which are True Positives (correctly predicted attacks), True Negatives (correctly predicted normal benign packets), False Positives (wrong predictions of attack), and False Negatives (wrong predictions of benign packets). A classification model's performance is usually illustrated using a special kind of matrix known as a confusion matrix. This matrix contains information about the accuracy of the model which it illustrates through the number of correct predictions and the number of wrong predictions. The matrix has 4 components which are True Positives (correctly predicted attacks), True Negatives (correctly predicted normal benign packets), False Positives (wrong predictions of attack), and False Negatives (wrong predictions of benign packets). False Positives will cause the IDS to raise false alarms thereby inhibiting the performance of the network. The wrong predictions will cause the system to drop actual benign packets because they were predicted as malicious by the system. There are many research works conducted to reduce the number of false positives and improve the accuracy of the model.

The Random Forest algorithm is categorized under the supervised ML algorithm which means the dataset is will have the input and output labelled. This algorithm is used to solve classification and regression problems. It is based on the concept of using the results of many classifiers (here trees) to resolve difficult and complex problems. This also enhances the efficiency of the model. The classifier contains numerous decision trees and uses the average to predict the result. The majority of the votes of prediction are used to obtain the result. The large number of classifiers that the algorithm utilizes leads to improved performance and reduced the risk of an overfitting problem which is common in classifiers. The reduced training time, improved accuracy, prevention of overfitting problems, and ability to accommodate missing values in the dataset are some of the reasons the algorithm is preferred.

Extreme Gradient Boosting algorithm is a supervised ML algorithm aimed at optimizing training for gradient boosting. In gradient boosting, the predecessor's errors are corrected by each predictor. The weights of each predictor are not changed but the training of predictors is done using the residual error of the previous predictors as labels. The XGBoost is an implementation of these boosted decision trees. The trees are created sequentially, and weights play a significant role in this algorithm. All independent variables are assigned their own weights and the weight of variables predicted wrong is increased. These are then fed to the second tree and so on. This ensemble gives a more precise model. XGBoost is predominantly used for ranking, regression and classification.

5. Machine Learning Model

The dataset contains 1188333 rows of intrusion and whitelisted or benign traffic with 79 features. To ensure that all the data is measured on the same scale standardization techniques are implemented. In this article, we make use of the z-score normalization technique. In this method, if a value is equal to the mean of all features it is normalized to 0. The range we have chosen is negative one to positive one. In the next step, we utilized the label encoder. The label encoder is used to convert the class labels into a numeric form so that the machine can read and understand the labels. The categorical value is assigned a value from 0 to the number of values minus 1. For this research, we used Python programming language to build the model. We used the sklearn library to import the random forest classifier, the confusion matrix, the label encoder, and the function to split the dataset and to calculate the recall score, precision score, and accuracy. From the xgboost library, we import the model. We used the seaborn and matplotlib libraries to visualize the output as a heatmap. After initializing the models and splitting the dataset, we fed the dataset into the model and the output is shown below.

BENIGN		798322
DDoS		383439
Web Attack	Brute Force	4550
Web Attack	XSS	1962
Web Attack	Sql Injection	60

Fig. 16.2 Dataset class labels and their counts

6. Results and Discussion

The XGBoost algorithm's precision, accuracy, recall, and F1-score are 0.901, 0.898, 0.891, and 0.878 respectively. The classifiers used 10 n_estimators and training to the testing ratio of 8:2. A heatmap using the confusion matrix was printed to measure how the model has performed and is illustrated below.

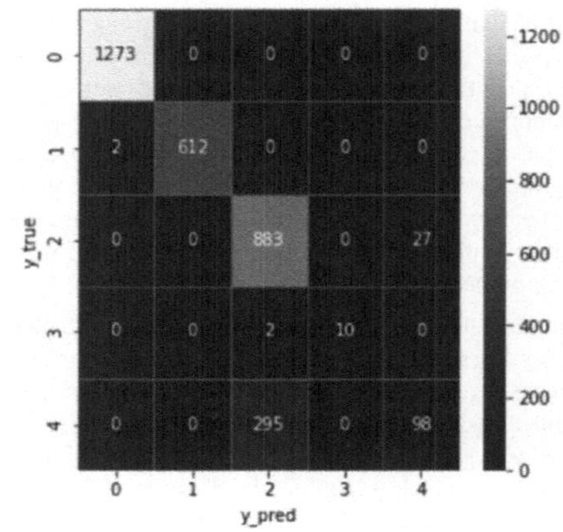

Fig. 16.3 XGBoost confusion matrix

The Random Forest algorithm's precision, accuracy, recall, and F1-score are 0.980, 0.979, 0.979, and 0.9793 respectively. The classifier was built with the random_state value set to 0 and the rest of the attributes as default values. The test-train split was 8:2. A heatmap using the confusion matrix was printed to measure the performance of the model and is shown below.

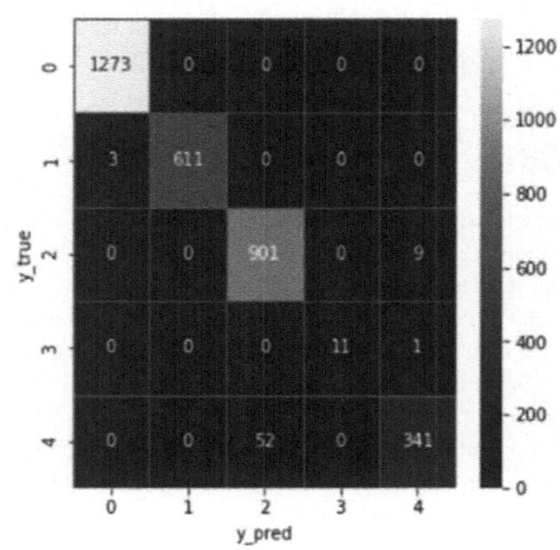

Fig. 16.4 Random forest confusion matrix

7. Conclusion

In the above two models, the Random Forest classifier performed better than the XGBoost classifier. This paper discussed the efficacy of 2 machine learning supervised algorithms in predicting malicious traffic. Future research may seek to broaden he variety of attacks in the training set, increase the volume of the data used, resolve class imbalances, reduce power consumption, improve the accuracy of the model, or focus on predicting zero-day and unknown attacks.

References

1. Yang, L., Li, J., Yin, L., Sun, Z., Zhao, Y. and Li, Z., 2020. Real-time intrusion detection in wireless network: A deep learning-based intelligent mechanism. IEEE Access, 8, 170128–170139.
2. Lee, S. J., Yoo, P. D., Asyhari, A. T., Jhi, Y., Chermak, L., Yeun, C. Y. and Taha, K., 2020. IMPACT: Impersonation attack detection via edge computing using deep autoencoder and feature abstraction. IEEE Access, 8, 65520–65529.
3. Elhenawy, M., Masoud, M., Glaser, S. and Rakotonirainy, A., 2020. A new approach to improve the topological stability in non-linear dimensionality reduction. IEEE Access, 8, 33898–33908.
4. Kim, A., Park, M. and Lee, D.H., 2020. AI-IDS: Application of deep learning to real-time Web intrusion detection. IEEE Access, 8, 70245–70261.
5. Al-Abassi, A., Karimipour, H., Dehghantanha, A. and Parizi, R.M., 2020. An ensemble deep learning-based cyber-attack detection in industrial control system. IEEE Access, 8, 83965–83973.
6. Gamage, S. and Samarabandu, J., 2020. Deep learning methods in network intrusion detection: A survey and an objective comparison. Journal of Network and Computer Applications, 169, 102767.
7. Zhou, J., Cao, Z., Qin, Z., Dong, X. and Ren, K., 2019. LPPA: Lightweight privacy-preserving authentication from efficient multi-key secure outsourced computation for location-based services in VANETs. IEEE Transactions on Information Forensics and Security, 15, 420–434

Note: All the figures in this chapter were made by the author.

Disruptive Technologies for Sustainable Development – Dr. G. Nagappan et al. (eds)
© 2024 Taylor & Francis Group, London, ISBN 978-1-032-55000-8

Decentralized Pharmaceutical Drug Supply Chain Management and Recommendation System Based on Blockchain and Machine Learning

17

M. Vijayanand[1]
Professor, Department of Computer Science and Engineering,
Saveetha Engineering College, Chennai, Tamilnadu

Kishore R.[2], Dinesh T.[3]
Department of Computer Science and Engineering,
Saveetha Engineering College, Chennai, Tamilnadu

Abstract: The top preeminence of the current healthcare system is the direct supply of medications amongst the manufacturer supply unit to the consumer. The pharmaceutical medication supply chain mechanism embraces a huge number of stakeholders, inclusive of distributors, product suppliers, retailer, medicine dealers and consumers. Today, forecasting ahead of time is also compelling and essential. Although it has becoming harder to track down the origin of raw materials and keep track of commodities as they move through the supply chain network, this has created chances for fake goods to join the real supply chain. Thus, to elevate and parlay the trackable goods in the pharmaceutical drug business logistics management using blockchain technology and machine learning. Blockchain is a sort of distributed ledger (DLT) that maintains an unchangeable record of all transaction data which is stored in Ethereum network that is transparent to all stakeholders and cannot be manipulated or changed. Implementing a blockchain-based machine learning system will provide the pharmaceutical industry with the tools necessary to improve drug traceability throughout the supply chain, leading to more efficient and secure healthcare in India.

Keywords: Ethereum, dApp, Sentimental analysis, Smart contract, Web3

1. Introduction

At its primitive level, Blockchain Network is a redistributed database of interrelated facts or an unclose record of all thought-of or modernized historical transactions that has been recorded which are shared among various parties. Data can not be expunged once the input has be delivered. Each and every exchange, regardless of when the transaction has been made, is explicitly and clearly chronicled in the blockchain network. In blockchain development and application, the most vital use case is bitcoin. The advantages of blockchain development completely annihilate the centralized administrative conundrum and traditional management hassles. A smart contract is a cluster of computer programs which performs and bring off self -double-checking, oneself -performing and with no action of forgery or changing the existing data in the block network .With incorporation of smart contracts into blockchain network technology made to have a capability for doing a assignment in actual time with low economical cost and provide a highest degree of reliability and safeguard . Pharmaceutical Recommendation systems is connected the supply chain network and blockchain realms, which can greatly profit from one another. It offers a foundation for the creation of decentralized recommender systems that use the blockchain database to safely keep the information about pharmaceutical products in order to prevent the supply of fake goods. In section 2, the literature review is present, followed by the methodology and submodule specifications system in section 3, which also explains the proposed system and its components, section 4 has the implementation, accompanied by the result in section 5 and the conclusion in section 6.

[1]Vijayananad@saveetha.ac.in, [2]kishorewolfeik@gmail.com, [3]droiddinesh.tamil@gmail.com

DOI: 10.1201/9781003428473-17

2. Literature Review

A novel device developed by C.ANTAL (2021), To enable the knowledge of all network peers, Ethereum smart contracts are designed for monitoring and measure the appropriate distribution and delivery of vaccines must adhere to the safe handling protocols set by the drug manufacturers. I. A. Omar (2021) suggests a cutting-edge system, a blockchain network-based catalogue allocation strategy based on smart contracts using a privatized Ethereum network to connect suppliers and merchants. A new system called CLOUDITY and R. Asyrofi (2020) to address a number of issues, including resource equipping, service choice, verification and validation of identity and regulation of data accessibility. One of the major method was suggested by Y. Madhwal (2015). The theory of Transaction Costs is used to assess the financial effects of implementing smart contracts for the management of the PoD. Results point to a possible decrease in processing costs by using a blockchain-based performance assessment system. The significant promise of blockchain is that the supply chain are working and activities for improved recital, distributed apportionment, and business automation, according to an innovative system put forth by S. E. Chang and Y. Chen (2020). This research offers a framework for these implementations from the perspective of interpretation, which aids in the understanding of blockchain use cases in real time supply and demand management. Salah (2019) offered an approach for monitoring and tracing soybeans throughout the agribusiness supply chain that effectively performs business operations by utilizing the Ethereum's digital distributed ledger system and smart contract. W. Alkhader (2021) suggests a cryptographic ledger-grounded approach for the supply of medical equipment as well as their decentralized digital production. To manage and monitor activities in a more organized decentralized methodology, transpicuous, ascribable, inspectable, and safe way, we create Ethereum smart contracts. The Merits of availing blockchain services deployed in supply-chain supervision system often outweigh the risks of doing so.

3. Methodology and Submodules Specification

The Entire Recommendation system based Web3 Supply chain is based upon, various modules in order to predict and suggest the pharmaceutical drug demand then manufacture module, retailer module and lastly customer module to validate the authenticity of the product after which one can take ownership of his medication, our prospective notion does not include this kind of period of time or technique. There is no central administration in the blockchain network; instead, each block taking part in the process verifies the transaction. Furthermore, since our blockchain-based structure utilizes a timestamped approach that enables end-user to track any record alteration, there is no chance of data exploitation. The goods stored in the Ethereum Blocks will help you identify them, The primary owner of the object will be the maker. While the QR code is being given to that product, the maker will ask the manager to add the item to the network. If the applicant is the real maker, the regulator will record both the goods and the producer on the network. The distinctive QR code for the product, which contains the product's information in encrypted text, will be used to generate a smart contract once the product is registered on the network. There is a Copy Sensitive digital picture embedded in the QR code to prevent duplication. The product is then forwarded by the manufacturer to the distributor, and its status is changed to "shipping"; ownership of the product does not change until a request from both parties for the purchase and sale of the product is accepted. The stake of a joint venture will automatically be transferred in the form of a Ethereum contract once money has been received in Ethereum blockchain network as soon as both parties agree to it. Clients will receive the Web app at this time. Then consumer can utilize the Web application to scan the QR code positioned to each item. Scanner examines the object, decrypts the encoded text using the given technique, and obtains details about the product's current manufacturer and owner, allowing the user to decide whether or not to purchase the item.

3.1 Sub Module Specification

There are five different modules which empowers the pharmaceutical drug supply chain mechanism they are namely supply-demand, manufacturer retailer and customer module With Supply and demand module, we develop a system for recommending medications stationed on patient reviews and ratings that predicts sentiment using a variety of vectorization techniques such as The Bow, TFIDF, Word2Vec, and manual data feature analysis. In other components, The Ethereum token allows the transactions to be kept in each block after the production of the drug is completed Once smart contracts have been formulated and implemented into blockchain networks. It is then given to the transportation division, enabling the delivery of medicines to the appropriate clients or institutions. Data stored in these areas cannot easily be accessed or changed. A QR code can be used to identify fraudulent goods.

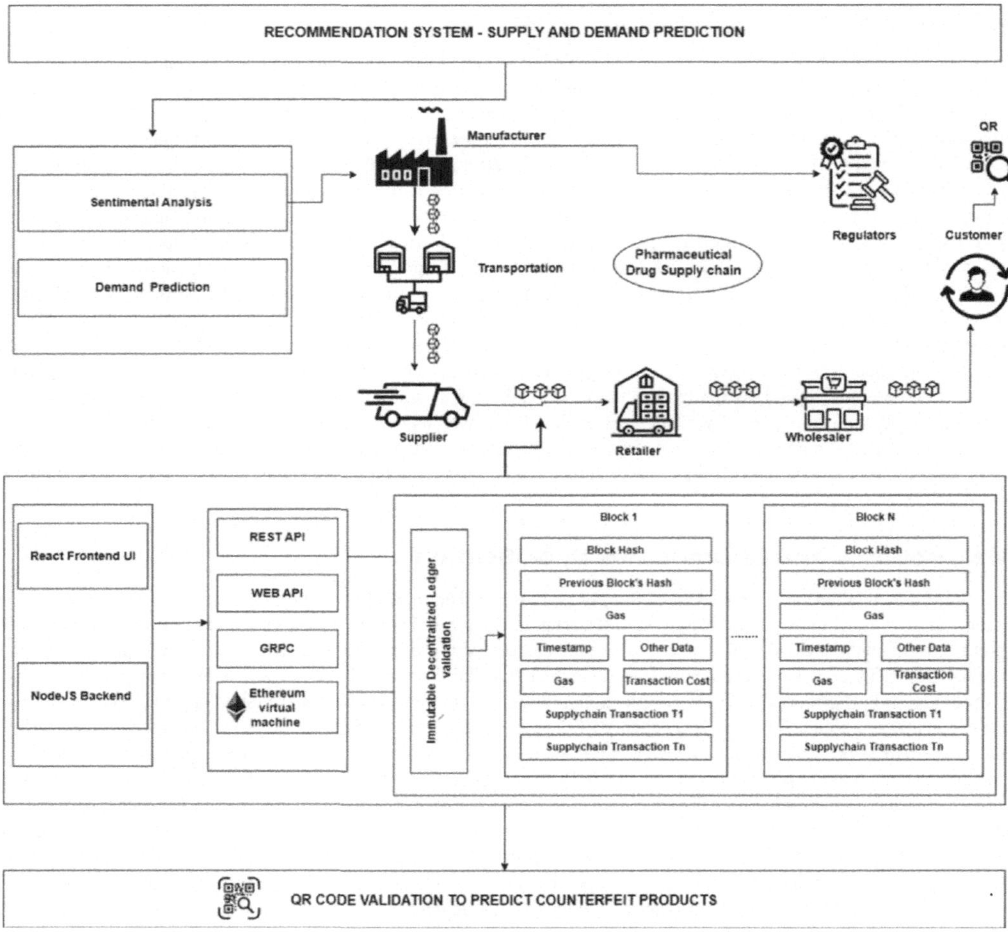

Fig. 17.1 System architecture

3.2 Consensus Algorithms Proof of Work in Ethereum

A smart contract contains all the transactional details and executes automatically. Ethereum contract is implemented via an programing called Solidity on several decentralized platforms. Smart contracts are distributed. In a distributed decentralized environment, consensus mechanism is a approach that sanction users or computer machines in the established network to cooperate. A agreement method known as proof of the work (PoW) calls for a lot of computing power from a highly secure connected network consist of numerous devices. Bitcoin and many other coins are built on the proof of work principle, which enables safe agreement. A decentralized acceptance method called proof of the testimonial or work (PoW) needs multiple network contributor to put forth effort to decipher an encrypted hexadecimal or Base16 number, It makes it possible to conduct private peer-to-peer transactions without the help of a reliable intermediary.

4. Implementation

The first stage in successful implementation of demand management is demand planning. It serves as the foundation for creating demand strategies. The reliability and viability of the demand strategy increase with precision. Its precision is based on whether the business can comprehend the industry and customer standards. A business also needs to be able to carry out its marketing and sales plans. The are various fundamental components in every recommender system. The initial stage of recommender models, which relates to data and gives suggestions to users, is the most crucial one. A post-processing is required to remove some repetitive suggestions. The manufacturer of the drugs receives the recommender model that is created using sentimental analysis, and the products are validated and added to the blockchain network to help remove counterfeit products.

4.1 Ethereum Web3Dapp Implementation

The information regarding the manufactured drug conformity, requirements and various vital attributes are stored into smart contracts with the help of Solidity. Unique address has been generated for each product and the Ethereum node has been deployed into the highly reliable cloud server with the help of Api endpoints the decoded stored facts has been delivered to the product owner via frontend which is completely developed using NextJS. For each new transaction new blocks has been formed in the Ethereum network Both manufacturer and consumer made aware of the products originality and legitimacy with the help of Quick response code .the verification is done over REST protocol between Client user and the blockchain peer to peer environment

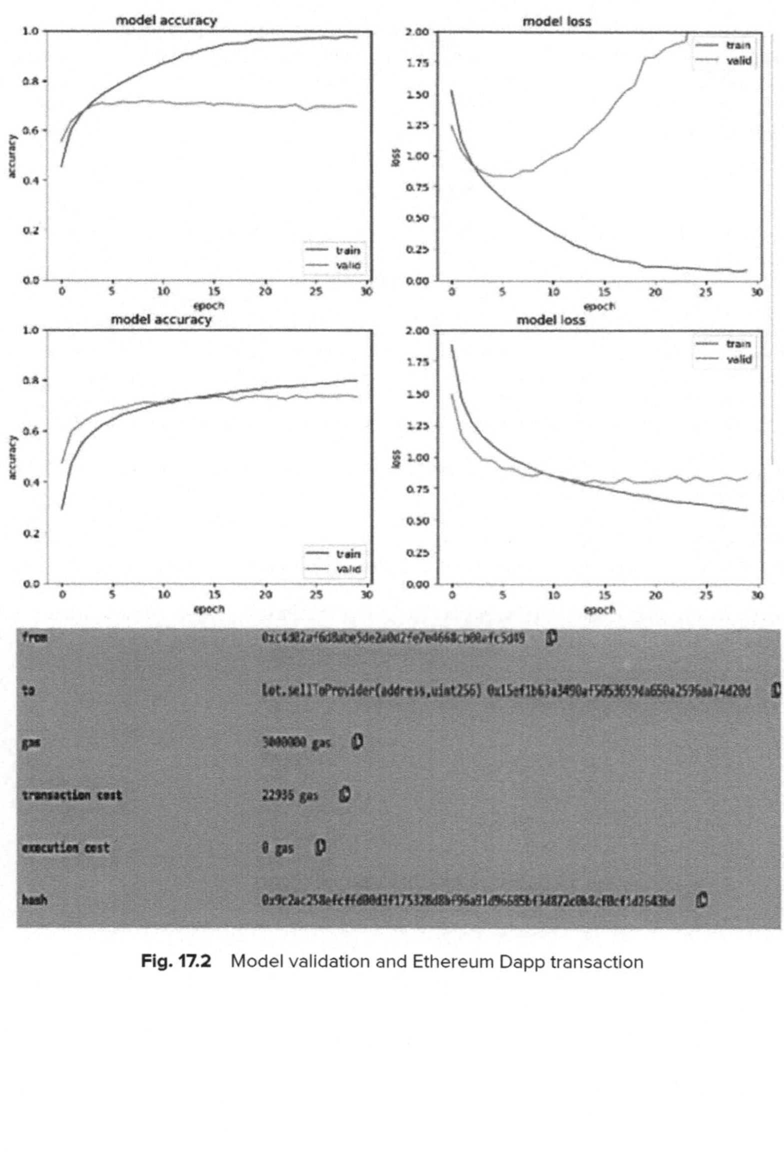

Fig. 17.2 Model validation and Ethereum Dapp transaction

5. Results and Discussion

Organizations use evaluations and ratings to forecast demand, and as a result Drug production can be sped up, allowing for the construction of a Realtime supply chain technique that includes blockchain technology binomial , Bayesian Network, Logistic and Linear Regression, SVM , Linear Multi SVC based on TF-IDF, Bow, and classifiers like LGBM, CART or Classification and Regression Trees, and Random Forest were among the machine learning methods used to build the recommender system.

The Linear SVC using TF-IDF beats all other models with 94 percent accuracy, according to our analysis of models using the five key measures of f1score, validity, recall, precision, and AUC score. Every pharmaceutical drug and its different category in the supply chain can use blockchain p2p Solution the producer, merchant, wholesaler, and seller. it presently contains info appropriate for Medicines and location, For every step in the logistics system Ethereum blocks were created and informational facts were stored

6. Conclusion

Counterfeit of essential products affects the business as well as the reliability of the manufacturers so with the assist of blockchain implemented supply chain the metadata and the logistics information regarding the paramedical drug cannot be manipulated or altered since smart contracts are deployed into every products, thus the authenticity of the medicinal drugs can be verified by using QR code embedded into the product. In order to force consumers to purchase the verified goods, Blockchain-based E-Pharma applications to be developed. Future improvements may include cloud services blockchain as a service by Amazon or Azure's Quorum Blockchain Service

REFERENCES

1. C. Antal, et al. (2021). Blockchain Platform For COVID-19 Vaccine Supply Management,in IEEE Open Journal of the Computer Society,.Vol (2): 164-178
2. I. A. Omar, et al., (2021). Supply Chain Inventory Sharing Using Ethereum Blockchain and Smart Contracts, in IEEE Access, vol. (10): 2345–2356.
3. R. Asyrofi et al, (2020). CLOUDITY: Cloud Supply Chain Framework Design based on JUGO and Blockchain.
4. Y. Madhwal et al., (2015). Proof of Delivery Smart Contract for Performance Measurements, in IEEE Access, vol (10): 69147–69159.
5. S. E. Chang and Y. Chen, (2020). "When Blockchain Meets Supply Chain: A Systematic Literature Review on Current Development and Potential Applications," vol (8): 62478–62494.
6. K. Salah, N. et.al, (2019). Blockchain-Based Soybean Traceability in Agricultural Supply Chain, in IEEE Access, vol. (7): 73295–73305.
7. W. Alkhader, et al. (2021).Blockchain-Based Decentralized Digital Manufacturing and Supply for COVID- 63 19 Medical Devices and Supplies, in IEEE Access,
8. R. Z. Rasi, et al (2022). A Literature Review on Blockchain Technology: Risk in Supply Chain Management. in IEEE Engineering Management Review vol. (50): 186–200
9. P. Gonczol, et.al (2022). Blockchain Implementations and Use Cases for Supply Chains-A Survey, in IEEE Access, vol (8): 11856–11871
10. S. Wang, D. Li, et al (2019). Smart Contract-Based Product Traceability System in the Supply Chain Scenario. in IEEE Access, vol. (7): 115122–115133.

Note: All the figures in this chapter were made by the author.

Disruptive Technologies for Sustainable Development – Dr. G. Nagappan et al. (eds)
© 2024 Taylor & Francis Group, London, ISBN 978-1-032-55000-8

A Study of Machine Learning Techniques for Supply Chain Management

18

N. Thilagavathi[1]
Assistant professor, Department of Computer Science and Engineering,
Rajalakshmi Engineering College

Arumugam Madan[2], **Vishal B.**[3], **Boobalan D**[4]
Department of Computer Science and Engineering,
Rajalakshmi Engineering College

Abstract: The crucial factor that influences the creation and distribution of products is known as Supply chain. While supply chains were previously sufficient to meet consumer demands, with the changing world landscape, it's critical for modern-day SCM processes to evolve accordingly. The contemporaneous era requires new models of supply chains due in part by exponential data collection rates. This article focuses on using innovative technologies such as machine learning within a current global context which will create an efficient Model suited for our times. Our central area of focus are three main components integral to all businesses involved; supplier selection, manufacturing optimization and forecasting among other related aspects involving distribution management. The introduction machine learning algorithms have potential benefits towards improving efficiency while offering insights into valuable predictions. This study seeks first not only analyze today's state-of-the-art technology in these areas but also discuss both real-world challenges when implementing this tech suite. Final results demonstrate positive outcomes if applied right whereby machine-learning significantly enhances performance throughout various stages from production through logistics thus generating informative information useful across industry sectors ranging from practitioners or researchers alike.

Keywords: Supplier, Manufacturer, Distributor, Machine learning, Web application

1. Introduction

Supply chain management (SCM) is rapidly utilising machine learning (ML) to boost productivity in processes including demand planning, inventory management, and transportation optimisation. To enhance predictions and speed up decision-making, machine learning techniques such as neural networks, decision trees, and clustering algorithms are utilised. The production, purchase, manufacture, and distribution of items all depend on the supply chain. End-to-end monitoring, predictive analytics, and risk management are being prioritised by businesses to strengthen resilience in the face of supply chain disruptions. To reduce costs and raise customer satisfaction, ML is utilised in demand forecasting, inventory management, transportation efficiency, quality control, and fraud identification. Companies can also increase efficiency by automating procedures and gaining real-time analytics. In the next two years, 87% of supply chain professionals want to practise resilience, according to a new Gartner study. This is a result of the rising number of supply chain interruptions brought on by calamities, political upheaval, and other occurrences. Automation and increased efficiency in SCM are also achieved via the use of digital technologies like AI and ML.

[1]thilagavathi.n@rajalakshmi.edu.in, [2]arumugam.madan.2019.cse@rajalakshmi.edu.in, [3]Vishal.b.2019.cse@rajalakshmi.edu.in, [4]Boobalan.d.2019.cse@rajalakshmi.edu.in

DOI: 10.1201/9781003428473-18

2. Literature Review

Using data from a well-known food industry, paper [1] developed a gradient boosted decision tree method to effectively handle service level failures in supply chains. Although it was 93% efficient, real-time or online supply chain management may be difficult to execute. Based on Canadian foundry orders, paper [2] employedneural networks to estimate product demand. Decision Tree algorithms were used in paper [3] to examine historical cost data and identify trends that conventional cost-control methods could miss. Based onvariables like source pricing, shipment schedules, and inventory levels, the algorithm was trained to forecast costs. The study [4] weakness is that RiskVis necessitates significant data availability and quality, which may not be possible for supply chain systems. In [5], genetic algorithms were applied to supply chain inventory optimization to determine the optimum inventory amounts, lowering costs and limiting stockouts. However, GA effectiveness can be impacted by data quantity and quality, which is a frequent problem in supply chain management systems. The study [6] suggested Conditional Generative Adversal Networks (CGAN), which assisted in lowering operating costs and enhancing responsiveness to market demands. However, the constraint can affect CGANs; if the requirement is not met, the model won't produce accurate findings.

A review paper [7] examined the taxonomy and evaluation of procurement, demand management, logistics, and transportation. They looked at back-propagation neural networks. (BPNN). The Cross Validation method. DT: Decision Tree algorithms that are the variables played a significant role, but BPNNs are susceptible to the setting of the weights; if the initialization is not suitable, the model may not converge. In article [8], transactional larceny was detected using logistic regression, but this technique relies on independent input traits. SARIMAX models and ACF/PACF were used in [9] to predict demand for agricultural products, but they have drawbacks when used with non-stationary time series. For forecasting vegetable expiration, Paper [10] suggested the ARIMA, SARIMA, LSTM, and XGBOOST models; however, these models may have trouble with complicated patterns in the data. In research [11], SVM, logistic regression, and Naive Bayes supervised machine learning algorithms were used to classify suppliers based on their performance, assisting businesses in finding high-performing vendors. With restricted data, supervised learning in supply chain management raises concerns about overfitting. To predict retail product sales and prevent overproduction and overstocking, study [12] suggested using past data and Random Forests; however, Random Forests can cause overfitting due to the large number of decision trees. The research mentioned in [13] was primarily concerned with handling inventory inventories using IoT RFID devices to track current stock levels. This job was aided by the use of machine learning -methods like Support Vector Machine, K-nearest Neighbors, and Bayes. The method's effectiveness in handling sizable databases is constrained, though.

Regression models such as Linear Regression, Ridge Regression, Lasso Regression, and Decision Tree Regression were used in a research by [14] to predict the order lead time for just-in-time (JIT) manufacturing systems. This was carried out in order to assist companies in lowering the risk of stockouts and enhancing their production and inventory planning. However, the approach's limited interpretability may be an issue in circumstances where transparency and interpretability are essential. Inbound logistics' effects on output from different sources were the subject of research in [15]. Inventory preparedness and plant planning were taken into account when improving transportation planning using machine learning techniques like KNN, Random Forests, and SVM. However, using this method to handle big datasets can become extremely costly. To enhance warehouse operations that manage product transportation and storing, research was conducted in [16]. According to information gathered from the storage, DEA and a pallet efficiency model were used to attain a 97% success rate. However, this approach did not take into account the supply chain's changing character, which is essential for long-term planning and decision-making. ML apps for supply chain activities, including supervised, unsupervised, and reinforcement learning, were compiled in [17]. SVMs were emphasized, but the choice of kernel function may have an impact on their accuracy. In [18], supply chain management distribution problems were discussed, and ML was suggested as a means of streamlining transportation and lowering costs. Bi-level code and an iterative, tiered supply chain strategy were essential for success. However, the model's presumptions might not correctly represent actual circumstances, which would produce less-than-ideal solutions. The BadNets technique, which creates hostile instances to reveal weaknesses in the machine learning models' supply chain, is covered in [19]. The research concentrates on CNN-based image identification algorithms, which might have trouble with pictures that are of poor quality or from odd perspectives. Backorders in the supply system were addressed in [20], and solutions including Distributed Random Forest and Gradient Boosting Machine were proposed. However, the Distributed Random Forest model needs a significant quantity of training data, which may be challenging to acquire for some tasks.

2.1 Scope of the Review

Role of demand forecasting in supply chain

In order to prevent stockouts and overstocking, demand forecasting is essential to supply chain management. The accuracy of demand forecasting is influenced by the calibre of the data, the model employed, and the underlying assumptions, it is vital to increase the accuracy of their demand forecasts and enhance the efficiency of their supply chains by utilising cutting-edge approaches like machine learning-based forecasting. Demand forecasting may assist companies in identifying market trends and opportunities, allowing them to maintain their competitiveness in a setting that is changing quickly.

Role of supplier selection in supply chain

The process of choosing a supplier is vital for businesses because it has a direct bearing on the reliability and quality of their goods and services. Businesses may minimise the risk of choosing the wrong suppliers and spend less time and money on the assessment process by automating and optimising the supplier selection process using machine learning algorithms. Machine learning may also aid in seeing patterns and trends in supplier data, allowing firms to make better decisions and bargain better contracts with their suppliers. In the end, using machine learning to supplier selection may boost the supply chain's overall efficacy and efficiency. Utilising relevant data from the past and machine learning, supplier selection may be automated. Additionally, businesses can employ dimensionality reduction or clustering strategies. Specifying needs, locating suitable suppliers, assessing, negotiating, and monitoring are the five processes in the supplier selection process.

2.2 A Review of Published Studies

The use of Supply chain management use machine learning to enhance a number of processes, including risk analysis, inventory management, and demand forecasting. The studies use several ML techniques with accuracy rates ranging from 84 to 97.8%. ML may be used to overcome issues with supply chain management and boost efficiency.

Table 18.1 ML Models used in the papers

Authors	Year of Publication	Methodology Used	Domain
Apurva Narayan [1]	2020	GBDT has 93% accuracy rate	Service level failures
Real Carbonneau [2]	2011	SVM,RNN, Neural Network	Application of ML for Demand Forecasting of products
Pan Yongming [3]	2009	Decision Tree has accuracy of 90.91%	Cost Control
Rick Siow Mong GOH[4]	2013	Riskvis Tool	Risk of Suppliers
R. Perumalsamy [5]	2010	Genetic Algorithms	Inventory Cost Control
Haifeng Lin [6]	2022	CGAN has accuracy rate is 96.27%	Operating cost control of Enterprises
Xuan Vy Pham [7]	2020	BPNN	Taxomony of Techincal issues
Yiyang Dong [8]	2021	LRC has accuracy is 97.80%	Fraud Detection in Transactions
Pramod K. Mishra [9]	2021	SARIMAX has accuracy is 90%	Demand Forecast and Pricing of agro products
Hashini Warnakulasooriya [10]	2020	ARIMA is 15% & XGBOOST is 70%	Forecasting Expiration of Vegetables
Ramkumar Harikrishnakumar [11]	2019	SVM has accuracy was at 84%	Selection of Best Suppliers
Archika Sharma [12]	2020	RF has accuracy is 88.84%	Predicting the sales of products
Ayaskanta Mishra [13]	2020	SVM has accuracy is 84%	Inventory Management of Stocks
Saurabh Singh [14]	2019	LR,has 96%RR has 97% accuracy rates	Predicting the Order Lead time to avoid Stockouts
AbdullahAlbadrani [15]	2021	KNN has accuracy of 96%	Time Efficient of Logistics
Francisco A. Diaz [16]	2009	DEA resulted in 97%	Improving Warehouse Management
Wenzel [17]	2019	SL,UL,RL	Applications of ML in Supply Chain
Jianhua Yang [18]	2009	Bi-level Programming	Distribution of products (Supplier to Customer)
Tianyu Gu [19]	2017	CNN image based classification	Identifying vulnerabilities of ML in Supply Chain
Samiul Islam [20]	2020	DRF,GBT increase of 20% when compared	Predicting likely supply-chain backorder situations

Papers that use machine learning to supply chain management are shown in the Table 18.1. For demand forecasting, inventory management, and logistics effectiveness, algorithms like decision trees, SVM, CNN, and RNN were employed.

3. Conclusion

This research demonstrates how machine learning (ML) has the ability to enhance performance by accurately optimizing different aspects of supply chain management. Future study possibilities include investigating cutting-edge ML methods, dealing with implementation issues, and handling environmental issues. Inventory management, production planning, transit scheduling, risk management, sustainability, teamwork, and real-time decision-making are a few future uses of ML in the supply chain. By utilizing the power of ML algorithms, the supply chain can improve efficiency, agility, and resilience, and make educated choices in reaction to unforeseen events.

References

1. Apurva Narayan,Pradeep K. Mahato,"Robust Supply Chains with Gradient Boosted Trees"2020 IEEE Symposium Series on Computational Intelligence (SSCI) https://ieeexplore.ieee.org/abstract/document/9308150.
2. Real Carbonneau,Kevin Laframboise,Rustam Vahidov."Application of machine learning techniques for supply chain demand forecasting"Received 3 August 2004, Accepted 1 December 2006, Available online 13 January 2011.byEuropean Journal of Operational Research. https://doi.org/10.1016/j.ejor.2006.12.004
3. Pan Yongming, Geng Xiaofei, Mi Guanxu."Study on Cost-control of Supply Chain based on the Decision Tree"2009 16th International Conference on Industrial Engineering and Engineering Management. https://ieeexplore.ieee.org/abstract/document/5344347
4. Rick Siow Mong GOH, Zhaoxia WANG, Xiaofeng YIN, Xiuju FU, Loganathan Ponnambalam, Sifei LU, Xiaorong LI"RiskVis: Supply Chain Visualization with Risk Management and Real-time Monitoring"2013 IEEE International Conference on Automation Science and Engineering (CASE). https://ieeexplore.ieee.org/abstract/document/6653910.
5. R. Perumalsamy and J. Natarajan, "Predictive analytics using Genetic Algorithm for efficient supply chain inventory optimization," 2010 Second International conference on Computing, Communication and Networking Technologies, Karur, India, 2010, pp. 1-8. doi: 10.1109/ICCCNT.2010.5591607. https://ieeexplore.ieee.org/abstract/document/5591607.
6. Haifeng Lin,JI Lin,Fang Wang."An innovative machine learning model for supply chain management".Journal of Innovation & Knowledge,Volume 7, Issue 4, 2022,100276,ISSN 2444-569X, https://doi.org/10.1016/j.jik.2022.100276.
7. Xuan Vy Pham,Angelika Maag,Sunthatalingam,Moshiur Bhuiyan."Predictive analysis of the supply chain management using Machine learning approaches: Review and Taxonomy"2020 5th International Conference on Innovative Technologies in Intelligent Systems and Industrial Applications (CITISIA). https://ieeexplore.ieee.org/abstract/document/9371842.
8. Yiyang Dong,Zhang Bohan,Keyu Xie,Lanchuan Lin."A Machine Learning Model for Product Fraud Detection Based On SVM"2021 2nd International Conference on Education,Knowledge and Information Management (ICEKIM). https://ieeexplore.ieee.org/abstract/document/9479632.
9. Pramod K. Mishra."Predicting Agri-Food Prices with Time-Series and Data-Mining based Methods"2021 5th International Conference on Intelligent Computing and Control Systems (ICICCS). https://ieeexplore.ieee.org/abstract/document/9432090.
10. HashiniWarnakulasooriya,JanangiSenarathna,Prasangika Peiris,Shaneli Fernando, Dr. Dharshana Kasthurirathna."Supermarket Retail – Based Demand and Price Prediction of Vegetables".2020 20th International Conference on Advances in ICT for Emerging Regions (ICTer). https://ieeexplore.ieee.org/abstract/document/9325451.
11. Ramkumar Harikrishnakumar, Alok Dand,Saideep Nannapaneni, Krishna Krishnan."Supervised Machine Learning Approach for Effective Supplier Classification".2019 18th IEEE International Conference On Machine Learning And Applications (ICMLA). https://ieeexplore.ieee.org/abstract/document/8999282.
12. Archika Sharma, M. Omair Shafiq."Predicting purchase probability of retail items using an ensemble learning approach and historical data".2020 19th IEEE International Conference on Machine Learning and Applications (ICMLA). https://ieeexplore.ieee.org/abstract/document/9356135.
13. Ayaskanta Mishra,Manaswini Mohapato."Real-time RFID-based item tracking using IoT &efficient inventory management using Machine Learning".2020 IEEE 4th Conference on Information & Communication Technology (CICT). https://ieeexplore.ieee.org/abstract/document/9312074.
14. Saurabh Singh,Umang Soni."Predicting Order Lead Time for Just in Time production system using various Machine Learning Algorithms".2019 9th International Conference on Cloud Computing, Data Science & Engineering (Confluence). https://ieeexplore.ieee.org/abstract/document/8776892.
15. Abdullah Albadrani,Faisal Alghayadh,Mohamed A. Zohdy,Esam Alouf,Richard Olawoyin"Performance and Predicting of Inbound Logistics Processes Using Machine Learning".2021 IEEE 11th Annual Computing and Communication Workshop and Conference (CCWC). https://ieeexplore.ieee.org/abstract/document/9376171.

16. Francisco A. Diaz - 2019."Supply Chain Management in the Logistics Industry"the Faculty of the WORCESTER POLYTECHNIC INSTITUTE June 21, 2009. https://core.ac.uk/download/pdf/212978589.pdf.
17. Wenzel, HannahSmit, Daniel Sardesai, Saskia."A literature review on machine learning in supply chain management".2019 [Publisher:] epubli GmbH [Place:] Berlin [Year:] 2019 [Pages:] 413-441. https://www.econstor.eu/handle/10419/209380.
18. Jianhua Yang,Zhangang Hao."The Study on Supply Chain Distribution Optimization Based on Bi-level Programming Model".2009 International Conference on Information Management, Innovation Management and Industrial Engineering. https://ieeexplore.ieee.org/abstract/document/5370298.
19. Tianyu Gu,Brendan Dolan-Gavitt,Siddharth Garg."BadNets: Identifying Vulnerabilities in the Machine Learning Model Supply Chain". Submitted on 22 Aug 2017 (v1), last revised 11 Mar 2019 (this version, v2)] Corneil University. https://arxiv.org/abs/1708.06733.
20. Samiul Islam,Saman Hassanzadeh Amin."Prediction of probable backorder scenarios in the supply chain using Distributed Random Forest and Gradient Boosting Machine learning techniques". J Big Data 7, 65 (2020). https://doi.org/10.1186/s40537-020-00345-2.

Disruptive Technologies for Sustainable Development – Dr. G. Nagappan et al. (eds)
© 2024 Taylor & Francis Group, London, ISBN 978-1-032-55000-8

Map Symbolic Gestures to Analogous Emoji by Hand Gesture Recognition System

19

N. V. Ravindhar[1]
Assistant Professor, Department of Computer Science and Engineering,
Saveetha Engineering College

G. Sujatha[2], S. Karpaga Kamatchi[3]
Master of Engineering, Department of Computer Science and Engineering,
Saveetha Engineering College

Abstract: Recent years have seen a phenomenal development in hand gesture recognition systems because of their ability to interact with machines efficiently. A webcam or built-in camera can be used to capture hand gestures. In the system, machine learning algorithms (convex hulls are used for pointing geometric objects) are coupled with the react JS framework. A variety of poses can be identified, and fingertip detection can be accomplished by analyzing the hand gestures. As part of the Proposed framework is the realm of hands subtracted from the foundation using geometric subtraction. A segmentation process is performed to identify and recognise the palm fingers and fingers of the hand. To predict the gesture, a rule classifier is applied based on predefined hand gesture labels. In our experiments, we demonstrate that our method is highly efficient and performs well on a small data set. Further, It performs a better than improvement approach to the adjacent hand gesture dataset. On top of that, emoji recognition and sign language detection are also included. A user can use the emoji recognition in interactive applications to express their reactions, such as thumbs up, thumbs down, etc. This application illustrates different sign languages used by deaf and dumb people by using American Sign Language.

Keywords: Gesture recognition, human computer interaction (HCI), and user-friendly interface

1. Introduction

The use of hand motions to control technology is becoming more popular and is an important way we interact with computers. In the past years, people have used keyboards and mice to control their computers, but computer technology has improved quickly. It has become important to find new ways to interact with computers. Technologies such as Voice Recognition and Sign Language Recognition. The body's posture or the movement of a hand can signify a range of emotions and actions. This form of communication is distinct from hardware-based methods and allows a computer to understand what people intend by recognizing their gestures or movements of different parts of their body. For many years, researchers have been studying this technology to facilitate human-computer interaction. Hand gesture recognition is highly useful in a variety of scenarios, such as recognizing sign language, creating virtual reality, providing interpreters for people with disabilities, and controlling robots. The research conducted by the authors focuses on how to identify the hand region from images, track its movements, and interpret the signs that are being made. This system can be broken down into three main stages: learning, detection and recognition.

[1]ravindharnv@saveetha.ac.in, [2]sujathagopinath1899@gmail.com, [3]anusethuraman1998@gmail.com

DOI: 10.1201/9781003428473-19

The other parts of the paper are organized like this. Section-2 looks at the existing research. Section-3 explains how we plan to do things and explains how different parts of our plan work. Section 4 tells us what happened in the study. Section-5 is the end of the story.

2. Literature Survey

Hasan used a fancy math tool to find hand signals using things other than shapes. The input hand image was divided into two distinct approaches - skin color segmentation through the HSV color model and clustering-based threshold methods. Although this research was focused on the four aforementioned classes, its primary limitation was that color changes were occurring very quickly due to differing lighting conditions, which could lead to inaccuracies or even failures. The system that recognizes hand movements has three important parts: segmentation, feature representation, and recognition techniques. The program uses different types of shapes to create models in space. It can create models that are 2D or 3D, and they are usually very accurate, about 86% of the time. However, the system does not take into account temporal gesture recognition due to insufficient light conditions causing Sometimes, areas that are not part of the skin might look like they belong to the hand because their colors are similar. There are three steps in this process: preparing the data (pre-processing), dividing the data into sections (segmentation), and performing some additional analysis and so forth. We need to pick the right tool to get the information we want. There are different tools that can be used to collect data like special gloves, markers or pictures of hands taken by cameras. This research had some problems with how bright the light was, turning the object around, making it bigger or smaller, and needing expensive equipment.

To put a system into action, we need to do three things.1. Utilizing a kinetic camera to identify hand motions, 2. Applying algorithms to detect and recognize hand movements, and 3. Identifying hand gestures with the help of edge detection and segmentation algorithms which may not be as effective as neural networks. The data set considered for this purpose is minor in size and can only differentiate between a few sign language gestures. The System architecture includes: 1. Capturing images, 2. Isolating the region with the hands, and 3. Utilizing distance transforms methods for gesture recognition. The drawbacks of this system include the limited number of recognizable gestures and the fact that these gestures are not used to operate any applications. Three main algorithms are employed in this implementation: Viola and Jones Algorithm, Convex Hull Algorithm, and an AdaBoost based learning Algorithm. The work was achieved by training a feature set comprising local contour sequences. However, two sets of images are required for categorization, which is another limitation.

This method uses a computer program called Java and special software called artificial neural networks to identify signals made by people's hands. There are two parts: 1. The module uses Java and can detect hands by looking at the background and turning video into a different format to find skin pixels. The system has another part called the predictive module that uses a type of computer program called a convolutional neural network. "The computer program gets an image from Java, and then it checks if it looks like any of the pictures it already knows." One problem with this system is that it requires special programming to make Java and Python work together.

3. Proposed Methodology

3.1 System Architecture

A system was made to sense when someone moves their hands and use that action to control a computer. Previous systems generally used spatial models rather than temporal models to recognize movements and were not able to work in real time. To address these shortcomings, a fresh architecture was designed which strives to create a system that can recognize hand gestures using cameras. It works really well and can keep up with quick movements. It works quickly and can be used right away. A system that people can use with computers was made. Instead of typing or clicking, people can use hand movements to control it.

The system receives input from a web camera or previously recorded video and uses an adaptive algorithm to detect the skin color of the user in the initial frames. Once it is established, a histogram clustering method is employed to localize the hand. Afterward, a machine learning algorithm is used to recognize hand gestures between consecutive frames for input into various applications.

Learning involves two components: the training dataset and feature extraction. The training dataset consists of various hand gestures which are used to teach the system how to respond. Feature extraction determines the center point of an image, splitting

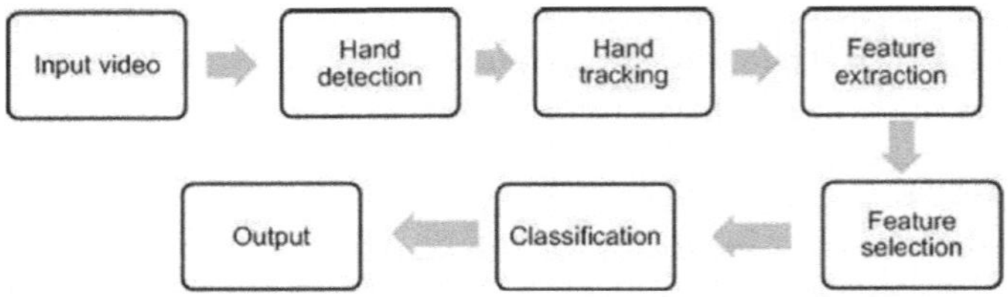

Fig. 19.1 Architecture diagram for hand gesture detection

it into two halves at its geometric center. The process of recognizing valid hand motions necessary to achieve a desired outcome starts with the input of an image from a webcam. This image is then compared to a dataset for detection. The hand must be kept at an approximate distance between 30 and 100 centimeters (about 3.28 ft) from the camera for reliable detection results. Gesture Identification: The number of digits on display in the hand movement is ascertained by utilizing defect points present in the gesture. When someone does a specific movement, the computer will do what the person wants. We used two sets of hand movements to test if our idea worked well. The first set of pictures shows different hand gestures. The symbols were given numbers in order from left to right and top to bottom, starting with 0 and ending with S3.

3.2 Modules Description

1. Emoji recognition
2. Sign language detection
3. Mouse functions

Emoji Recognition

This module explains how emoji are created through the use of voice and facial expressions. The process of identifying emoji involves tracking hand gestures, which are then mapped to a range from 0 to 20 based on the predicted landmarks. Once the coordinates have been established, they are matched with predefined points to generate an output emoji. Commonly used techniques for capturing input include hand belts, data gloves and cameras. Our system uses a cheap built-in camera to find both still and moving signals. This part makes pictures look better. The camera takes a picture. The picture goes through some changes like changing colors, removing fuzziness, and setting a limit. Then, software finds the lines and shapes in the picture.

Fig. 19.2 Emoji Recognition

Sign Language Detection

Sign Language is a type of nonverbal communication that utilizes body movements to express ideas, either as an alternative or in addition to verbal words. We use our hands, face, and body to move. Other ways of communicating without words using your body, like making facial expressions and using personal space and joint attention displays may differ from gestures which have specific meanings and can vary depending on social context. This initiative is to educate a Machine Learning system capable of distinguishing pictures of diverse sign language, such as an alphabet letter and numeric hand movement. An investigation of the proposed and present algorithms reveals that the precision is high.

Mouse Functionality

This module explains that a computer mouse is an input device used to pinpoint and interact with what it's pointing at. A virtual AI mouse is employed to avoid the need for any durability time, unlike a regular hardware mouse. This type of mouse provides the same functionality as a traditional one, but it works through hand gesture recognition which is tracked by the system's webcam.

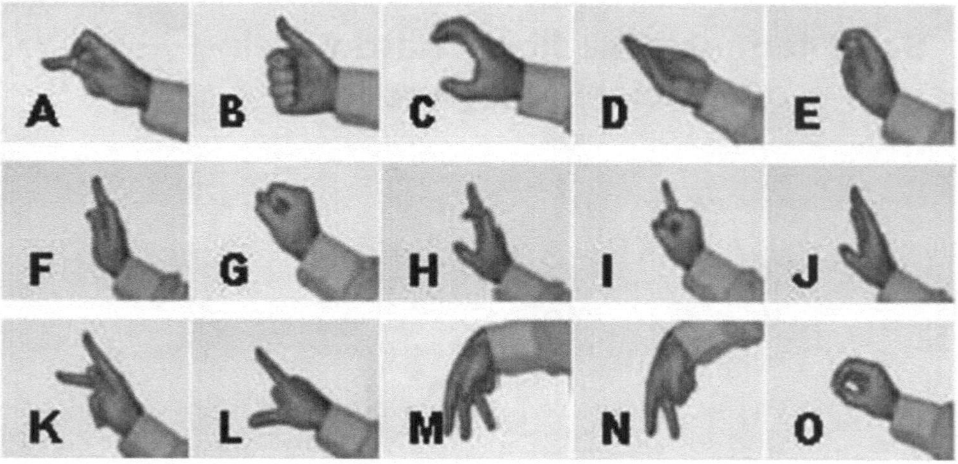

Fig. 19.3 Sign language detection

4. Results

The proposed system achieved its mean accuracy, and two centroid distance datasets were constructed: the first dataset consists of 7848 records for the initial seven hand postures, and the second dataset contains 2170 records for Portuguese Sign Language vowels gathered from four users. RapidMiner (Miner) was utilized to identify the most suitable kernel based on the datasets being examined. The best setting found for the kernel was one that's straight and has a cost of one.

5. Conclusion

Gesture recognition is vital for constructing efficient human-computer communication. This article explains how the system is put into effect through the pictures that are taken. Hand detection is achieved using Viola and Jones facial detection, along with convex Hull and AdaBoost algorithms. The system can be further improved to interpret signals from humans, such as page turning, scrolling up or down a page, recognizing sign language and emoji symbols.

References

1. Monika B. Gandhi, Sneha U. Dushane, and Ashwini M. Patil (2018). "Cursor Control System Using Hand Gesture Recognition."
2. Vinay Kr. Pasi, Saurabh Singh, and Pooja Kumari (2016). "Cursor Control using Hand Gestures" in the IJCA Journal.
3. Chaithanya C, Lisho Thomas, Naveen Wilson, and Abhilash SS (2019). "Virtual Mouse Using Hand Gesture" S.Ibrahim M., Shehata M. and Badawy W., (2013). "Image Recognition: A State-of-the-Art Review", IEEE Transactions on Circuits and Systems for Video Technology, Vol. 23, No. 2, pp. 311–325.
4. D. Liu Vinay Kr. Pasi, Saurabh Singh, and Pooja Kumari (2016). "Cursor Control using Hand Gestures" in the IJCA Journal.
5. MAUNG, T. H. H.(2019). Real-Time Hand Tracking and Gesture Recognition System Using Neural Networks. Proceedings of World Academy of Science: Engineering & Technology, 50, 466–470.
6. VATAVU, R.-D., ANTHONY, L. & WOBBROCK, J. O.(2012). Gestures as point clouds: a $P recognizer for user interface prototypes. 14th ACM international conference on Multimodal interaction. Santa Monica, California, USA: ACM.
7. VIJAY, P. K., SUHAS, N. N., CHANDRASHEKHAR, C. S. & DHANANJAY, D. K.(2012). Recent Developments in Sign Language Recognition: A Review. International Journal on Advanced Computer Engineering and Communication Technology, 1, 21–26.
8. WOBBROCK, J. O., WILSON, A. D. & LI, Y.(2007). Gestures without libraries, toolkits or training: a $1 recognizer for user interface prototypes. Proceedings of the 20th annual ACM symposium on User interface software and technology. Newport, Rhode Island, USA: ACM.
9. WU, Y. & HUANG, T. S.(1999). Vision-Based Gesture Recognition: A Review. Proceedings of the International Gesture Workshop on Gesture-Based Communication in Human-Computer Interaction. Springer-Verlag.

Note: All the figures in this chapter were made by the author.

Disruptive Technologies for Sustainable Development – Dr. G. Nagappan et al. (eds)
© 2024 Taylor & Francis Group, London, ISBN 978-1-032-55000-8

Self-Diagnosis Medical Chatbot Using Artificial Intelligence

20

Ragul Krishnan G.[1]
Department of Computer Science and Engineering,
Saveetha Engineering College, Chennai

G. Nagappan[2]
Professor & Head of Department, Department of Computer Science and Engineering,
Saveetha Engineering College, Chennai

Abstract: 2016 was clearly the year of the chatbot. The use of websites and apps will be replaced, according to some of the critics. The development of chat bots or Intelligent Conversational Agents using AI or machine learning techniques is a fascinating challenge realm. Various scrutiny and development initiatives, conversation/dialogue agents are being developed using AI, ML approaches. They are still advancing and experimenting with their research and development. Businesses of various sizes, from very small start-ups to major companies, are using these virtual agents. Both code-based and interface-based chatbot development frameworks are widely accessible on the market. Nevertheless, they lack the adaptability and value needed to create genuine discussions. Before contacting a doctor, the goal was develop a health care chatbot utilising AI able to identify illness,give a brief description of it. The medical chatbot is created to lower healthcare expenditures and provide curative information. Several chatbots performs curative encyclopaedias.

Keywords: Artificial intelligence, Natural language processing, Identify illness, Promoting better health

1. Introduction

A chatbot is a software tool that engages in verbal or written communication with people. By communicating with users in a human-like way, a medical chatbot enhances the work of a health practitioner and aids in enhancing their performance. Intelligent medical chatbots have a wide range of applications that could benefit doctors, nurses, clinicians, sufferers, or their families.Even if we are becoming acclimated to the most recent alterations brought on by the Covid19 pandemic, the worry of contracting the disease is rising as new varieties start to emerge. Despite the stringent lockdown regulations, the instances have not yet been found, disrupting our regular lives. Applying for a doctor's appointment is one of the challenges. Although hospitals adhere to Covid-19 regulations to the letter, there is a common misperception that hospitals are a magnet for Corona patients. We developed the concept of at-home self-diagnosis in order to lessen this dread and discourage individuals from frequently leaving their homes. The user can review their health records to identify any serious difficulties with Cancer, Diabetes, Heart, Liver, Kidney, Malaria, or Pneumonia top health problems in India. Users can input their health report details and receive a diagnosis for the same through a medical web page that uses Flask to carry out this procedure. Web development techniques and machine learning were used to create this page. The chatbot service also provides responses to other health-related questions. When a user types in symptoms like cold,vomiting, fatigue, etc., the chatbot will ask them to describe them. The model will

[1]rahulkrish4761049@gmail.com, [2]hod.cse@saveetha.ac.in

DOI: 10.1201/9781003428473-20

then predict that the user has dengue. It will also go on to list more potential symptoms and estimate the likelihood that the sickness will actually manifest. The users will be able to assess the severity of their ailments and take the necessary action as a result, which is the key advantage of this research. They can then decide whether a clinic or hospital visit is required or whether a self-diagnosis and telephone consultation would be sufficient. You can save time by doing this task rather than wasting time waiting and making several journeys to the hospital.

2. Literature Review

Simon Hoermann looked at the current research on the efficacy of synchronisable text-based online psychic health intrusions. Artificial intelligence (AI) as a means of teaching complex new ways to surviving and facilitating a ongoing cycle. Artificial intelligence (AI), which describes the phenomenon of the engine from which the competition is to feel and act and achieve results, maintains that when using simulation, a machine accurately mimics the expression of the human intellect, the development of the role of the congestion, to change, and methods of problem-solving. This study scout, a nutshell. AI used to performs some tasks associated with technology that connects to other people in order to "exercise" and "solve problems".The best capacity mimic is provided by artificial intelligence. As AI chatbots converse with experts in various fields, researchers are laying the groundwork for their security. Also, a change occurs and gives trustworthy chatbot are recommended. One particularly cutting-edge area of the industry is used to illustrate the effects of artificial intelligence.

Taking things to the point being replaced with AI, where they were only to have escalated. So curative AI can aid give excellent medical prognosis, despite a fact that the modifications did not happen quickly. Human intellect will eventually reach a point where it is unaffected by a safety margin that is shorter than it is now. There is therefore time for a dialogue with the AI bot to recognise and appreciate the health advantages.

Additionally a computer, AI fine matching was compatible with super-robots that respond to chat. This is a criteria turn; it is not merely a change. We give a brief description of how discussion revolutionaries will help you comprehend how this technology is used and its effects.

If a person is regularly evaluated, it may be possible to identify potential issues before they impair the body. Other issues that are not included in the study are analytical charges, government policies that ensure the effective use of tailored drugs.

3. Proposed System

The user conversation is organised in a linear fashion that begins with symptom extraction, continues with symptom mapping to identify the matching symptom, and concludes with a diagnosis to establish whether the patient has a major or minor ailment; They will be directed to a suitable doctor if they do. By using the login details that are saved in the database, the user can be located. State graph is used to depict the dialogue design of chatbots. The logic for state transitions is created, templates for natural language production are used, and the system takes the initiative to reach out to the user and solicit feedback for prognosis. In addition to its hello and good-bye states. Our bot begins by requesting the user's email address and password in order to log in, and it then enters a loop of symptom extraction states until it gathers enough data to provide a diagnosis. The moment they were given their initial prognosis, patients have of returning to the loop to discuss new symptoms with the doctor. They also have the option of viewing their chat history to see what they have previously discussed.

4. Implementation

4.1 System Architecture

This flow diagram depicts how the system will establish itself once the user's text chat is open. Following that, a series of questions based on the user's responses will be asked. User will receive symptom ideas from the system for system clarification, and the system will automatically map the precise symptoms. Finally, the disease was determined to be either severe or minor, and the user was provided treatment options and the doctor's contact information.

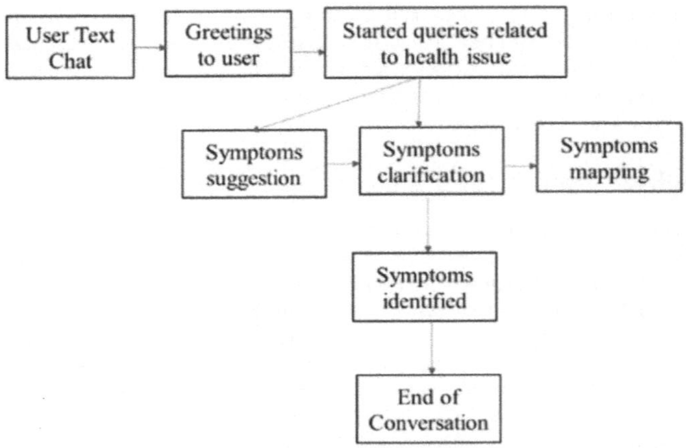

Fig. 20.1 Architecture diagram

5. Modules Description

5.1 List of Modules

- Dataset
- Users
- Webchat
- Suggestion and clarification
- Prediction

Dataset collection

Collections of data from the medical profession are made up of illness types and their corresponding symptoms. Further names and phone numbers of doctors who are suggested for patients with severe health issues have also been included. Dataset is gathered and saved in an.xml file that is kept on the local drive.

Users

These are the system's users. Patients and physicians can both be users. People can communicate with the system to learn about their medical conditions, schedule appointments with specialists, and other things. Physicians can use the system to view their scheduled appointments on a specific date.

Web text chat

The actual discussion takes place in this module. It is through this module that users communicate with the system. This page displays all of the data that was pulled from the database.

Suggestion and clarification

This module will use the string search algorithm to process the user's queries. This module will evaluate each symptom that the patient has stated and categorise them as either major or minor. If the condition is serious, a specialist will be recommended for the patient to see. If the condition is mild, it will recommend a suitable treatment.

Prediction

The system will determine what type of ailment you might have and provide the contact information for a doctor if you have serious health issues; otherwise, it will advise you to take some time off.

6. Conclusion

The utilisation of Chatbots either the desktop or mobile app is user-friendly and open to anyone who understands how to type in their own language, according to the evaluation of numerous publications. Based on your symptoms, a medical chatbot offers individualised diagnoses. Other details about the site, potency, and span of the symptoms, as well as detailed rendition of the symptoms, are medical features, could be supported in the future, which would considerably improve the bot's performance in terms of symptom recognition and diagnosis. Customized Medical Assistant implementation mainly relies on AI algorithms and training data. Ultimately, Many lives could be saved by customised medicine, and it would increase public awareness of health issues. As was previously stated, messaging apps will dominate the future since consumers will spend the most time using them. .People can have this medical dialogue wherever they are. However it has certain drawbacks, such as the fact that occasionally the patient may receive an irrelevant response or the bot may offer a solution that the user finds hard to trust.Performance will be enhanced further by adding more high-quality data. For additional testing, Alternate hyper-parameters and datasets should be used to train the training model. This was an investigation into the use of deep neural networks for dialogue generation in an intelligent chatbot. By extending the utilisation of the database and adding new word combinations, the chatbot's effectiveness can be increased so that it can manage all forms of ailments

References

1. Saurav Kumar Mishra, DhirendraBharti, Nidhi Mishra. (2017). Dr.Vdoc: A Medical Chatbot that Acts as a virtual Doctor. Journal of Medical Science and Technology. 6(3): 16–20.
2. Vineth, R., R. Rithish, D. V. S. N. Sai Varma, and B. V. Ajay Prakash. (2020). Smart Health Care Chatbot for Prognosis of Treatments and Disease Diagnosis Using Machine Learning. Journal of Computational and Theoretical Nanoscience. 17(9-10): 3947–3951.
3. Mishra, S. K., Bharti, D., & Mishra, N. (2018). Dr.Vdoc: A Medical Chatbot that Acts as a Virtual Doctor. Journal of Medical Science and Technology. 6(3): 16–20.
4. Divya, S., V. Indumathi, S. Ishwarya, M. Priyasankari, and S. Kalpana Devi. (2018). A self-diagnosis medical chatbot using artificial intelligence. Journal of Web Development and Web Designing. 3(1): 1–7.
5. V. Manoj Kumar. (2016). Sanative Chatbot For Health Seekers. International Journal of Advanced Trends in Computer Science and Engineering. 5(3): 16022–16025.
6. B. E. Comendador, B.M. Francisco, J.S. Medenilla, S.M. Nacion, T.B. Serac, (2015). Pharmabot: a pediatric generic medicine consultant chatbot. J. Autom. Control Eng. 3(2): 137–140.
7. P. Kaur, M. Sharma. (2017). A survey on using nature inspired computing for fatal disease diagnosis. Int. J. Inf. Syst. Model. Des. 8(2): 70–91.
8. Doina Dragulescu,Adriana Albu. (2015). Medical Predictions System, International Journal of Engineering Research and Applications. 2(3): 1988–1996.
9. Fontil V, Radcliffe K, Lyson HC, Ratanawongsa N, Tuot D, Yuen K, et al. (2019). Testing and improving the acceptability of a web-based platform for collective intelligence to improve diagnostic accuracy in primary care clinics. JAMIA Open. 2(1): 40–8.
10. Laranjo L, Dunn AG, Tong HL, Kocaballi AB, Chen J, Bashir R, et al. (2018). Review conversational agents in healthcare: a systematic review. J Am Med Informatics Assoc. 25(9): 1248–58.
11. Palanica A, Flaschner P, Thommandram A, Li M, Fossat Y. (2019). Physicians' perceptions of chatbots in health care. J Med Internet Res. 21(4): e12887.
12. Braun V, Clarke V. (2006). Using thematic analysis in psychology. Qual Res Psychol. 3(2): 77–101.
13. Lyles CR, Ph D, Harris LT, Le T, Flowers J, Tufano J, et al. (2011). Qualitative evaluation of a mobile phone and web-based collaborative care intervention for patients with type 2 diabetes. Diabetes Technol Ther. 13(5): 563–9.

Note: The figure in this chapter was made by the author.

Disruptive Technologies for Sustainable Development – Dr. G. Nagappan et al. (eds)
© 2024 Taylor & Francis Group, London, ISBN 978-1-032-55000-8

Sensors Based Laptop Controller Using Arduino App

21

E. Christo Raymonde[1]
Department of Networking and communications,
SRMIST, Kattankulathur, Chennai

K. Kalaiselvi[2]
Associate Professor, Department of Networking and communications,
SRMIST, Kattankulathur, Chennai

Abstract: Controlling a laptop utilizing a set of movements is the process of detecting and interpreting gestures made by a human hand in real-time. This can be achieved through various sensors, which is a field of study that focuses on how computers can interpret and understand visual information from the real world. The control strategy lies in the choice of proper sensors that can be connected to Arduino board using additional circuitry. The Arduino board is then programmed to read the sensor data and to send appropriate command to the laptop. The present work focuses on integrating an Ultrasonic sensor, an IR sensor, a Motion sensor and a sound sensor to act as an integrated sensor system and to coordinate different control actions in a laptop. The Python libraries such as Pyserial PyautoGUI, are used to get the desired behavior.

Keywords: Arduino, hand gesture, Motion sensor

1. Introduction

In the modern society, everyone is surrounded by machines and gadgets at all times. The use of mobile phones, tablets, computers and smartwatches has increased manifold due to creative new technology and Internet use. Human computer interface (HCI) combines the hard circuitry with the programming to communicate between a user and a computer. Communication can be easier with a touchless user interface, which can be controlled using voice or gestures. Sensors can be used to detect the presence of people in a room, and the Arduino board can be programmed to turn on or off the appliances accordingly. For example, sensors can be used to detect the movement of the user hand, and the Arduino board can be programmed to control the game or multimedia application accordingly Gergely Sziladi (2016).

In addition to these applications, there is also ongoing research in the field of controlling a laptop using Arduino and sensors.. On this task, we have got implemented an easy control with Arduino interlinked with various sensors to perform actions in the system monitor to move tabs or pages or control playing of a video with the assist of hand gestures.

The project focuses on integrating IR sensor, motion sensor, and sound sensor with Arduino. The serial Python and Pyautogui library enable the control of actions in a computer. Arduino dispatches the commands using the serial port to the laptop. With the aid of python that is running on the pc and the statistics from the various sensors, the preferred motion will be executed. The paper is structured as with surveying the literature, then proposing the work, followed by the requirement analysis for hardware and software. The implementation procedure is then discussed and presented.

[1]ec7719@srmist.edu.in, [2]kalaisek2@srmist.edu.in

DOI: 10.1201/9781003428473-21

2. Existing System

An elaborate analysis of the previous works reports on the use of computer motion control by different techniques. Kale et al. focusses on the manipulating of the actions of a robot through gestures Kale (2016). Oudah et al refers how to move a robot using a mobile phone. Oudah (2020). Yasen et al in his paper has investigated the different methodologies in using gestures Yasen (2019). Mukherjee in his paper states that gesture is an expression Mukherjee (2018). A comprehensive study describing the fundamental techniques of computer vision and the constraints is given by Murthy et al. Murthy (2009) A system of recognition referring with recognition classification and extraction of features are put forward by Khan et al. Khan (2012). Suriya et al. intakes a discussion on the methods, algorithms used for interaction of machine with humans Suriya (2015)

3. Proposed Methodology

3.1 Case Diagram

The present study intends to transform hand gestures, sounds, and force induced motion into appropriate signals to control various actions in a laptop with the utilization of Arduino integrated with IR sensors, motion and sound sensors. The integration of the sensors with Arduino is done using C++ and Python programming. It is possible to use an Arduino board with motion and sound sensors to control a laptop. We can use the sensors to detect different movements and sounds and map them to specific actions on the laptop, such as keyboard shortcuts or mouse clicks. Use the Serial communication interface to send the mapped actions to the laptop. we can use the Arduino Keyboard library to simulate keyboard inputs or the Arduino Mouse library to simulate hand gestures, mouse movements and clicks as in Fig. 21.1

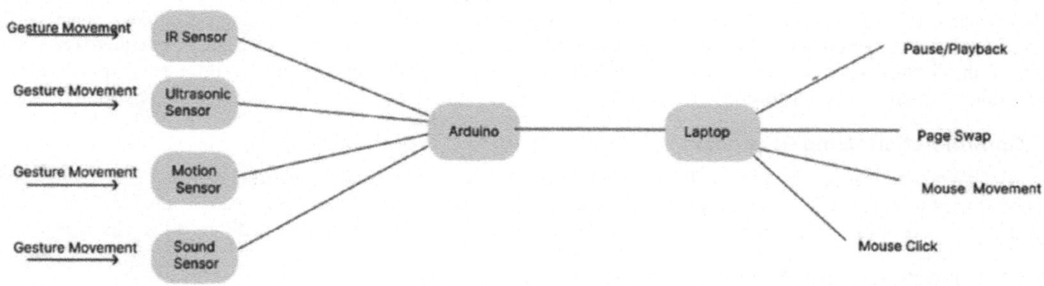

Fig. 21.1 Case diagram

3.2 Hardware Requirement

The basic requirements regarding the hardware includes Arduino UNO, the sound sensor (KYO38), the ultrasonic sensor (HCSRO) and IR sensor which is commercially available.

3.3 Software Requirement

Arduino IDE - The Arduino platform includes an Integrated Development Environment (IDE) for programming the board, which allows users to write code in C/C++ or other programming languages. This IDE is compatible with the Arduino board. Python IDE – A Python IDE (or Integrated Development Environment) utilizes codes written in Python to perform any operation. PySerial library links serial ports that are remote through RFC 2217 and PyAutoGUI libraryI takes up programs to control the input devices.

3.4 Circuit Diagram

See Fig. 21.2 on next page.

3.5 Control Action Using Ultrasonic Sensors

The working of ultrasonic sensors lies in the transmission of sonic waves at ultrasonic frequencies that on obstacle is reflected back to a receiver. The ultrasonic sensor is wired to the Arduino board. The pulseln() function is used to measure the time taken

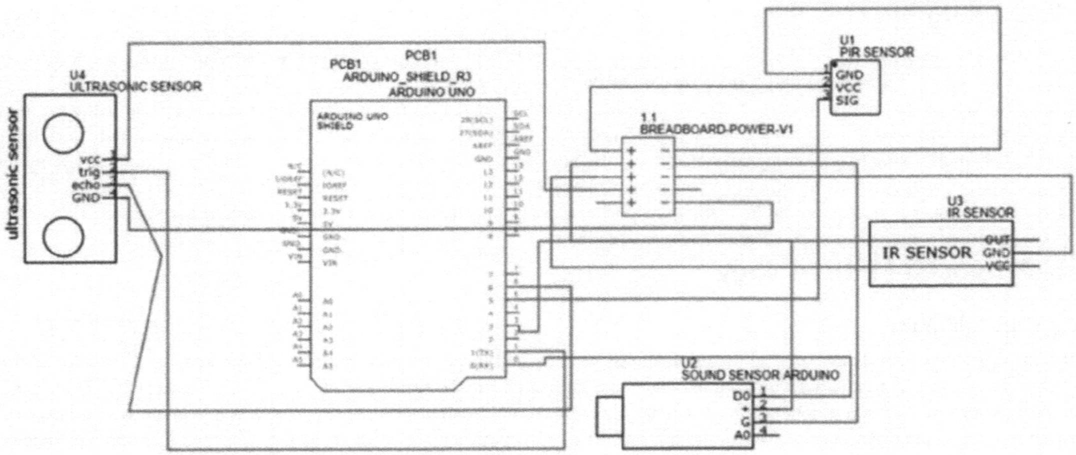

Fig. 21.2 Circuit diagram

for the ultrasonic sensor to send and receive an ultrasonic pulse, and is converted into distance using the formula: distance = (T/2) * S where T is the time and S is the speed of sound. The gestures are interpreted by writing codes. For example, if the distance is less than a certain threshold, it could mean that the hand is close to the sensor, which could be interpreted as a "fist" gesture. If the distance is greater than a value meant to be the threshold, it could mean that the hand is far away from the sensor, which could be interpreted as an "open palm" gesture.

3.6 Control Action Using IR Sensors

An additional control was provided by an IR sensor. The infrared sensor is circuited to the Arduino board. The analogRead() function is used to measure the voltage output of the sensor, which is proportional to the infrared radiation detected. A code written is used to read the voltage measurements from the infrared sensor and interpret them as different hand gestures.

3.7 Control Action Using Motion and Sound Sensors

The motion sensor detects a change in its motion perpendicular to it upwards, it could be mapped to a mouse movement based on the hand movement(if the movement is right direction, then the mouse will move to the right direction), and if the sound sensor detects a tapping sound, it could be mapped to a mouse click. Use the Serial communication interface to send the hand gesture information to a computer or a display module The entire working module can be implemented in Fig. 21.3.

Fig. 21.3 Interfacing with the laptop

4. Results and Discussion

4.1 Algorithm Steps

Step 1: The sensors are interfaced with the Arduino UNO R3 board

Step 2: The Ultrasonic sensor perform action like the play or pause of the video

Step 3: The swapping of pages on the monitor screen is performed using IR sensor.

Step 4: The motion sensor like that of an accelerometer is used as a mouse

Step 5: Any sound consequent to a tapping action is detected by a sound sensor like that of a piezoelectric used to control the mouseclicks Left and RIGHT

4.2 Programming Arduino in Gestures

The motive of this work lies in programming the Arduino in the conversion of actions and gestures perceived by the sensors that are transduced to suitable commands. The Serial data from Arduino has to be read and certain keyboard key presses needs to be invoked. To achieve this a special Python Module called PyAutoGUI is installed. The implementation of the control of laptop moves like controlling the mouse,clicking the mouse right and left, pressing the key multiple times, scrolling of a page in the monitor using PyAutoGUI exempting the physical movements. The system is tested and refined by adjusting the sensor thresholds and mapping parameters to achieve the desired level of sensitivity and accuracy.

4.3 Simulating the Sensor Actions with Arduino and Python Libraries

To detect the commands relating to the gestures and movements through the serial port Arduino ends out commands like UP/DOWN/NEXT/PREVIOUS/CHANGE. The functions PyAutoGUI.moveTo() are further used for the implementation of actions like controlling the mouse movement, mouse right and left click,moving the pages up and down. The pyAutoGUI library is primarily used for automating keyboard and mouse actions on a computer.

We can process the distance measurement using an ultrasonic sensor to interpret it as a hand gesture. For example, we can define a set of thresholds to classify distances as "close", "medium", and "far", and use those classifications as hand gestures. We can use the pyautogui.hotkey() function to to play or pause a video. And the pyautogui.moveTo() function to move the mouse

The pyAutoGUI library can be used in conjunction with an Arduino board and an IR sensor to control a laptop using infrared signals. The IR sensor is connected to an Arduino board and a sketch is uploaded that reads the IR signals and sends them to the computer via serial communication. The IRremote library can be used to interface with the IR sensor and the Serial library to send data to the computer. The pySerial library allows to communicate with the Arduino board via serial communication. The Python script written listens for incoming data from the Arduino board and uses the pyAutoGUI library to simulate keyboard or mouse actions based on the received IR signals. We can use the pyautogui.press() function to simulate a key press or the pyautogui.moveRel() function to move the mouse for swapping of pages.

The motion sensor is connected to the Arduino board. We can use any motion sensor that can be connected to the Arduino board, such as an accelerometer or a gyroscope. A program for the Arduino board is written that reads the data from the motion sensor and sends it to the laptop. We can use the Serial library in Arduino to communicate with the laptop. The Python script written receives the data from the Arduino board using the pySerial library and uses the pyAutoGUI library to control the laptop. For example, we can use the motion sensor data to move the mouse cursor, simulate keystrokes, or control the volume. The Python script reads the data and uses the moveTo() function from the pyAutoGUI library to move the mouse cursor on the screen based on the x and y values received from the motion sensor.

The sound sensor is wired to the Arduino board. A program is written for the Arduino board that reads the data from the sound sensor and sends it to the laptop. We can use the Serial library in Arduino to communicate with the laptop. A Python script is used to receive the data from the Arduino board using the pySerial library and pyAutoGUI library is used to control the laptop. Also we can use the sound sensor data to simulate mouse clicks, open applications, or play media. For example, the Python script reads the data and uses the click () function from the pyAutoGUI library to simulate a left-right click on the laptop whenever the sound sensor detects a sound.

5. Conclusion

The present work highlights the laptop control with a conglomeration of Arduino and four different sensors utilizing Python libraries The work has a futuristic scope in exploring the use of sensors to detect the emotions of the user. For example, sensors can be used to detect changes in the user heart rate or skin conductivity, which can be used to predict the emotional state in which the user is in. It also paves way to explore the use of Arduino and sensors to create immersive augmented reality experiences. Sensors can be used to track the movement of the user head, and the Arduino board can be used to control the display of virtual objects on the laptop screen. Overall, controlling a laptop using Arduino and sensors has many potential applications and is an area of ongoing research and development.

References

1. G. Sziládi, T. Ujbányi and J. Katona (2016). Cost-effective hand gesture computer control interface, 7th IEEE International Conference on Cognitive Infocommunications *(CogInfoCom)*, Wroclaw, Poland, pp. 000239–000244
2. Kale, Pradeep , Yeshi, Mayur. (2016). Hand gesture recognition for human-computer interaction. International journal of scientific development and research. 1. 8–14
3. Oudah, M.; Al-Naji, A.; Chahl, J.(2020) Hand Gesture Recognition Based on Computer Vision: A Review of Techniques. J Imaging, 6, 73.
4. Yasen M, Jusoh S. (2019). A systematic review on hand gesture recognition techniques, challenges and applications. PeerJ Computer Science 5:e218R. Mukherjee, P. swethan, R. Pasha and S. s. Rawat (2018), Hand gesture-controlled laptop using arduino," International Journal of Management, Technology and Engineering, p. 7
5. Murthy, G.R.S.; Jadon, R.S.(2009) A review of vision based hand gestures recognition. Int. J. Inf. Technol. Knowl. Manag.. 2, 405–410.
6. Khan, R.Z.; Ibraheem, N.A(2012). Hand gesture recognition: A literature review. Int. J. Artif. Intell. Appl., 3, 161.
7. Suriya, R.; Vijayachamundeeswari, V.(2014) A survey on hand gesture recognition for simple mouse control. In Proceedings of the International Conference on Information Communication and Embedded Systems (ICICES2014), pp. 1–5.

Note: All the figures in this chapter were made by the author.

Disruptive Technologies for Sustainable Development – Dr. G. Nagappan et al. (eds)
© 2024 Taylor & Francis Group, London, ISBN 978-1-032-55000-8

Detecting Phishing Websites Using Machine Learning

22

V. Uma Rani[1]
Associate Professor, Department of Computer Science and Engineering,
Saveetha Engineering College, Chennai, India

R. Pradeep[2]**, N. Jayanathan**[3]**, V. P. Muhammed Rashid**[4] **and S. M. Shrinikkesh**[5]
B.E Scholar, Department of Computer Science and Engineering,
Saveetha Engineering College, Chennai India

Abstract: Phishing is a type of criminal activity that involves the creation of bogus websites that closely mimic actual ones in order to fool users into disclosing personal information like as passwords, account details, and ATM numbers. Once a user logs into the phishing website, their sensitive information is then captured by the scammers behind the scheme. Phishing sites that anticipate to get hold of sufferers' touchy information, redirecting them to view a fake website that seems like a legitimate website is any other type of online crook pastime and one of the particular problems in many regions, such as e-authorities. Blended and wholesale. The detection of a hacked site is a truly indistinct and complicated hassle with many components and standards that are not solid. Because of this, and also the ambiguity inorganizing websites due to the intelligent systems that programmers use, some proactive strategies can be beneficial and powerful tools that can be used, which include neural systems and metalworking strategies. In this paper, phishing web page recognition mechanism uses machine learning algorithm, random forest classifier to locate phishing pages. From this experimental results, Random Forest provides 83.8% accuracy which is higher than other classifiers.

Keywords: Phishing websites, Machine learning, Random forest

1. Introduction

Phishing is a major threat to Internet users in which malicious internet websites act as an original website and try to steal or misuse the user's information and data's like passwords, debts, or master card numbers (Alsenani,2023). Since many peoples try to get resources provided by government and financial institutions on official websites, remarkably this had increased phishing attacks in recent years. Phishers established this has a vast network to earn money and also started to earn from it. Typically, this attack is carried out by sending a false e-mail or message that pretends to be a real request from a well-known business organization. A link which is received usually sends a potential victim to a corrupt website, which is often looks like the assumed entity's legitimate website. In the phishing websites, victim enters sensitive information, which the phishers can then misuse.

Although there are many anti-phishing programs and other methods to find phishing attempts in messages and to stumble on phishing content on websites, phishers are coming up with new and more strategies to avoid the programs and systems to be had. However, phishing involves social engineering, getting the same results with phishing websites is more challenging. Phishing poses a significant risk to the trust and financial security of both individuals and organizations. In some cases, it serves

[1]umaranibharathy@gmail.com, [2]pradeep070302@gmail.com, [3]jayanathan71@gmail.com, [4]vpmuhammedrashid9@gmail.com, [5]shrinikkeshsm@gmail.com

DOI: 10.1201/9781003428473-22

as the starting point for more significant attacks that can result in substantial economic losses. In fact, millions of dollars are lost daily due to this form of crime. According to the survey of Anti-Phishing Working Group's (APWG) on the year 2020, the amount of reported fake URL incidents has more than doubled, indicating a worrying trend compared with earlier years.

Many research has been taken place to prevent phishing attacks with the help of different organizations all over the world. The key to preventing phishing attacks is to detect phishing websites and educate users on how to identify them to prevent the theft of their data. Various methods can be employed to detect such attacks, but machine learning algorithms are considered to be one of the most effective. This study explores three kinds of Machine Learning (ML) algorithms namely Random Forest (RFC), Support Vector Machine (SVM) and Decision tree (DTC) used for detecting fake websites and discusses their effectiveness.

2. Literature Review

Several research works is carried out for detecting phishing or fake websites using machine learning. In this section discuss some work carried out to recognize phishing website detection and datasets used for their study.

Menaka (2023) have utilized the UCI dataset to investigate spam and phishing groups using six distinct machine learning algorithms and ensemble learning. In addition, the researchers conducted three different tests in which first phase uses ML with original dataset, the second phase utilize the feature selection mechanism namely Best First and CfsSubsEvaluation ,along with ML and the third phase uses Ranker and Principal Components Optimization along with ML. The experiments conducted by WEKA tool and it shows the RFC outperformed the other machine learning algorithms across all three trials, achieving higher accuracy scores above 90%.

Hussain M (2023) investigates randomly collected phishing websites using the RFC algorithm. After testing seven different algorithms, the researchers found that the NLP along with RFC yielded the best results, achieving high accuracy above 90%. Jha, A. K(2023) have worked with the five different supervised ML models, including SVM, MLP, NB, RF and DT, on a 1,00,315 data sets. The ten different characteristics and two new attributes recommended by the authors are taken as input. They use hyperparameter tuning to select best four attributes for final training.

Shirazi H (2023)have proposed an intra-URL application that utilizes information from Yahoo and Google to create a URL search engine for fake URL detection. The proposed search engine utilizes ML to identify phishing URLs within a real-world data collection. They use lexical and domain attributes collected from websites for fake website detection. Safi A (2023) have employed a genetic algorithm (GA) to optimize the parameters of multiple ML algorithms, including RFC, AdaBoost, Gradient Boost(GB), and Bagging. The optimized models were ranked according to its accuracy, and the top three ML algorithm were selected as primary classifiers and tested using two distinct phishing datasets.

From this literature survey, RFC is popular and good for phishing website detection. In this research utilize the three ensemble ML RFC, DTC and SVM for learning (Ganesh P ,2023) and it shows the accuracy of RFC is better compared with others.

3. Proposed Methodology

In this research, The phishing websites or blocked URLs are detected using RFC and give alert to end users while browsing or accessing a phishing site(Al-Tamimi,2023). This system can be utilized for identification and authentication purposes, serving as an authorized tool to prevent individuals from being misled. The overall flow of proposed methodology is shown in Figure1. It has three stages

1. Data set creation and preprocessing
2. Training
3. Prediction

3.1 Data Set Creation and Preprocessing

The data is gathered from Google search operators and the Phish tank archive. Initially the features are extracted from collected URL for detecting phishing domains. Some parameters used for detection is

- Attributes of URL
- Attributes of Domain

- Attributes of Page
- Attributes of Content

After feature extraction, use several natural language processing is applied to clean and structure the dataset.

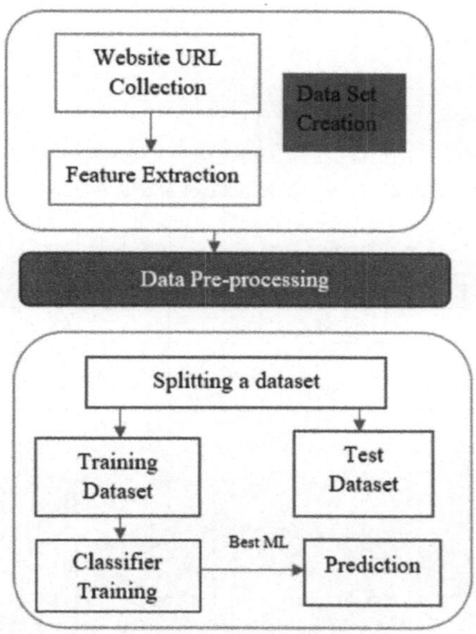

Fig. 22.1 Proposed methodology

3.2 Training

This research includes three machine learning algorithms for training.

1. RFC
2. SVM
3. DTC

The ML classifiers learn the model using training data set.

3.3 Prediction

The best model is selected based on accuracy and it is deployed for website for final prediction.

4. Results and Discussion

The experiment is carried out using a Python-based web program for detecting phishing websites, with the selected best ML algorithm being utilized. Initially data are collected from phish tank archives. The collected dataset has 2456 instances and 18 features in it. Integer values 0 and 1 are used as attribute values, with 0 denoting authentic content and 1 denoting phishing. To conduct the experiment for phishing website detection, the dataset is first processed to obtain mature data in the required format, and then split into two halves - 75% for training and 25% for testing. The subset of the dataset used to train the machine learning model.

In this research three algorithms were used to detect the phishing sites. Based on three algorithms RFC algorithm has given more accuracy about 83.80%. This is shown in Table 22.1. The RFC is deployed and check the current URL is legitimate or fake. This is illustrated in Fig. 22.3 and Fig. 22.4.

1	Domain	Having_@_	Having_IP	Path	Prefix_sul	Protocol	Redirectic	Sub_dom	URL_Leng	age_dom	dns_recor	domain_r	http_toke	label	statistical	tiny_url	web_traffi
2	www.liquidgeneration.com	0	0	/	0	http	0	0	0	0	0	1	0	0	0	0	2
3	www.onlineanime.org	0	0	/	0	http	0	0	0	0	0	1	0	0	1	0	1
4	www.ceres.dti.ne.jp	0	0	/~nekoi/s	0	http	0	1	0	1	0	1	0	0	0	0	0
5	www.galeon.com	0	0	/kmh/	0	http	0	0	0	0	0	0	0	0	0	0	0
6	www.fanworkrecs.com	0	0	/	0	http	0	0	0	1	1	1	0	0	1	0	1
7	www.animehouse.com	0	0	/	0	http	0	0	0	0	0	1	0	0	1	0	1
8	www2.117.ne.jp	0	0	/~mb1996	0	http	0	1	0	1	0	1	0	0	0	0	2
9	archive.rhps.org	0	0	/fritters/y	0	http	0	2	0	0	0	1	0	0	0	0	2
10	www.freecartoonsex.com	0	0	/	0	http	0	0	0	0	0	1	0	0	0	1	2
11	www.cutepet.org	0	0	/	0	http	0	0	0	2	0	0	0	0	0	0	2
12	www.taremeparadise.com	0	0	/	0	http	0	0	0	2	0	2	0	0	0	0	2
13	www.internetdump.com	0	0	/users/po	0	http	0	2	2	0	0	1	0	0	0	0	1
14	darkkaminari.net	0	0		0	http	0	0	0	1	1	1	0	0	1	0	1
15	www.iei.net	0	0	/~bkos1/v	0	http	0	2	0	0	0	1	0	0	1	0	1

Fig. 22.2 Phishing dataset and its features

Table 22.1 Accuracy of ML classifiers

S. No	Algorithm	Accuracy
1	RFC	83.80%
2	DTC	80.89%
3	SVM	80.78%

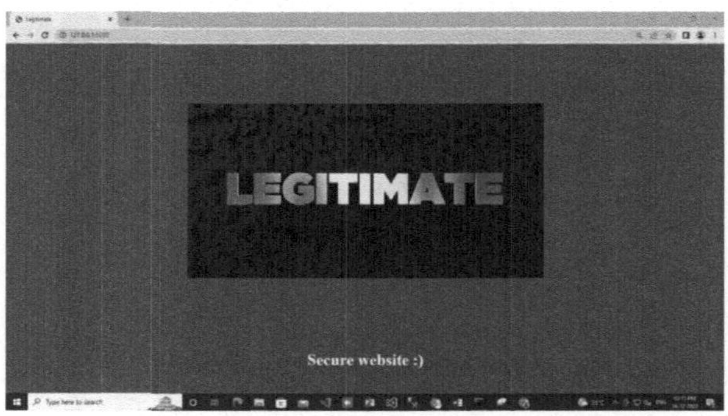

Fig. 22.3 Leigtimate URL detection

Fig. 22.4 Fake URL detection

5. Conclusion

Phishing refers to a type of cybercrime that involves the use of social engineering and fraudulent tactics to obtain sensitive personal information. In addition, phishing is any other not unusual sort of rip-off. In this research, experiments have been done with collected real time phishing datasets and tested the three ML classification algorithms based on accuracy. The cause of this studies work is to predict whether or not a given URL is a hacked website or no longer. From this analysis, it seems that the RFC is the pleasant classifier with excessive class accuracy of 83.5%. In future, this research is extended to collect more features and use optimization along with feature selection to improve the accuracy and other performance metrics.

References

1. Alsenani, T. R., Ayon, S. I., Yousuf, S. M., Anik, F. B. K., Chowdhury, M. E. S. (2023). Intelligent feature selection model based on particle swarm optimization to detect phishing websites. Multimedia Tools and Applications, 1–33.
2. Menaka, S., Harshika, J., Philip, S., John, R., Bharathiraja, N. and Murugesan, S., 2023, February. Analysing the Accuracy of Detecting Phishing Websites using Ensemble Methods in Machine Learning. In 2023 Third International Conference on Artificial Intelligence and Smart Energy (ICAIS) (pp. 1251–1256). IEEE.
3. Hussain M., Cheng, C., Xu, R., Afzal, M. (2023). CNN-Fusion: An effective and lightweight phishing detection method based on multi-variant ConvNet. Information Sciences, 631, 328–345.
4. Jha, A. K., Muthalagu, R., Pawar, P. M. (2023). Intelligent phishing website detection using machine learning. Multimedia Tools and Applications, 1–26.
5. Shirazi H, Muramudalige S. R, Ray I, Jayasumana, A. P, Wang, H. (2023). Adversarial Autoencoder Data Synthesis for Enhancing Machine Learning-based Phishing Detection Algorithms. IEEE Transactions on Services Computing.
6. Safi A, Singh S (2023), A Systematic Literature Review on Phishing Website Detection Techniques, Journal of King Saud University-Computer and Information Sciences.
7. Ganesh P, Kalaiarasi S (2023), SVM and Random Forest Algorithm are used well to detect phishing attacks for enhanced accuracy, Journal of Survey in Fisheries Sciences, 10(1): 2982–2988.
8. Al-Tamimi Y,,Shkoukani, M. (2023). Employing cluster-based class decomposition approach to detect phishing websites using machine learning classifiers, *International Journal of Data and Network Science*, 7(1): 313–328.

Note: All the figures and table in this chapter were made by the author.

Disruptive Technologies for Sustainable Development – Dr. G. Nagappan et al. (eds)
© 2024 Taylor & Francis Group, London, ISBN 978-1-032-55000-8

Wearable IoT Device Enabled with Geo-Fencing for Monitoring Mentally Retarded Child

23

J. Sivasankari[1]
Associate Professor, Velammal College of Engineering and Technology,
Madurai, Tamilnadu, India

T. Maha Priya[2], **U. Sakthi Malavika**[3],
S. Yoga Pandeeswari[4]
Student, Velammal College of Engineering and Technology,
Madurai, Tamilnadu, India

Abstract: Approximately 89% of girl children and 26% of boy children are subjected to abuse. A child goes missing somewhere in the globe every 46 seconds. Children are the foundation of a country; if their future was threatened, it would have an effect on the development of the whole country. Children who experience abuse lose their emotional and psychological stability, resulting in a negative impact on their career and future. The things that happen to these vulnerable kids are not their fault. However, due to the state of the economy and a desire to concentrate on the future and profession of their child, parents are obligated to earn money. Consequently, it becomes challenging for them to constantly adhere to their kids. We offer a setting in our system where this issue may be effectively handled. It enables parents to keep an eye on their kids in real time without having to intervene manually, almost as they were standing next to them, while concentrating on their own careers. This motivates us to develop an autonomous IOT wearable device to monitor the children's activities and their location without any intrusion.

Keywords: IoT, Child safety, Geo-Fencing, Wearable device

1. Introduction

Children with autism struggle to communicate with others, especially other children their own age. Children struggle emotionally to express themselves and communicate with others. The prevalence of autism has greatly increased during the past 40 years. Tens of millions of people worldwide, including two million Americans, have autism. Almost 2 % of the kids suffer from autism as per the recent report. The development and actions of children with autism may be better understood by specialists with the aid of diagnostic instruments. Also, it offers crucial data outlining the advantages and disadvantages of kids. As a result, paediatricians will have more opportunities to implement effective therapeutic approaches. The child abuse issue has also become more frequently active issue and it still a challenging issue for the parents to monitor their children. Currently, girls make up approximately 89% of all incidents of child abuse recorded globally, with boys making up the remaining 26%. Every 46 seconds, a child is reported missing somewhere on the earth. Abused children lose their emotional and psychological stability, which has a detrimental effect on their future and career. These fragile children are not to blame for the repulsion that occurs to them. Parents are compelled to longing for money, nonetheless, due to the status of the economy and a desire to focus on the future and career of their child. As a result, they find it difficult to consistently be there for their children.

[1]jss@vcet.ac.in, [2]mahapriyat@gmail.com, [3]malavika180202@gmail.com, [4]yogapandeeswari123@gmail.com

DOI: 10.1201/9781003428473-23

We provide a configuration in our system where this problem can be dealt with successfully. It enables parents to focus on their own careers while keeping an eye on their kids in real time without having to actively intervene, almost as if they were standing next to them. This inspires us to create an autonomous Internet of Things wearable device to discreetly track the children's whereabouts and activities. Skin's electrical conductivity can be detected using galvanic sensor, assisting carers in making an autism diagnosis, suggested in (Saude and Vardhini.2020). As a result, a GSR sensor can be used to measure how autistic children respond to a variety of emotional states, including happiness, hunger, melancholy, and anxiety. Face-to-face interaction is the most popular therapy strategy for kids with autism, insisted by (Rahman et al. 2022). Although expensive treatments are effective, folks with limited incomes or those who live in rural areas cannot easily acquire them.

The proposed approach enables distant specialists to conduct data analysis and appropriate therapeutic actions. Despite their high cost, effective therapies are difficult for those on low incomes or living in rural locations to access. 15,042 cases of missing children were recorded between 2011 and 2019, according to a figure made public by the Royal Malaysian Police in 2020, and 363 children have still not been located. As the number climbed, there were 1523 incidences of parental neglect documented.

Making matters worse, parents were unable to monitor their children's whereabouts or even receive notifications when their kids left a secure location, particularly their house or school. With smart location tracking software, parents can track their children's whereabouts in real time thanks to the development of smartphone technology. Location of the children and time synchronisation are provided through a sophisticated location tracking application that makes use of Differential GPS technology to ensure with Geo-Fencing based app development (Rameesa et al. 2018)This was done to make sure that the parents and the police would have some background information or hints about any position of the kids may have had prior to going missing.

2. Problem Statement and Related Work

Current technologies for kid security systems work to detect a child's presence in a school or on a school bus by using RFID chips, GSM modules, and GPS devices (Gupta and Harit 2016, Jabbar et al. 2019). Other than the child's location, this kind of technology offers no security features. The majority of the user interface will take the shape of a website or be tied to an unfriendly external monitoring system (Moodbidri et al. 2017). Consider that the current method only produces the child's position under specific situations when the child is believed to be in unsafe settings. Objective of the proposed work is to develop an autonomous system, offers comprehensive child security, is tracked in real-time, and reports its findings to a parent application. For speedy results, data is sent between the system's many components using the elastic search technique.

Farman Ullah (2021) proposed a geo fencing enabled wearable device for COVID-19 patients to monitor their vital parameters and to get keep posted of the patient's location in 24*7 manner and the measurements was analyzed using RSSI level based Geo-fencing technique. A smart wearable device system for women's safety is implemented using GPS and GSM technologies by DhirajSunehra's (2021).

With embedded modules, a smart child safety wearable can be created. Wearable technology adoption extends beyond the consumer sector. Its effects are also becoming more and clearer in the business world. With the proliferation of smart, connected devices, there are now more ways than ever to equip workers to perform their jobs, enabling faster, safer, evidence-based decision-making and fostering greater business agility and collaboration.

3. Proposed Work

The proposed methodology relies on a variety of functionalities, including child application layers for visualization and analysis, wearable sensor layers, IoT gateway layers, cloud server layers, and so on. IoT environment enabled as a wearable device will collects the information based on Bluetooth and Wi-Fi. Geo-Fencing enabled IoT device will monitor the location and circumference area of a child within the updated database. Cloud servers make use of Raspberry Pi equipped with deep learning neural network algorithm to analyze data and do remote monitoring with user authorization. Child abuse and monitoring the mentally retarded child can be reduced by using geo-fencing enabled real-time sensor data, by monitoring the susceptible and abnormal circumstances, depicted in the Fig. 23.1.

3.1 Methodology

This project includes three stages of indulgence namely, data acquisition, data classification, alert notification and updation of current location.

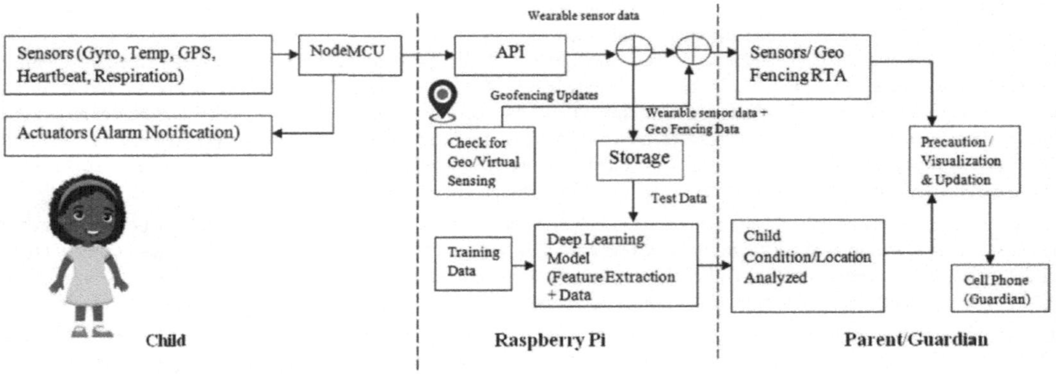

Fig. 23.1 Functional architecture of proposed methodology

3.2 Data Acquisition

Various wearable sensors are utilized to monitor the vitals of child's body. And the data collected will be transferred from sensors to Raspberry Pi 3B+ model.

3.3 Data Classification

(a) Raspberry pi 3B+ model has been used for classifying the abnormality concerning the vitals collected from the various sensors.

(b) Based on the literature survey of various classification algorithm for the topic entitled **"Autonomous geo-fencing enabled IOT device for ensuring child safety",** best classification techniques has to be identified

(c) Existing well known classification algorithm using machine learning was examined for accuracy calculation. As the decision of child safety system is sensitive, accuracy plays an imperative role. Hence the deep learning based classification model developed in python language will be loaded into and used for prediction.

The virtual geo-fencing system will be set up utilizing the GPS module. Geofencing will be turned on in our project for any certain geographic location. A warning message will be delivered to the parent smart phone if the youngster leaves the geo-fencing. The pi camera that comes with the Raspberry Pi 3B+ model may be used to improve security measures.

3.4 Alert Notification

After successful classification, the alert notification will be generated for two cases, such as of distress situation and also if the child moves out of the geo-fencing and sent to the parent's mobile via SMS. This project focuses mainly on the autonomous operation of child safety system. And it helps to reduce security threats to the children and gives relief for the parents.

4. Results and Discussion

The proposed Activity Monitoring and Geofencing application's functionality is tested in an experiment. A youngster is given a GPS tracker to carry during the experiment so that the child's whereabouts may be monitored in real time.

Evaluation of the suggested application Longitude and latitude coordinates are obtained from satellites by the GPS tracker, which then sends the information to a cloud database and shown in Fig. 23.2. The location information was then successfully collected; by retrieving the information from cloud and parent's smart device will displays the location.

Application that shows the where abouts of the child will update the live location of the child and the instant image was captured using Pi Camera, as they enter or leave the geo-fence region while wearing the GPS tracker, shown in Fig. 23.3.

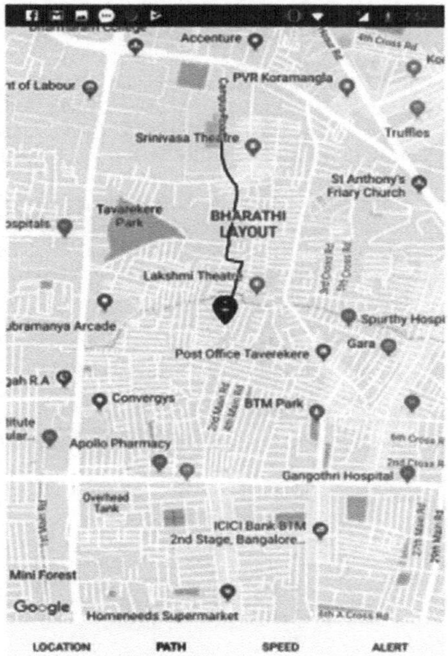

Fig. 23.2 Location of mentally retarded child—Geo fencing enabled with GPS

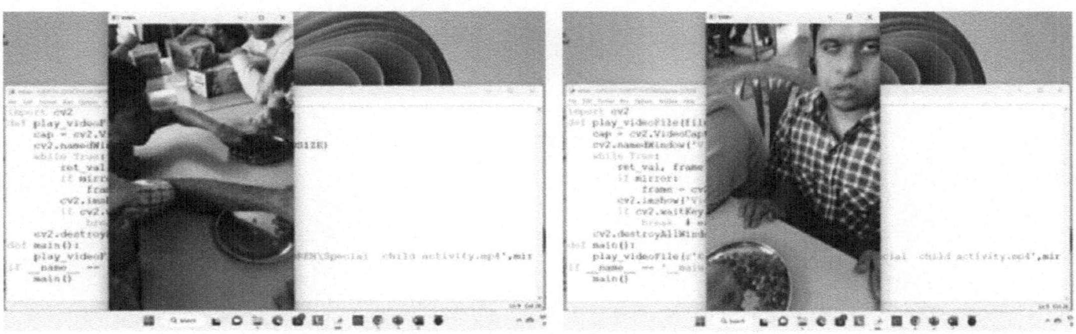

Fig. 23.3 Images captured using Pi-Camera connected with Rasperry-Pi

5. Conclusion and Future Scope

The mentally retarded child will be continuously monitored by this technology, and the parent will receive real-time information. Parents can use an android application to follow their child's movements in real time and to assist in monitoring their health using this approach. With the use of this technology, more precise values may be obtained, and real-time applications can exchange data quickly and with minimal memory usage. This system can also be equipped with an alarm system that will sound in risky situations and can be set to call the parent's phone or a nearby police station in case of emergency. In the future, deep learning based algorithms will be used to analyse images engaged by the Pi-Camera to decide with the distress situation.

References

1. Rahman, S., Ahmed, S. F., Shahid, O., Arrafi, M. A., & Ahad, M. A. R. (2022). Automated detection approaches to autism spectrum disorder based on human activity analysis: A review. Cognitive Computation, 14(5), 1773–1800.
2. Ullah, F.; Haq, H.U.; Khan, J.; Safeer, A.A.; Asif, U.; Lee, S. (2021). Wearable IoTs and Geo-Fencing Based Framework for COVID-19 Remote Patient Health Monitoring and Quarantine Management to Control the Pandemic. Electronics 2021, 10, 2035.
3. D. Sunehra, V. S. Sreshta, V. Shashank and B. U. Kumar Goud.(2020).Raspberry Pi Based Smart Wearable Device for Women Safety using GPS and GSM Technology. IEEE. INOCO*N*, pp. 1–5.
4. N. Saude and P. A. H. Vardhini.(2020). IoT based Smart Baby Cradle System using Raspberry Pi B+ , ICSIDEMPC, pp. 273–278.
5. W. A. Jabbar, H. K. Shang, S. N. I. S. Hamid, A. A. Almohammedi, R. M. Ramli and M. A. H. Ali. (2019). IoT-BBMS: Internet of Things-Based Baby Monitoring System for Smart Cradle. *IEEE Access*, vol. 7, pp. 93791–93805.
6. Rameesa. O, C Periasamy and Priyanka. M. T. (2018). Real Time Child Monitoring System Based on Raspberry Pi And Beacon Technology Using Android App. Int. J. of Adv. Res. **6** (Mar). 1410–1416.
7. Moodbidri and H. Shahnasser, (2017). Child safety wearable device. International Conference on Information Networking (ICOIN*)*, Da Nang, Vietnam,pp. 438–444.
8. A. Gupta and V. Harit.(2016).Child Safety & Tracking Management System by Using GPS, Geo-Fencing & Android Application: An Analysis.*Sec.Int.Con.on CICT,* pp. 683–686.
9. Wang, Yixin, et al.(2015).RSSI-based bluetooth indoor localization. Int. Conf. *MSN, 1–5.*
10. Ninad Tanksale, Ajay Vedpathak, Amey Panse.(2015). Cloud Based Child Tracking System Using Raspberry Pi, Int. Journal of Science and Research.

Note: All the figures in this chapter were made by the author.

Disruptive Technologies for Sustainable Development – Dr. G. Nagappan et al. (eds)

Building Real Estate AntiForge Platform Using Blockchain

24

K. Suresh Kumar[1]
Associate Professor, Department of Information Technology,
Saveetha Engineering College, Chennai, India

Sanjai Kumar N.[2]
Department of Information Technology,
Saveetha Engineering College, Chennai, India

Abstract: The value of real estate to a nation's economy cannot be overstated. Despite the lengthy history of real estate transactions, fraud continues to thrive owing to the industry's lack of transparency and immutability. From the time a person begins looking for a home to the time they actually buy one, there are several places where the present system may be gamed. However, purchasers may end up paying more than they should if middlemen are involved. Transparency and illiquidity issues, high selling costs, specific impulses, and sluggish sale procedures are all examples of inefficiencies within the real estate market that might lead to a bubble bursting. A new software paradigm for controlling digital resources in low- or no-trust environments has been made possible by blockchain technology. Commemorative events are used for business transactions, the dissemination of empirical data, and the promotion of online and offline cooperation. Due to the immutability of Blockchain data, this project intends to create a decentralised housing product in the Real Estate Marketplace in order to avoid title disputes that may be resolved with the help of the Ethereum Blockchain. Each asset is represented by an ERC721 token. The property owner must provide proof of ownership prior to the minting of tokens representing the title of the properties. To show ownership, one must thus create a pre-knowledge proof using zokrates and zk-SNARKs. The use of zk-SNARKs allows for the development of a verification system that can provide proof of ownership without disclosing sensitive information. After the token's validity has been established, it will be listed on OpenSea, making it possible to buy and sell the tokenized resources.

Keywords: ZK-SNARKs, NFT, OpenSea, Decentralized, Ethereum, Blockchain, Zokrates, Smart contracts, Rinkeby

1. Introduction

In today's world, property is vulnerable to scammers and thieves. The project's primary goal is to promote openness to the real estate industry and ensure that no fraudulent activities are possible as a result of phoney contracts. The best course of action is to develop tamper-evidence methods and to bar the liability of third parties from financial dealings. Because of this, the immutable, tamper-evidence Blockchain technology has begun to spread its roots outside the realm of cryptocurrencies. By eliminating the need for all parties to agree on a common third party as part of a transaction's verification process, blockchain technologies constitute a new class of distributed systems (Karl Wust and Arthur Gervais, 2018).

The RoR of the Department of Revenue and Land Reforms is making the right decision as it enables more transparent fiscal operations and negotiations. The land register uses blockchain technology to enable secure property transfers (Arati Baliga et al. 2018).

[1]sureshkumar@saveetha.ac.in, [2]sanvettu@gmail.com

DOI: 10.1201/9781003428473-24

The idea of smart contracts allows for automated record streamlining, to the buyer through an operation form. By using blockchain technology and smart contracts, ownership rights may be established with absolute certainty (Vishnu Chandra and Baladevan Rangaraju, 2017). Trust in business transactions has been bolstered by this technology, which facilitates the automated monitoring and enforcement of agreements. It strengthens client trust in the government, makes information easier to obtain for clients, and makes land records more secure (Gavin Wood et al. 2014). The sales process is open, secure, and immune to manipulation and censorship. After agreements are permanently recorded. Blockchain technology is a revolutionary innovation that has the potential to displace current business models reliant on third parties to establish and maintain trust (Stefan Tai, Jacob Eberhardt and Markus Klems, 2017).

Second-generation blockchains (like Ethereum) were established in 2014, and they enable users to write and run their own code, known as "smart contracts," across the whole network of nodes. As a result, anybody may develop and release software on a universal network (Alexandru Oprunenco and Chami Akmeemana, 2018).

2. Literature Review

Several research projects are done to prove who owns a piece of property. In this section, talk about some of the work you did before to show that you own the property.

The work of X. Zheng et al. (2019) considering both Efficiency and Safety, Blockchain's properties of decentralisation make it an ideal match for a wide range of industries. it provides a concise overview of the investigation's performance and security problems. Moreover, it conducts tests on Ethereum and conducts a survey of other prominent blockchain platforms in order to understand more about the scalability characteristic of blockchain, before moving on to talk about its possible uses and the lessons it has taught. In the end, it tries to pinpoint the unanswered questions and promising avenues for further study.

Study of Marcus Walshea et al. (2019) by taking into account energy consumption and computational overheads, modern authentication algorithms aim to create authorised sessions via insecure channels with a minimal resource footprint. Methods using just public/private key architecture need more time and effort to compute and maintain. Interactions inside the protocol to pinpoint problem show that our NIKP method works well for authentication even in low-resource IoT settings.

The empirical evidence of Chao LIN et al. (2020) addressing operational issues associated with high-density sensor and actuator deployment in a smart city requires a scalable and reliable crowd management infrastructure. The use of crowdsourcing in crowd management is common, however most existing systems rely on a third-party platform for security and reliability. A few of incentive-based blockchain-based crowdsourcing systems (BCSs) have been developed, however they are not built with security in mind. As a result, we investigate the necessary safeguards for a private and reliable BCS and provide a workable alternative.

3. Methodology

Figure 24.1 Shows the architecture of the proposed methodology

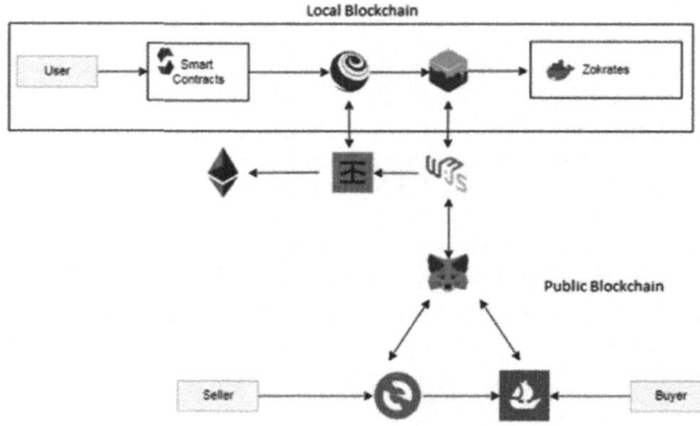

Fig. 24.1 Proposed architecture

It has been deployed locally using truffle and ganache and subsequently to a network using rinkeby. The objective is to create a decentralised home product for sale on the Real Estate Marketplace. In this case, the ERC721 Tokens represent the individual properties. A property's owner must provide proof of ownership before tokens representing the properties may be minted. To show ownership, one must thus create a pre-knowledge proof using zokrates and zk-SNARKs. zk-SNARKs are implemented in a verification system to show ownership of real estate without disclosing private details about the property. It talks to the deployed Rinkeby smart contract using web3/metamask. After the token has been approved, it will be listed on a trading platform like OpenSea, where its buyers and sellers will be able to transact in the tokenized version of the underlying asset. There are five parts to the prototype's implementation:

1. Local Deployment
2. Zokrates Setup
3. Network Deployment
4. Minting Tokens
5. Opensea Storefront

Code from Visual Studio is used to create the smart contracts. Truffle helped us create the legal documents. The solidity-based smart contracts were then deployed using ganache for local migration and a web3 provider. The zokrates prototype runs on a local Windows 11 machine equipped with an Intel i5-1135G7 processor running at 2.4 GHz and 8GB of DDR4 memory.

3. Results and Discussion

3.1 Local Deployment

1. *Ganache:* At first, "truffle compile" is used to write and compile contracts for smart contracts". It will display the list of available accounts with public and private keys. Figure 24.2 shows the result of start of local blockchain called ganache.

Fig. 24.2 Ganache local blockchain

2. *Implement Zokrates:* There are only two persons involved, the Prover and the Verifier, therefore the Verifier has faith in his own pre-verification preparations.

3.2 Network Deployment

Figure 24.3 Shows the output of network migration

Five Properties has been purchased and collected in the opensea marketplace as shown in Fig. 24.4.

Figure 24.5 Shows the empirical analysis of ERC token standards such as ERC20, ERC721 and ERC1155

```
2_deploy_contracts.js
=====================

   Replacing 'Verifier'
   --------------------
   > transaction hash:    0xfd7646de73686f71985e4f45df7fd3c32bcc81a60295dbab26
02fb8457ad67ef
   > Blocks: 1            Seconds: 9
   > contract address:    0x224fc04949DA11834Ea1Cc87e1561777C5885042
   > block number:        10806035
   > block timestamp:     1654536602
   > account:             0x1D4396D3DfdFFFB57e2Ae580f175E7DFCf08759A
   > balance:             1.07415333334098163
   > gas used:            1394595 (0x1547a3)
   > gas price:           10 gwei
   > value sent:          0 ETH
   > total cost:          0.01394595 ETH


   Replacing 'SolnSquareVerifier'
   ------------------------------
   > transaction hash:    0x4fa93b9397f8f7a61f47a08642dc50fc445ce441e138c90fdf
3d533415bb1f65
   > Blocks: 1            Seconds: 9
   > contract address:    0xBaFDC0Ca2ca2Dd160CdCE5dDe032Dd02a2e57C7D
   > block number:        10806036
   > block timestamp:     1654536617
   > account:             0x1D4396D3DfdFFFB57e2Ae580f175E7DFCf08759A
   > balance:             1.04077437334098163
   > gas used:            3337896 (0x32eea8)
   > gas price:           10 gwei
   > value sent:          0 ETH
   > total cost:          0.03337896 ETH

   √ Saving migration to chain.
   > Saving migration to chain.
   > Saving artifacts
   -------------------------------------
   > Total cost:          0.04732491 ETH
```

Fig. 24.3 Network migration

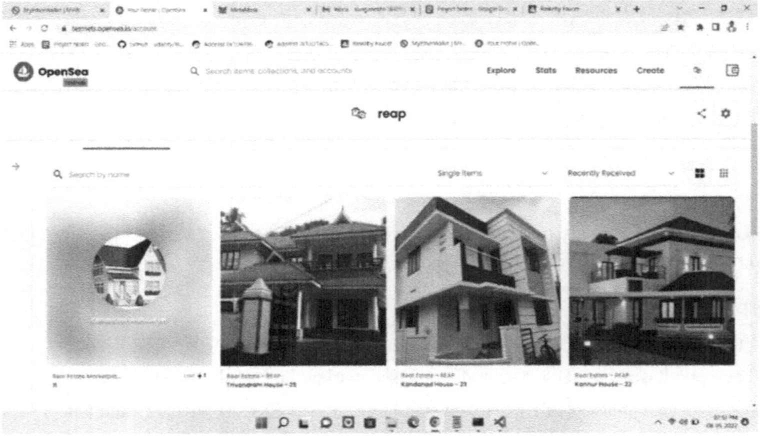

Fig. 24.4 Property listed in opensea marketplace

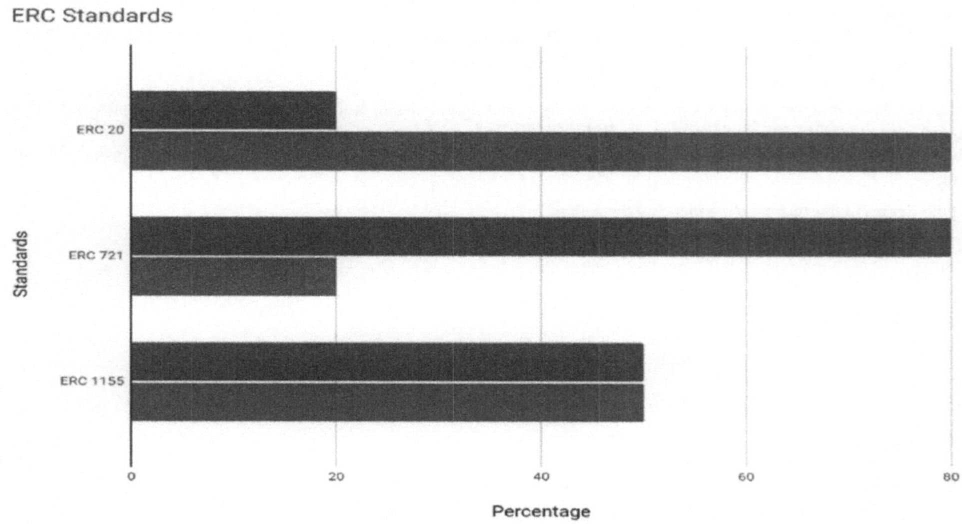

Fig. 24.5 Empirical analysis of token standards

ERC 20: Organizations and governments are the most common users of ERC-20, since it is a set of principles for regulating certain blockchain activity. You probably already know that ERC-20 tokens may be exchanged for other tokens of the same kind.

ERC 721: ERC-721 was designed specifically for tokens that cannot be exchanged for other tokens. In other words, digital assets are created in accordance with these token's standards that are both unique and irreplaceable.

ERC 1155: Under a single smart contract, users were able to create tokens that were both interchangeable and unique. Simply put, it addressed a previously unmet need.

4. Conclusion and Future Work

There is transparency regarding the usage of Blockchain technology for land registration and the part it plays in maintaining the accuracy of property records. The land registration procedure is made more transparent, straightforward, and open with the usage of blockchain technology. This programme is helpful at the land register for addressing inquiries such as "how," "when," "where," "which," and "why" regarding a property's title. Also, it enables us to determine whether any actions were made in a certain nation. Every deed and land title is shown. Indeed, this programme will lead us to progress and easy access to life, not only for us but for future generations as well. In future, the system can be integrated with artificial intelligence to make predictive models for the real estate system and can use to mint ERC1155 token.

References

1. X. Zheng, Y. Zhu, and X. Si, "A survey on challenges and progresses in 1205 blockchain technologies: A performance and security perspective," Appl. 1206 Sci., vol. 9, no. 22, pp. 1–24, 2019
2. Walshe, M.; Epiphaniou, G.; Al-Khateeb, H.; Hammoudeh, M.; Katos, V.; Dehghantanha, A. Non-interactive zero knowledge proofs for the authentication of IoT devices in reduced connectivity environments. Ad Hoc Netw. 2019, 95, 101988.
3. Karl Wust and Arthur Gervais. Do you need a blockchain? In ¨ 2018 Crypto Valley Conference on Blockchain Technology (CVCBT), pages 45–54. IEEE, 2018.
4. Alexandru Oprunenco and Chami Akmeemana. "Using blockchain to make land registry more reliable in india. LSE Business Review", 2018.

5. Arati Baliga, I Subhod, Pandurang Kamat, and Siddhartha Chatterjee. Performance evaluation of the quorum blockchain platform. arXiv preprint arXiv:1809.03421, 2018.
6. Lin, C.; He, D.; Zeadally, S.; Kumar, N.; Choo, K.-K.R. Secbcs: A Secure and Privacy-Preserving Blockchain-Based Crowdsourcing System. Sci. China Inf. Sci.2020,63, 130102. https://doi.org/10.1007/s11432-019-9893-2
7. Vishnu Chandra and Baladevan Rangaraju. Blockchain for property, 2017
8. Stefan Tai, Jacob Eberhardt, and Markus Klems. Not acid, not base, but salt. In Proceedings of the 7th International Conference on Cloud Computing and Services Science, pages 755–764, 2017.
9. Gavin Wood et al. Ethereum: A secure decentralised generalised transaction ledger. Ethereum project yellow paper, 151(2014): 1–32, 2014.

Note: All the figures in this chapter were made by the author.

Disruptive Technologies for Sustainable Development – Dr. G. Nagappan et al. (eds)
© 2024 Taylor & Francis Group, London, ISBN 978-1-032-55000-8

Music Recommendation by Emotion Detection Using SVM

25

Vijayanand M.[1]
Professor, Saveetha Engineering College,
Chennai, Tamilnadu

Prem chand T.[2]**, Vedhadharshan B.**[3]**, Varunkumar K.**[4]
Final Year Student, Saveetha Engineering College,
Chennai, Tamilnadu

Abstract: Music is connected with human emotion, and it has a significant impact on human routine, daily life, and current advanced technologies. Typically, the user's job is to manually browse the playlist or song according to their mood. As part of this study, we propose a robust and nuanced model that can generate a selection based on the user's present mental state and expression. Manual searching of songs take more time and sometime we get bore of it. By using real time emotion detection and music recommendation model the user can enjoy music based on their feelings. Existing techniques for music recommendation and emotion recognition are computationally slow, less precise, and even call for further features. This essay makes practical recommendations for music. Time extraction of facial expressions and gestures, as well as classification into a certain emotion, will automatically create a playlist with a reasonably cheap computational cost. This will make it easier for music listeners to appreciate and interact with music than in the past.

Keywords: Machine learning, Support vector machine, Convolution neural network

1. Introduction

Music assumes a fundamental part in lifting a singular life and it is a craftsmanship, music helps humans to keep calm and overcome from the stress, depression also music plays role as amusement for music lovers and artist. In today world there are many music players are in internet world wide to entertain people they all have unique characters like web design, features. People choose best player which will suit for their everyday use. Music players have highlights like converse, quick forward and backward, variable playback speed, volume balance etc. Although these music player provides better user experience these music players have some disadvantages, music lovers have to search for the songs manually.

While considering emotion based music player the manual searching of song is avoided, while emotion based music player provide feature like real time emotion recognition and music recommendation. Emotion based music player uses machine learning algorithm to extract and recognize emotion using facial gestures. Using support vector machine algorithm facial expression extraction is done and using convolution neural network algorithm expression recognition is done.

2. Literature Review

Devansh Shukla et al. (2022) studies on emotion detection and music recommendation system the emotion of user is captured using facial landmarks and the drawbacks in this system is that the proposed model generates playlist from the online websites.

[1]vijayanand@saveetha.ac.in, [2]premchandt333@gmail.com, [3]vedhadharshan2001@gmail.com, [4]kvarunkumar7266@gmail.com

DOI: 10.1201/9781003428473-25

Vinayak Bali, Shubham Haval et al.(2019) uses recurrent neural network (RNN) algorithm for this model and also uses grayscale images for emotion detection and also had disadvantages that While using RNN algorithm the training process becomes slow and complexity becomes high. Shlok Gilda, Husain Zafar et al. (2017). The author uses deep learning algorithm in this model also uses both image and audio as a input and they also realized that the accuracy percentage is low while using deep learning algorithms. Dhruvisha Bansal, Pinkal Bhatt et al. (2020). This programme recognises the user's facial expressions of emotion and plays music in accordance with those expressions. A machine learning technique called the Support Vector Machine (SVM) algorithm is used to identify the emotions. Te-Hsun- Wang, James Lien et al.(2009). In 2009 the invention basically describes the facial action units that can be used and algorithm for combining 2 facial action units for accurate detection of the emotion and disadvantage is that it describes about only two types of emotion and it is not accurate. Kamal Chandra Paul et al. (2021). The researcher uses viola-jones algorithm for face detection and local binary pattern histogram for processing the images and this research leads to increase the usage of viola-jones algorithm Kamal Naina Soni, Kushagra Agrawal et al. (2021). The currently available methods for automatically constructing playlists are numerically wasteful, lack precision, furthermore, at times require the utilization of subordinate hardware, for example, EEG or sensors. Mr. T. M. Hayath, Ms. Sulthana et al. (2012). The author uses (FER) facial expression recognition based frame works for this model. Banpreet Singh et al. (2020). A machine learning technique called the EMO algorithm is used to identify the emotions. Braja Gopal Patra et al. (2015). A machine learning technique called the EMO algorithm is used to identify the emotions.

3. Suggested System

In the suggested study, the model extracts various expression using web cam provided in system then it compares the captured image with the dataset imported to the model, if the tone of image matches with the data set then it is compared with the section of emotion that is sad, happy, happy or neutral. If it matches then it goes to music playlist where we have stored the songs. Then music is played. Algorithms from the field of machine learning, such as the support vector machine and the convolution neural network, are utilised in these works for the purposes of picture identification and extraction. The dataset is also "pre-processed". Data pre-processing is another step taken with the dataset, and it includes replacing any absent or erroneous values with the mean value or the most consistent value in that column. After the raw data has been cleaned and organised, the treated fields are split into a training group and a test set. We aimed to provide user-friendly music, emotion based music player is an idea of automating multiple interaction between music player and its user.

4. Proposed Architecture

Figure 25.1 is an architecture schematic that details the many moving parts of an audio suggestion system.

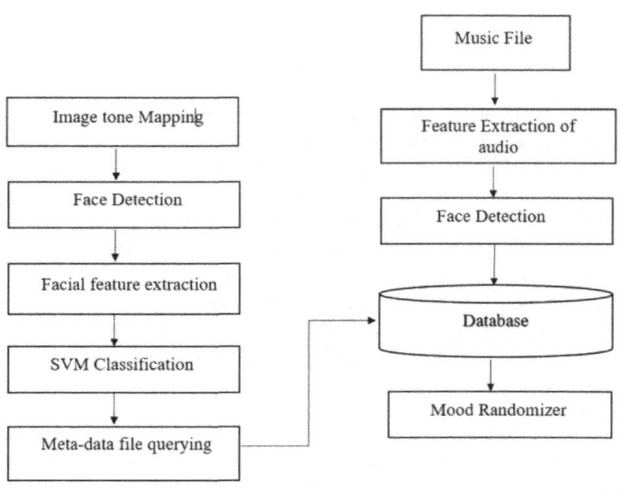

Fig. 25.1 Proposed music system architecture

5. Implementation and Analysis

5.1 Data Set

The data set consists of various facial expression is collected from internet or we can take pictures of our own expression and is given to the model as input data, where emotion is classified. The input images are usually grayscale images.

5.2 Algorithm—Support Vector Machine, CNN

The suggested model was created using the support vector machine algorithm. An example of a guided machine learning method, support vector machines are typically applied to issues involving categorization and regression. In our case the algorithm used for classification of feelings. It is done by facial extraction feature and through query. The extracted image is compared with the grayscale images in data base if it matches then the song based on current feeling is played. In detail first stage is the input image is taken by the model through webcam and in next phase image tone mapping is done following it face detection is done by support vector machine algorithm. Usually support vector machine algorithm uses n dimension to point the surface. By doing so the facial expressions are marked in n dimensional space then it is classified and as mentioned above the input image is compared with database then music system is played. Support vector machine algorithm makes sure of vectors which creates hyperplane in between dimensions which represent extreme instances.

Another algorithm used in this model is convolution neural network. The vital work of this algorithm is image tone mapping. Image tone mapping usually refers to giving adjustments to image such as brightness. The convolution neural network algorithm always deals with grayscale images it gives adjustments to grayscale images nothing but dataset. After the adjustments done to grayscale images the dataset will give effecting matching accuracy.

6. Result and Analysis

The emotion based music player output is represented in Fig. 25.2.

Fig. 25.2 Music player detects happy mood of user

The below graph compares the accuracy of previously used RNN and recently used SVM Algorithm in Fig. 25.3.

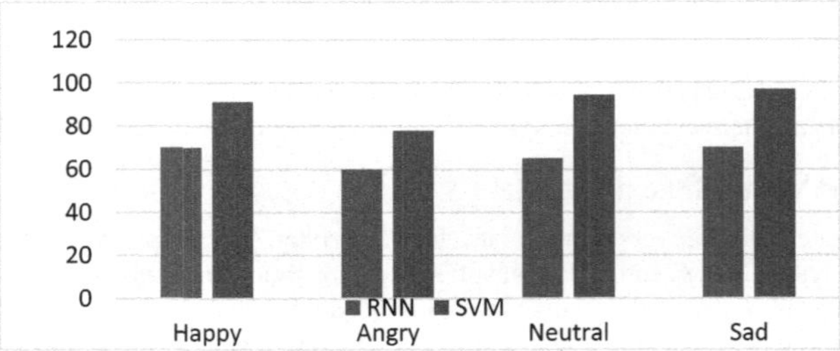

Fig. 25.3 Graphical representation for accuracy comparison between RNN and SVM

The results discovered that the input image given to train the model is less and sometimes the emotion detected varies. Usually, the Gray scale images are provided as dataset to match the real time face seen in web cam. Using the less dataset results in wrong recognition of emotion. Also new dataset is not added frequently. Most of them uses different technique to implement the model but they failed to produce the accurate model also the feeling-based music player gets some margin to handle the picture. The implementation cost is high where it requires the good pixels of image and good quality of camera. The cost of the model may rise further if an expert is enlisted to assist with the process of selecting and operating the necessary tools.

7. Conclusion

The accuracy percentage becomes better while using support vector machine algorithm. After training the model with support vector machine the accuracy of detecting sad mood becomes 96%, happy is 91%, angry is 77% and neutral becomes 94%. And the studies found that accuracy of SVM is better than any other machine learning algorithm. Since for detecting emotion with good accuracy the good pixels of image is needed with good webcam facility. In future we have planned to add datasets with better pixels of image for increasing accuracy. we have planned to reduce the complexity to build the model so that it will be easier for everyone in the society to use the model.

Reference

1. Devansh Shukla, Shivam Singh, Shubham Sawant, Shubhangi Chavan.(2022), Emotional Based Music Player, International Journal of Creative Research Thoughts(IJCRT), ISSN: 2320–2882.
2. Vinayak Bali, Shubham Haval, Snehal Patil, R. Priyambiga.(2019), Emotion Based Music Player, Journal of Software Engineering & Software Testing, ISSN: 2457-0516. Doi: 10.5281/zenodo.264769.
3. Shlok Gilda, Husain Zafar, Chintan Soni, Kshitija Waghurdekar.(2017), Smart music player integrating facial emotion recognition and music mood recommendation, International Conference on Wireless Communications, Signal Processing and Networking (WiSPNET), doi:10.1109/WiSPNET.2017.8299738.
4. Dhruvisha Bansal, Pinkal Bhatt, Megha Dusane, Avneet Saluja, Kushal Patel.(2020), Emotional Based Music Player Device, International Research Journal of Engineering and Technology(IRJET), e-ISSN: 2395-0056. p-ISSN: 2395–0072.
5. Te-Hsun- Wang, James Lien.(2009), Facial expression recognition system based on rigid and non-rigid motion separation and 3D pose estimation, International Conference on Intelligent Computing and Intelligent Systems(IEEE), 42(5): 962–977. doi: 10.1016/j.patcog.2008.09.035.
6. Kamal Chandra Paul, Semih Aslan.(2021), An Improved Real-Time Face Recognition System at Low Resolution Based on Local Binary Pattern Histogram Algorithm and CLAHE, Optics and Photonics Journal, doi:10.1109/CVPR.2001.990517. I-511-I-518.
7. Kamal Naina Soni, Kushagra Agrawal, Navni Pandya, Nupur Agrawal.(2021), Web Application for Emotion-based Music Player, Ijraset Journal For Research in Applied Science and Engineering Technology, doi: 10.22214/ijraset.2021.39143.
8. Mr. T. M. Hayath, Ms. Sulthana, Mr. Kampli Shabbir. (2012), A Novel Method To Design Emotion-Based-Music-Player, International Journal of Engineering Development And Research, ISSN - 2321-9939.
9. Madhuri Athavle. (2021), Music Recommendation Based on Face Emotion Recognition, Journal of Informatics Electrical and Electronics Engineering, DOI: 10.54060/JIEEE/002.02.018.
10. Braja Gopal Patra, Dipankar Das, Sivaji Bandyopadhyay. (2015), Music Emotion Recognition System, International Symposium Frontiers of Research on Speech and Music (FRSM).

Note: All the figures in this chapter were made by the author.

Disruptive Technologies for Sustainable Development – Dr. G. Nagappan et al. (eds)
© 2024 Taylor & Francis Group, London, ISBN 978-1-032-55000-8

Android Based Women Safety Application Using Accelerometer Sensor

26

S. Sasi Kumar[1]
Professor, Department of Computer Science and Engineering,
Saveetha Engineering College (Autonomous), Chennai, India

Aishwarya T.[2]
Department of M.E Computer Science and Engineering,
Saveetha Engineering College (Autonomous), Chennai, India

Edward DansoAnsong[3]
Associate Professor, University of Ghana, Accra, Ghana

Abstract: Women traveling alone in our world today is ballistic. And all over the world, there are so many happenings that question women's safety. This program is designed to guarantee protection for women and exceeds the proposed scheme with advanced functionality. Our project aims to develop women's security applications that have advanced site tracking technology to enable friends and family to receive longitude and latitude via short user message services. Location- sharing features through Short Messaging Services are automatically sent during shaking service from the mobile which is detected by the accelerometer sensor via human shake frequency. The shaking frequency must exceed the emergency threshold of short messaging services. In case the user of this application is running while shaking the phone which meets the threshold frequency then audio gets enabled automatically. The running motion of the user is identified using google fit. The recipient can receive the notification through short messages, and every communication is secured using a secure mechanism. Our system provides the capabilities to sync and send personalized messages to contacts in the local database. There is also a feature that updates the message about the location of the victim from time to time to the receiver.

Keywords: Accelerometer, Threshold frequency, Women security, Location frequency

1. Introduction

Women in our country have been given equal fundamental rights in comparison to all men; however, people have failed to remember this rule. Women predominately contribute to the evolution and developmental expansion of our country; still, they are living in distress. In general, people recognize the importance of women's security, but at the same time, we must also recognize the fact that there is still a need for a support system for any individual's protection while they are alone. Having this app on any individual phone can reduce their risk and assist when they need them, whether he/she is in an unexpected or dangerous situation. The following are the features of the app that set from other apps created to date:

1. Initially, we have to log in and enter our details such as name, gender, mail id, enter the contact numbers and mail id of family members, and friends into the application and allow access to location and media.
2. While traveling alone or whenever an individual senses its need, run the application, it will send an alert message to the synced contacts

[1]sasikumar@saveetha.ac.in, [2]aishuaishu1962@gmail.com, [3]edward@gmail.com

DOI: 10.1201/9781003428473-26

3. As soon as the accelerometer sensor is activated by shaking the phone, it sends the message contain the location link of the victim to all synced contact numbers.
4. Unique feature of our app is a message with a location link is sent continuously to the synchronized contact numbers frequently. In case the step count of the user keeps increasing, it automatically enables the camera and audio access.

2. Literature Survey

Many researchers have done a lot of work in the past using various technologies to improve women's protection. The authors of [13] Aditi Meshrama, Pushpa Choudhary B, Nagendra R Velagab focused on two aspects while creating this project. The first is self-defense, and the second is to give the victim's location to specific access numbers. In this project, we use the accelerometer, location sharing, and Google Fit. They allow for the prediction of the application's performance or achievement. To measure the sensitive nature of accelerometers, the Frequency Ratio Method (FRM) and the Minimum Point Method (MPM) are used. These two methods are modified here to make them suitable for testing accelerometers. To evaluate a commercial MEMS (Micro-Electro-Mechanical Systems) accelerometer's capability, its sensitivity was tested at various base frequencies. Accelerometers are sensors that detect how quickly something moves. The test results are in good agreement with the manufacturer's stated sensitivity of 24.2 mV/g [4] Laser Self - Mixing Grating Interferometer for MEMS (Micro-Electro-Mechanical Systems) Accelerometer Testing. By fine-tuning the circuitry and dynamics of the machine, more precise testing results can be obtained. Because of its high precision, simple mechanical structure, and stability, the technique is particularly well suited to industrial applications. An accelerometer has been implanted based on the above observations. The role of activity recognition requires the mapping of accelerometer data in time series to a single physical activity. Six basic activities cover a great deal of time on a normal day: hiking, jogging, climbing up and down, sitting, standing, and lying. We are looking at six common activities. The mobile is supposed to be in the user's pocket, bag or hands. The accelerometer calculates gravity speeds of approximately 9.8 m/s^2, which is included in the values of the y-axis for activities in which the user is in an upright position. Location sharing is a basic feature of location-based services (LBSs) that helps to share a user's present location information with other users. This project makes use of Google Maps to predict the user's position. Various map applications were tested, including HereWeGo, Waze Maps, and Google Maps. All sensor sharing messages created within the sharing range will be forwarded without redundancy control in the pure sensor sharing scheme.

3. Proposed Methoology: Architecture

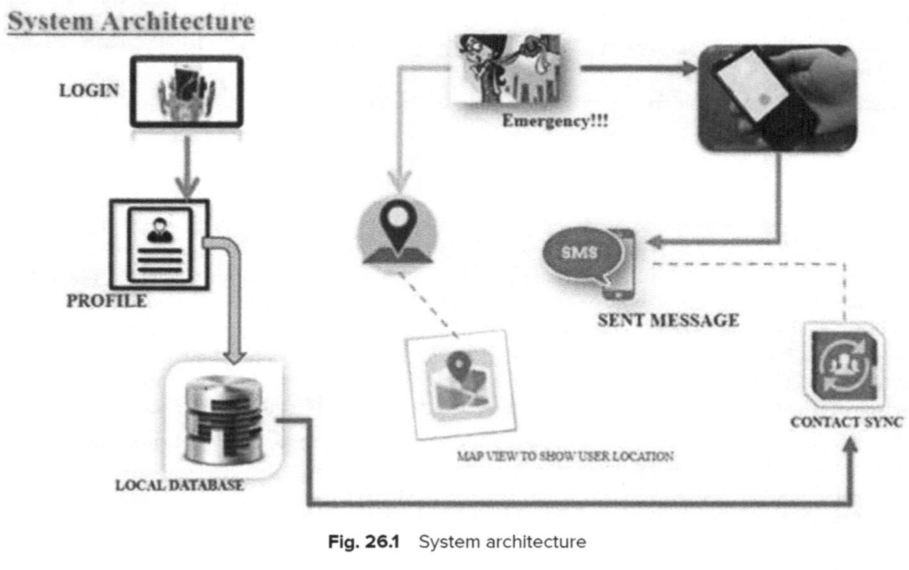

Fig. 26.1 System architecture

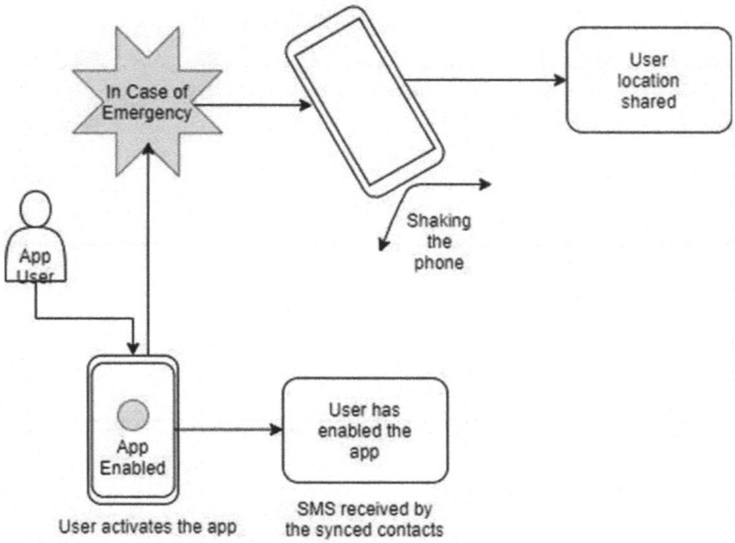

Fig. 26.2 Sub architecture

The application initially consists of a login module where in a profile for the user is created with username and password. The login process also includes the addition of contacts to which the emergency alert message is supposed to be going. Necessary contacts with their name and phone number are added. The contacts can be further selected, updated and deleted if necessary. The location of the user is detected by the GPS provider and the URL of the location along with the latitude and longitude of the location is sent as Short Service Message. Time to time location of the user is sent via Short Service Message so as to update the very precise location of the victim.

4. Modules

Login/Register: Initially, after the app is downloaded the user will register with the basic details like name, address, email, contact number, username, password. Along with this SOS contacts are added. These contacts are editable.

Shake Service: The Shake Service ultimately uses an accelerometer that is built in any android phone. The mobile's body posture is calculated by monitoring the shift in acceleration in three axes with a tri-axial accelerometer. The user enables the app in case of an emergency. Simultaneously, the step count is calculated. If the count keeps increasing constantly, the camera and audio are enabled

Location Sharing: The module is predicting the location, which is a built in user's side, uses the present location information. So there's a timely update on the location in case of travel. If the user is in an emergency, the shake service is activated which in turn creates an URL and sends an SMS/online message to the SOS contacts. It shares the very exact latitude and longitude.

Online/Offline Message Notification: When the user enables the SOS button in the application, it gives access to the accelerometer sensor. When the user shakes the phone and activates the accelerometer sensor SMS with a warning sign is sent to the synced contacts. These contacts are in the local database. Online message service using email and other messaging platforms like Whatsapp is activated when the internet supply is on.

5. Accelerometer Data

Let x y z axes are the accelerator meter in acceleration sensor. This acceleration data provide the magnitude of the acceleration and direction of the acceleration. if the device is moved from left to right, the axis generate a positive acceleration. if the device is moved from right to left, the axis generate a negative acceleration. The Fig. 26.3 show the following sequence:

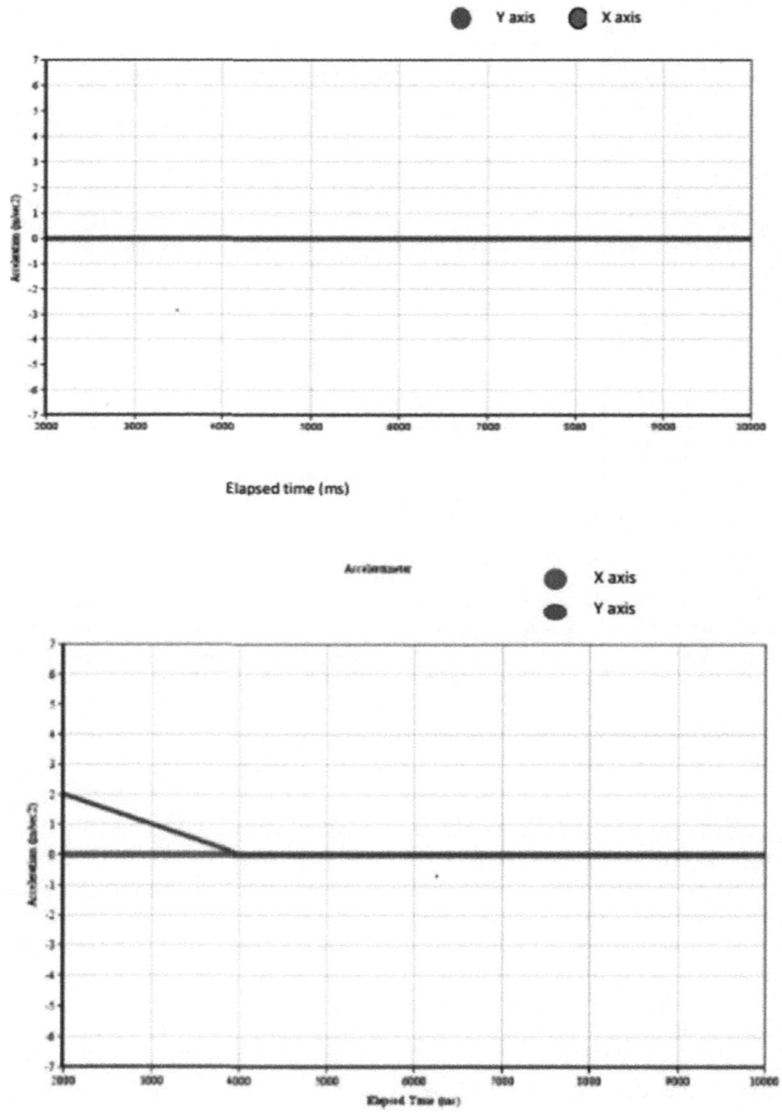

Fig. 26.3 Acceleration graph

1. The device should be laid flat back for 10 seconds.
2. The device can be move from left to right along the x axis.
3. The device can be motionless for 10 seconds.
4. The device can be move from right to left along the x axis.
5. The device can be motionless for 10 seconds.

6. The device can be move from left to right along the x axis.
7. The device can be motionless for 10 seconds.
8. The device can be from right to left along the x axis.
9. The device can be motionless for 10 seconds.

6. Result

This application has been created for woman safety where we exclusively inclueded sos contact upto any number. Accelerometer which inbuid in every smart phones. So based on this we added video call option and call option also to enable sos contact that we are in danger. This might help them in identifying the exact location using latitude and longitude of application.

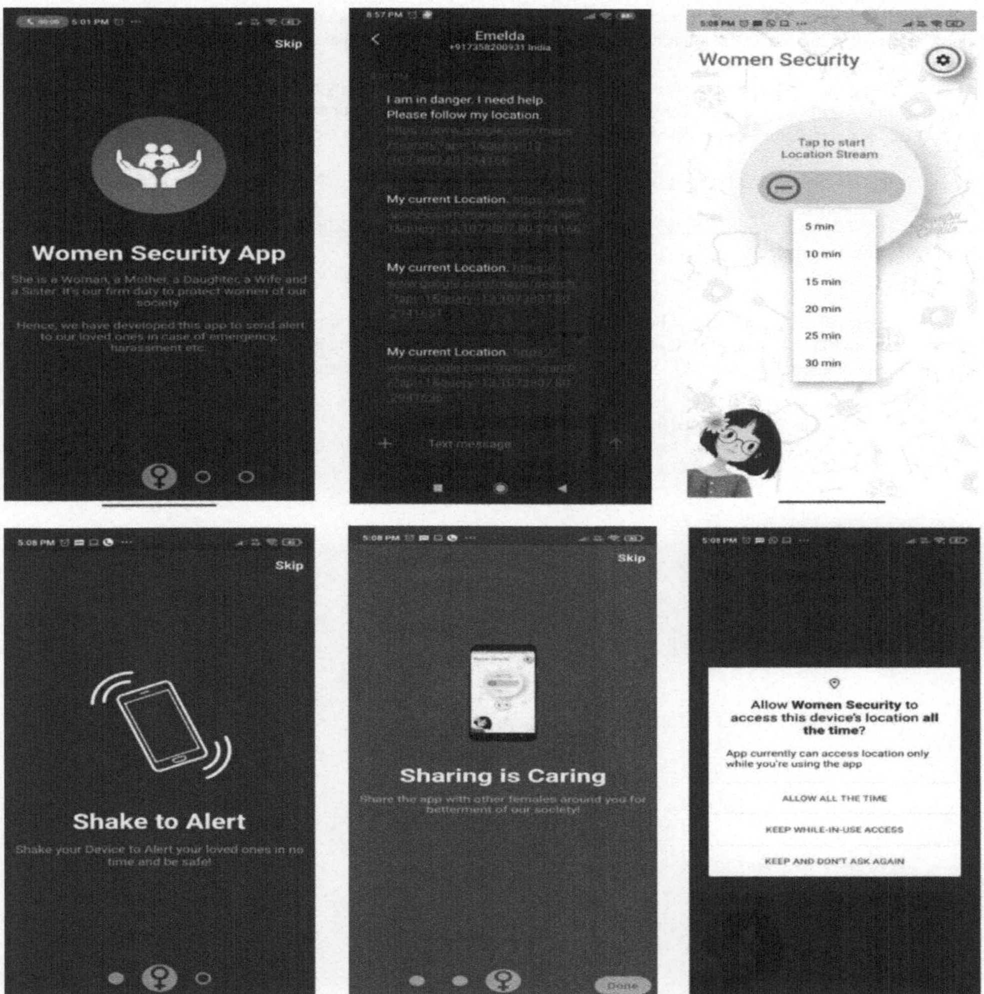

Fig. 26.4 Screenshot of woman security application

7. Conclusion

A nation might not make strides without women's advances, but we can see that women suffer a lot from abuse, which is an obstacle to their progress. That's why we are decided to work on this project which benefits millions of women. The biggest advantage of this protection mechanism is that women can feel confident as they head out as soon as possible to aid when in danger through this app. An additional objective of this app is not only sending messages through email and online messaging platforms but also access the camera and audio, it screens it to synced contacts which makes the situation less threatened.

Reference

1. Carolina Alvarez, Rebbecca Wilson, et al. 2020, The association of acculturation with accelerometer-assessed and self-reported physical activity and sedentary behavior: The Hispanic Community Health Study/Study of Latinos, 2–4.
2. Yidong Liu and Tieying Ma, 2018, Parasitic Resistance-Based High Precision Capacitive MEMS Accelerometer Phase Shift and Its Usage for Temperature Compensation, 32–34.
3. Bernard M. D. Vonck, Marc J. W. Lammer, et al. 2021, Cortical potentials evoked by tone frequency changes compared to frequency discrimination and speech perception: Thresholds in normal-hearing and hearing-impaired, 5–9.
4. Dongmei Guo Haiqing Jiang Liheng and Shi Ming Wang, 2018, Laser Self-Mixing Grating Interferometer for MEMS Accelerometer Testing, 3–6.
5. Zhongyi Shen, Xin Zhang, and Dacheng Yang, 2018, A Location-Based Extended Sensor Sharing Algorithm in Vehicular Ad Hoc Networks, 1–2.
6. Md Zakirul Alam Bhuiyan, Member, IEEE, Rahman Tashakkori, Senior Member, IEEE, et al, 2020, Privacy Enhanced Location Sharing for Mobile Online Social Networks, 8.
7. Wei Cherng Chenga, Masayoshi Aritsugia, 2018, A user sensitive privacy-preserving location sharing system in mobile social networks, 9–10.
8. Ajumal P A, Ananthakrishnan S, 2019, A Framework to Study Heuristic TSP Algorithms with Google Maps by API, 48.
9. A. Z. M. Tahmidul Kabir, Al Mamun Mizan, 2020, Safety Solution for Women Using Smart Band and CWS App, 56.
10. Aditi Meshrama, Pushpa Choudhary B, et al, 2019, Assessing and Modelling Perceived Safety and Comfort of Women during Ridesharing, 4–5.

Note: All the figures in this chapter were made by the author.

Disruptive Technologies for Sustainable Development – Dr. G. Nagappan et al. (eds)
© 2024 Taylor & Francis Group, London, ISBN 978-1-032-55000-8

Image Forgery Detection Using Machine Learning

27

R. Kaladevi[1]
Associate Professor, Department of Computer Science and Engineering,
Saveetha Engineering College, Chennai

Mary Emelda Florence F. X.[2]
Department of Computer Science and Engineering,
Saveetha Engineering College, Chennai

Alvaro Rocha[2]
Professor,Computer Science and Engineering,
University of Coimbra, Portugal

Abstract: In the world of the internet, image forgery has become a prolific method for nefarious users all around the globe. With the humongous amount of data surrounding the internet, it has become easy to use fake content, especially in journalism, social media, surveillance systems, and more. Digital image forensics has gained more attention as it has become easier for anyone to make forged images. Vast areas concern image manipulation: a treated idea can increase the reliability of false news, and impostors can use manipulated images to disguise to be someone else. Thus, a comparison of Transfer Learning and the CNN algorithm is made to solve the problem of forged data being used for unethical purposes while detecting the forgery made on images. The best algorithm is obtained to derive a solution in the future.

Keywords: Image forgery, Transfer learning, CNN algorithm

1. Introduction

Due to the emergence of social networks such as Instagram, Facebook, and WhatsApp, there has been a massive rise in the last decade of image data generation. Using video processing and image software like Adobe Photoshop and GNU Gimp for creating fake photos and videos is a vital alarm for social networking companies like Meta. Before action can be performed based on a questionable idea, authenticity is essential and should be verified. In this paper, a discussion has been made on this problem to bring out a solution. This challenge has 2 phases, dealing with the first phase in this project is to detect if it is a real or fake image. The project learnings are Image Handling and Processing. In this paper, a comparison between CNN and Transfer Learning algorithms is made using an Image data generator. Observing the performances of CNN vs. Transfer Learning, the best algorithm is used to detect the fake image. In section 2, we have the literature survey, followed by the proposed system in section 3, which also explains the algorithms used in the system. Section 4 has the model and evaluation graphs from the analysis, followed by the result in section 5 and the conclusion in section 6.

[1]kaladevi@saveetha.ac.in, [2]fxemelda@gmail.com, [3]amrrocha@gmail.com

DOI: 10.1201/9781003428473-27

2. Literature Survey

Zankhana Barad (2020) elaborated on deep learning techniques for detecting forged images. The study outcomes are findings, analysis, and details of publicly available image forgery datasets [1]. Liqiong Lu (2019). A fresh framework based on ResNet-20 for script identification integrates Global CNN and Local CNN. Many segmented images and patches concerning the aspect ratios of the images are obtained. Next, these segmented patches and images are given as inputs to Local CNN and Global CNN for training, respectively. Lastly, the AdaBoost algorithm combines the outcomes of Local and Global CNN for decision-level fusion to fetch the final result [3]. Praveen Gupta (2022) the study proposes developing a Deep Learning based model. For this, a dataset of forged and original are collected from a CASIA database. This is then preprocessed by utilising techniques like vision resizing and denoising. Images obtained after processing are used for training the DL model. The working of the model is tested using loss and accuracy as the major parameters. Lastly, it is examined that the model can produce 95% greater accuracy[4]. Mohit Baviskar (2022) the ultimate aim is to perform a relative examination of methods of forged images. By this an eight-layer CNN-based model is proposed with sigmoid for the last one and activation of relu function for the top seven layers. VGG-16 and VGG-19 pretrained models are used for comparing with the proposed model since it is the most widely used model for image classification [5]. Abhishek Thakur (2018)this evolves around two main algorithms to classify and localise image forgery. First algorithm uses convolution neural network based on deep learning to organise authentic and splice images. In the second algorithm, saliency algorithm based on machine learning detects and localizes forged images [6]. Anushka Singh (2021) this paper has two essential aspects of inculcating deep convolutional-based neural networks to detect image forgery. At first, an examination is done for different preprocessing methods and convolutional neural networks (CNN) architecture. Later the evaluation was carried out for the additional transfer-based learning of pre-trained ImageNet and implemented over the dataset CASIA V2.0[7]. Nagaveni Hebbar (2022)

3. Proposed System

Considering the previously used method, binary classification, for identifying the forged image, we have taken a new approach to determine the forged images. The process is depicted in the below flow diagram.

At first, the images from the datasets are obtained. They are segregated into pristine images, that is, the original image and fake images. These counterfeit images will have shape conflict, depth of images will be modified. Both sets of images are sent for data processing. An equal number of images are split between CNN and Transfer Learning algorithm. The CNN algorithm has hidden convolution and pooling layers. Images go through all these layers, and filtration is done based on the inputs. An F1 graph is obtained for accuracy between fake and pristine appearance. Similarly, in Transfer Learning, the images go through the layers of pre-training, training, and extraction of features. Finally, the F1 graphs of CNN and Transfer Learning are compared, and the most accurate score is obtained. The complete process can be abridged as follows:

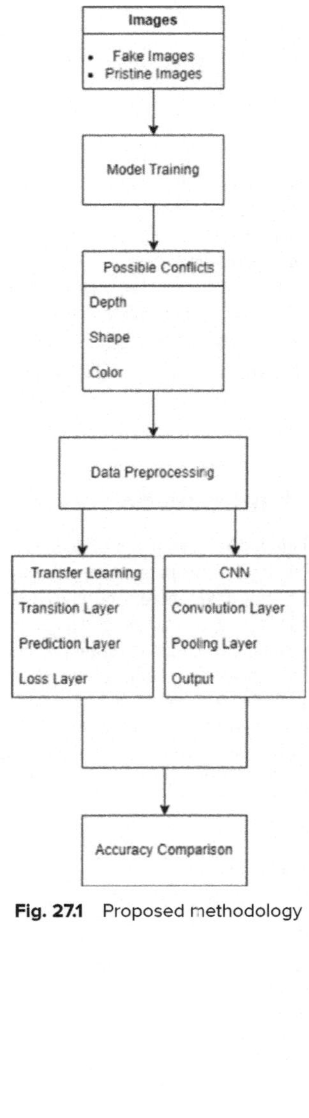

Fig. 27.1 Proposed methodology

3.1 Convolution Neural Network

A CNN (convolutional neural-based network) is called the machine learning's subset. It is one of the many artificial neural networks used for different data types and applications. Deep learning algorithms use this as a network architecture and are used explicitly for tasks involving pixel data processing and image recognition. Even though there are other varieties of neural-based networks in deep learning, CNN is the network architecture of choice for recognizing and identifying objects.

3.2 Transfer Learning Algorithm

The principle of transfer learning is as follows: a model is sent for training for a larger dataset and then applied to a smaller dataset. We halt the early convolutional-based

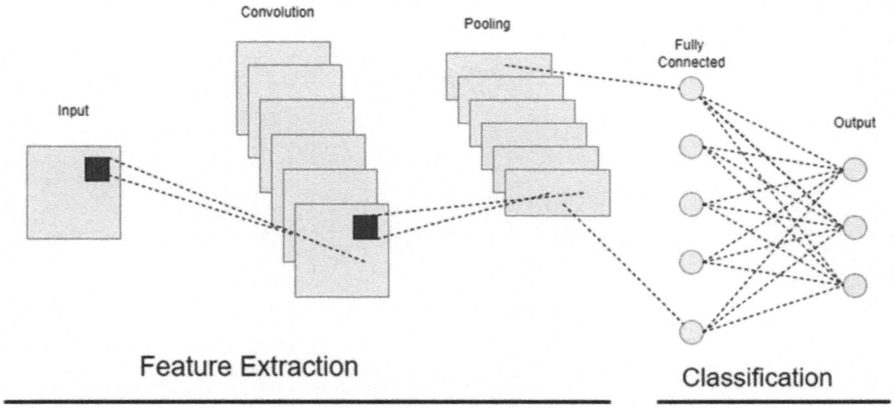

Fig. 27.2 Convolution neural network flow diagram

network layers for object recognition and introduce only a few layers in the last that allow a prediction. The convolutional-based layers extract low-level, general features that can be used throughout images.

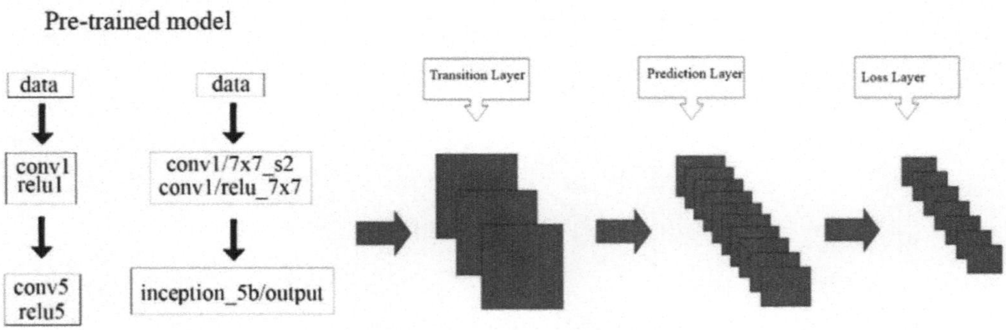

Fig. 27.3 Transfer learning flow diagram

3.3 Model Evaluation and Training

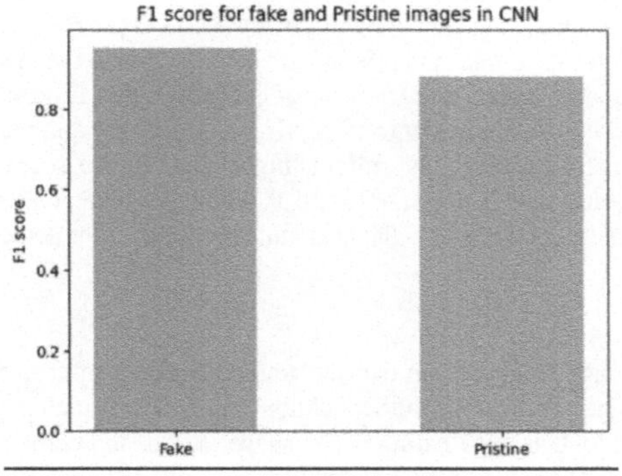

Fig. 27.4 F1 score graph for fake and pristine image in CNN

The F1 score is mapped between 0 and 1. The accuracy of the images in CNN is better if the F1 score is more. F1 score of fake image = 0.95 and F1 score of pristine image = 0.91

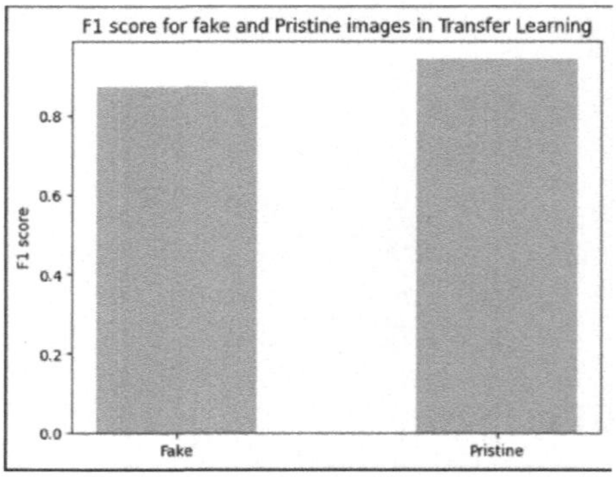

Fig. 27.5 F1 score graph for fake and pristine image in transfer learning

Figure 27.5 is the F1 score graph of fake and pristine images in the Transfer Learning. The F1 score is mapped between 0 and 1. The more the F1 score, the better the accuracy of the images in the Transfer Learning. F1 score of fake image = 0.8 and F1 score of pristine image = 0.95

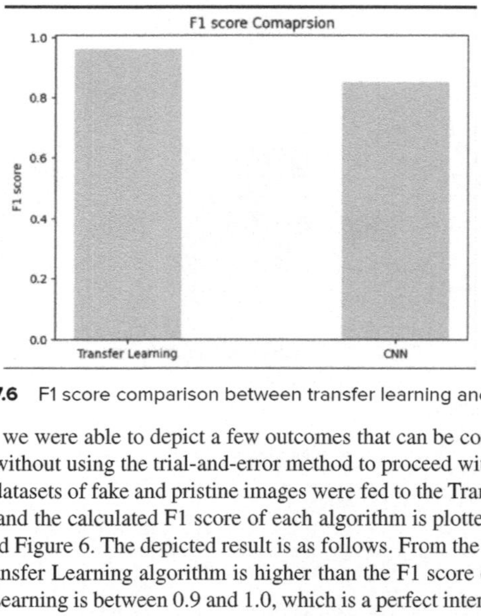

Fig. 27.6 F1 score comparison between transfer learning and CNN

Upon analysis and experimentation, we were able to depict a few outcomes that can be considered for actual time application. Further, this approach can be taken without using the trial-and-error method to proceed with the related works. As stated in the proposed methodology, pre-trained datasets of fake and pristine images were fed to the Transfer Learning and CNN algorithms. Levels of analysis were performed, and the calculated F1 score of each algorithm is plotted in the graph as a fake and pristine image, which is given in Figure 5 and Figure 6. The depicted result is as follows. From the calculated F1 score comparison, we can see that the F1 score for the Transfer Learning algorithm is higher than the F1 score of the CNN algorithm. As shown in Fig. 27.5, the F1 score for Transfer Learning is between 0.9 and 1.0, which is a perfect interpretation and fulfills the satisfaction level. From this, we conclude that Transfer Learning is the best-suited algorithm for detecting forged images.

4. Conclusions

This paper is focused on how the image from a given dataset (trained dataset) and the past dataset were used to predict the pattern of fake images or forged images used in the authentication or any social media networks. The datasets are trained in both Convolution Neural Networks and Transfer Learning. The shape, depth, and color conflicts are obtained and segregated as real and fake. Then upon training, the results of both algorithms are obtained. The F1 score is obtained by plotting graphs against each image and comparing them for accuracy. The best result is obtained by using the Transfer Learning algorithm. This algorithm can be used to find solutions for image forgery detection.

REFERENCES

1. Zankhana J. Barad (2020) " Image Forgery Detection using Deep Learning: A Survey."IEEE
2. Sudeep D. Thepade; Sanket Bhandari; Chirag Bagde; Rutuja Chaware; Krutik Lodha(2021) "Image Forgery Detection using Machine Learning with Fusion of Global and Local Thepade's SBTC Features." IEEE
3. Liqiong Lu; Yaohua Yi; Faliang Huang; Kaili Wang; Qi Wang(2019)," Integrating Local CNN and Global CNN for Script Identification in Natural Scene Images", IEEE, Vol 7
4. Praveen Gupta; Chour Singh Rajpoot; T. S. Shanthi; (2022), "Image Forgery Detection using Deep Learning Model", IEEE
5. Mohit Baviskar; Sheetal Rathod; Jay Lohokare (2022), "A Comparative Analysis of Image Forgery Detection Techniques, IEEE
6. Abhishek Thakur; Neeru Jindal (2018), "Machine Learning Based Saliency Algorithm For Image Forgery Classification And Localization", IEEE
7. Anushka Singh; Jyotsna Singh (2021), "Image forgery detection using Deep Neural Network", IEEE
8. Ismail Taha Ahmed; Baraa Tareq Hammad; Norziana Jamil (2021), "Effective Deep Features for Image Splicing Detection", IEEE
9. Nagaveni Hebbar; Ashwini Kunte (2022), "A Deep Learning Framework with Transfer Learning Approach for Image Forgery Localization", IEEE
10. Gokul Nair; Kaustubh Venkatesh; Dipankar Sen; Reena Sonkusare (2021), "Identification of Multiple Copy-move Attacks in Digital Images using FFT and CNN", IEEE

Note: All the figures in this chapter were made by the author.

Disruptive Technologies for Sustainable Development – Dr. G. Nagappan et al. (eds)
© 2024 Taylor & Francis Group, London, ISBN 978-1-032-55000-8

Smart Students Attendance System Using Face Detection

28

E. Sujatha[1]
Professor, Department of Computer Science and Engineering,
Saveetha Engineering College, Chennai, Tamilnadu

Manikandan M.[2]
Department of Computer Science and Engineering,
Saveetha Engineering College, Chennai, Tamilnadu

Abstract: Face detection technologies have made a significant change in the world. The daily monitoring of student attendance systems includes a fantastic accomplishment in the form of an intelligent attendance system with real-time facial detection. A face detection-based attendance system is a process in which the participant's face is captured for analyzing and recording their attendance using highly descriptive camera-based face biometrics. In this architecture facial detection system can accurately locate and capture human faces as images or videos captured by camera. Viola Jones' algorithms methods were included to enhance performance of facial detection It converts each frame into images so that students face can be recognized and their presence smoothly, and highly scalable database is used to store the information regarding each student.

Keywords: Face detection, Attendance system, Viola Jones

1. Introduction

This project aims to build attendance system that uses face detection to mark student attendance, on-time, and off-time. It also describes the iterative development process that was used, as well as how structural modelling and an architectural design were used to implement the needs that were identified during the requirements analysis. Being a highly important aspect of administration, attendance can sometimes turn into a tedious, repetitious task that pushes for errors. The traditional way of taking attendance is difficult to handle attendance book and also student average percentage of attendance difficult to calculate. Each organization has its own method of measuring student attendance. Some companies have adopted a document-centric approach, while others have adopted digital processes such as biometric and card exchange. Standard facial detection methods and techniques attempt and fail to address issues such as scaling, position, lighting, transformation, rotation, and occlusion. The proposed framework aims to address the shortcomings of existing systems. Face detection has advanced significantly, but the steps remain face analyzer and extraction. In addition, full descriptions of built-in features and steps to complete the prototype are covered. Enroll new students, add photos to training records, view attendance reports, and more. It can be used in corporate offices, schools and organizations where security is essential. It focuses on the implementation of the proposed system, i.e. a smart attendance system using OpenCV. Attendance is then recorded in the institution's Excel format, prefixing each student's name with the appropriate date and subject of the lecture.

[1]sujathae@saveetha.ac.in, [2]manikandanvalli098@gmail.com

DOI: 10.1201/9781003428473-28

2. Literature Review

The author of R. S. Bharadwaj, (2019) described a system that employed the Eigenfaces method for face recognition. They have worked on background subtraction for binary and greyscale images after performing face detection, face cropping, and background subtraction. The Eigenface approach was employed by the author because of its efficiency, quickness, and capacity for learning. The author claims that face identification was carried out in three steps, including face detection, feature extraction, and classification. For face detection, the Viola-Jones method was employed. The classification process uses the Ada boost algorithm. According to the author's research, Venkata Kalyan Polamarasetty et al, (2018) face recognition may be broken down into three steps: face detection, face recognition, and auto-tagging. For face detection, the Haar cascade technique was employed. The Ada Boost learning method is used to train this system. This algorithm uses machine learning. By comparing the face recognized by the Haar cascade algorithm with database data, face recognition was carried out. When a match is made, the tag name of the person is displayed. W. Zeng (2019) uses a technique to track attendance that combines face recognition and RFID card technology. Alex Net's convolutional neural network is used to recognize faces and extract features. In order to prevent visitors from outside the school from entering any classes, a live camera is employed to watch the situations. Describes a system for recording attendance that makes use of RFID, face recognition, and fingerprint scanning. The GUI was programmed in MATLAB so that student information and attendance could be updated and presented. The authors of Anusaya Tantak et al. (2017) described a system that uses the Eigen faces method for face identification. After performing face detection, face cropping, and background subtraction, they worked on background subtraction for binary and grayscale images. The Eigenface approach was used by the authors for its efficiency, speed, and learning ability.

3. Proposed System

3.1 Architecture of the Proposed System

The diagram above depicts the entire architecture of this face analysis -base attendance system. This smart attendance system's operation very straightforward and simple to comprehend. For our project, we will require several hardware components in order to make this system functional is shown in Fig. 28.1

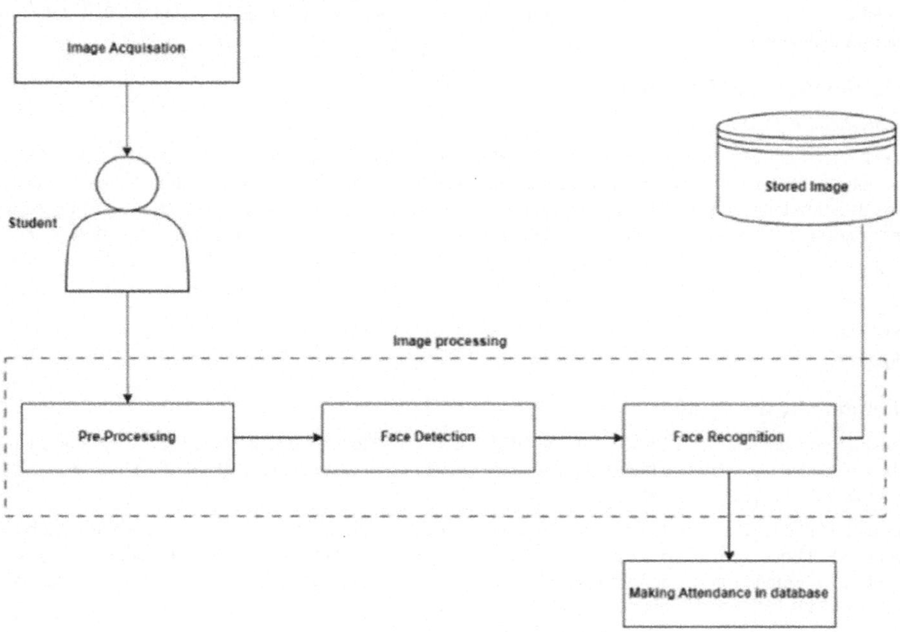

Fig. 28.1 System architecture

The suggested job is to take a picture of every student's face and record database. The student's must be photographed in such a way that all of the features on their face can be recognized. The system records the video, and additional processing identifies faces from the video, eliminating the need for teachers to manually record attendance for recognized classes and updating the attendance database. This system is developed with Python OpenCV. We suggested the entire attendance management system in the lecture hall in order to gather the attendance, positions, and face images.

Building a solution that can be implemented in existing IoT setups. Making the solution highly robust and scalable. The video footage will be obtained via a physically linked camera system, from which facial detection frames will be taken. The acquired images will then be pre-processed first, then fed to a facial detection algorithm for the determination of facial features. The captured face's identity will then be determined using an existing database and the identification algorithm.

3.2 Methodology

There is various cause of action that must be performed in order to construct the smart attendance management system.

- Enrolment
- Pre-Processing
- Face exploration and Detection
- Confirmation by the system
- Attendance records into centralized database systems
- Attendance marking

3.3 Enrolment

The student added the data base includes the person's name, admittance ID, class information, and section. The image of the student's face shown in the camera window is likewise stored in the database of student records together with all of this metadata.

3.4 Pre-Processing

Based on facial detection, our model. To create 128-d vectors from the training dataset, we used the face detection package, which is a pre-trained network.

3.5 Face Exploration and Detection

Viola-Jones Method for face detection

The first object detection framework that produced workable outcomes in actual circumstances. Paul Viola and Michael Jones proposed the concept in 2001. It was designed focus on the issue of face detection, it may also be trained to recognise many object classes. It is implemented as CV HaarDetectObjects in Open CV (). The vision approach has been used to build this algorithm in the software "Matlab". The Viola - Jones contains of 4 techniques for the facial parts detection:

- Haar Feature
- Integral Image
- AdaBoost learning
- Cascading

3.6 Confirmation by the System

A smooth and effective operator called Local Binary Pattern (LBP) functions by thresholding the area around each pixel in an image and then evaluates the result as a binary number. When used in conjunction with LBP, For face detection, LBP and HOG are combined to create the LBPH algorithm.

- The four factors that LBPH considers for face detection are Radius: Radius is used to create a circular local binary pattern. Neighbors: Neighbors are used to set cyclic local binary patterns and are often set to 1. Usually 8. Grid X: Shows the number of cells arranged horizontally. Normally a set to 8. Grid Y: Displays the number of cells arranged vertically. Normally set to 8.
- The algorithm is trained on a database of student headshots that need to be recognized. Student detection is enabled by a unique ID that is set when the record is created.

- Using the LBP technique, produce accurately represents source enhancing face features. The algorithm makes use of the sliding window concept and bases its decisions on factors like radius and neighbors.
- Before extracting the histogram, the photo is divided into several grids using the Grid X and Grid Y parameters. Since this is a grayscale image, each histogram can only display 256 digits (0 to 255). only Reveals the presence of intensity for each pixel. Face detection is performed by first training a histogram generation algorithm. Each histogram represents one image of her from the training dataset. Repeat the above steps for the input image to create a histogram of that image. The histogram of the original image is compared to the histogram of the dataset image, and the closest histogram is used to select a matching image from the dataset. You can compare histograms using a variety of techniques including absolute value, Euclidean distance, and more. To compare the histograms, use Equation 1 to determine the Euclidean distance. HistD - Image Histogram from Dataset HistR - Image Histogram in Real Time Due to the slight difference between the histogram of the student's image and the photo of the dataset, the algorithm provides a unique ID for the student.

3.7 Attendance Records into Centralized Database System

- In order to begin, we first identify the faces and compare each one to the encodings we have from our training set.
- Each face in our database is used to determine the Euclidean distance:
  - To assume that the faces were similar, the gap must be less than a certain threshold.
  - The face is regarded as unknown if the distance exceeds the threshold.

3.8 Attendance Marking

When a lecture is in Excel format, the presence of a student is marked in front of their name and registration number to create trailing facial detection reports. If a match is found in the database, student attendance is automatically marked by lecture name and time in an Excel spreadsheet. Otherwise, an error will be displayed and attendance will not be marked.

4. Result

The smart attendance management system works simply and efficiently. The Existing system takes more manual effort to perform attendance tracking but in the proposed system works automatically as soon as an individual student registration is created by the administrator.

- Student enrollment
- Face analyser
- Along with their IN and OUT time the corresponding subject class hours is also recorded
- Dynamic Attendance sheet report generation by Excel (xlsx) format.

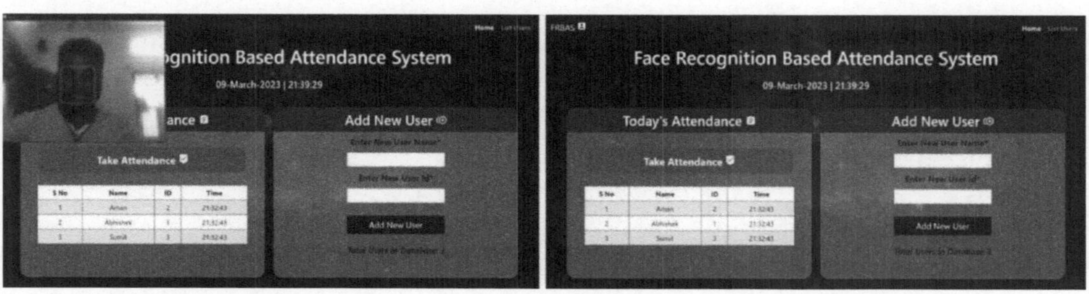

Fig. 28.2 Final output of real time face detection based on image dataset

5. Conclusion

The paper offers a Viola-Jones technique-based method for face detection and detection. The suggested approach's performance is far better when to compared to that of other face detection techniques that is available nowadays, shown a higher level of detection accuracy. The the rate of identification accuracy is high and sought in the majority of applications, the proposed approach can be implemented in various educational or enterprise organizational at an ease.

References

1. R. S. Bharadwaj, T. S. Rao, and T. R. Vinay, (2019), "Attendance management using facial detection," Int. J. Innov. Technol. Explor. Eng., vol. 8, no. 6, pp. 1619–1623.
2. Aayushi Bansal (2018) "A factors affecting face recognition" IJAMTES volume 8, issue II, pp. 669–672.
3. Venkata Kalyan Polamarasetty et al. (2018), "Attendance System based on Face Recognition", International Research Journal of Engineering and Technology (IRJET), Volume: 05 Issue: 04 | Apr-2018, pp. 4606–4610.
4. Anusaya Tantak et al. (2017), "Face Recognition for EAttendance for Student and Staff", IOSR Journal of Computer Engineering (IOSR-JCE), Volume 19, Issue 2, Ver. IV (Mar.-Apr. 2017), pp. 89–94.
5. Hapani, Smit, et al. (2018) "Automated Attendance System Using Image Processing." 2018 Fourth International Conference on Computing Communication Control and Automation (ICCUBEA). IEEE.
6. W. Zeng, Q. Meng and R. Li, (2019) "Design of Intelligent Classroom Attendance System Based on Face Recognition," 2019 IEEE 3rd Information Technology, Networking, Electronic and Automation Control Conference (ITNEC), Chengdu, China, pp. 611–615.
7. Anusaya Tantak et al. (2017), "Face Recognition for EAttendance for Student and Staff", IOSR Journal of Computer Engineering (IOSR-JCE), Volume 19, Issue 2, Ver. IV (Mar.-Apr. 2017), pp. 89–94.
8. X. Han and Q. Du, "Research on face recognition based on deep learning, (2018)" 2018 Sixth International Conference on Digital Information, Networking, and Wireless Communications (DINWC), Beirut, pp. 53–58.
9. Mathana Gopala Krishnan et al. (2015), "Implementation of Automated Attendance System using Face Recognition", International Journal of Scientific & Engineering Research, Volume 6, Issue 3, pp. 30-33.

Note: All the figures in this chapter were made by the author.

Disruptive Technologies for Sustainable Development – Dr. G. Nagappan et al. (eds)
© 2024 Taylor & Francis Group, London, ISBN 978-1-032-55000-8

Internet Accessibility for Visually Impaired Using Automatic Speech Recognition and Speech Synthesis

29

S. Sasi Kumar[1]
Professor,Department of Computer Science and Engineering,
Saveetha Engineering College (Autonomous), Chennai, Tamil Nadu, India

Kiruthika S.[2]
Department of Computer Science and Engineering,
Saveetha Engineering College (Autonomous), Chennai, Tamil Nadu, India

Abstract: There are several issues that the blind and visually impaired face daily. Accessing the internet for instant up-gradation is challenging for them. These were noticed via connecting with them, researching about them on internet and social media, and doing surveys. Most of the content that is available on the internet is inaccessible to them. As a result, people ought to struggle to access information without using any digital technology. The purpose of the plan is to enable blind people to examine all of the information that is accessible on the internet. The work's novel technology is to make digital technology approachable through internet browsing using a tailored inquiry that the consumer has chosen. Voice recognition and synthesis has brought the technology to a more suitable environment for the visually impaired. They will use voice-to-text and text-to-speech conversion technology to access extensive personal web content through surfing. Using the Google API to recognize the voice and synthesize text, the blind and visual impairment (BVI) can be benefited.

Keywords: Automatic speech recognition, Automatic speech synthesis, Google application programming interface, Visually impaired, Web surfing

1. Introduction

Visually impaired deal with a variety of problems in their daily life. These have been discovered by interacting with them and reading about them in social media and polls. They are unable to access the majority of the knowledge that is available on the internet. They face challenges to learn, update knowledge and surf the internet. The voice feature in the live environment enables accessing the internet for all individuals as it concludes limited information. People are therefore having difficulty accessing the material in-depth with the aid of digital technologies.

The work's novel technology is to make digital technology approachable through internet browsing using a tailored inquiry that the consumer has chosen. A dynamic social platform urges to implementation of all access to the disabled person. Google API is one of the most used application interfaces that the user makes benefit from the feature. Voice recognition and synthesis has brought the technology to a more suitable environment for the visually impaired. Existing systems such as Alexa, Siri, and OK Google perform a vital role in today's lifestyle. The proposed system enhances internet accessibility which shows or iterates customized results by surfing through the internet by using speech recognition and speech synthesis approach in websites, youtube, and Wikipedia contents.

[1]sasikumar@saveetha.ac.in, [2]kiruthika0096@gmail.com

DOI: 10.1201/9781003428473-29

In the upcoming chapters, we have literature survey, followed by the proposed system and algorithms used to identify and output preferred content. After that, operating principle with experiments is explained followed by result analysis and conclusion.

2. Literature Review

The concept of providing internet accessibility for visually impaired people using AI can be proposed via digital technologies. But lately, the focus of research has been on applying deep learning to speech-related problems.

For people with restricted vision, vision loss has a significant negative influence on spatial cognition and object recognition. In order to aid BVI's spatial cognition, Stereo Pilot, a wearable target localization device, is introduced in this (Zhikai Wei et al. 2022). The objective of this (Yu Shi et al. 2022) is to transform the ambiguous and temperamental ASR output into intuitive text for individuals and potential factors. The approach is described as a genomic text creation challenge.

TTS systems that produce speech from beginning to end have recently produced great speech. This (Sakriani Sakti et al. 2022) suggests a different TTS strategy that is more in line with the Lombard effect in people. Using two different kinds of user research, this (Jia Liu et al., 2021) examines and analyses the implications of the research participants' perception of visual grooves in different pneumatic display modes. The study of voice audio signals has seen a great deal of investigation recently. This evaluation of voice recognition has been provided by (Muna Alrazgan et al. 2021) to help other researchers with the most important subjects enclosed in the past six seasons of research.

The objective of this (Shah Nazir et al., 2021) is to examine the performance of sampling frequency with varied data ranges, calculation speeds, and efficiency depending on the dataset type. In this paper (Shinnosuke Takamichi et al., 2021), the dynamic text-to-speech (TTS) approach is clarified with synthesizer that divides speech into absorbable lexical pieces while maintaining the output's naturalness.

This work designs, assesses, and enhances the suggested link building functionality for Vision Impeded (VI) users to efficiently conduct web search operations. The results show that WCAG 2.0 criteria and Nielsen's methodologies helped identify a lot of usability issues that any one of them could have missed. if employed independently (Kamran Khowaja et al., 2020).

The perspective divides modification is required into three categories: data augmentation, model parameter modification, and embedding. We propose a contextual of the performance of voice recognition adaption classifier upon the relative computational complexity reductions that have been demonstrated in the (Joachim Fainberg et al. 2020).

Machine learning is a relatively new pitch, yet it has already outperformed others in a variety of applications, including voice recognition. A thorough analysis of the research on deep learning for vision applications is included in this (Ismail Shahin et al., 2019). The integration of machine learning to voice application services, particularly voice recognition, has garnered a lot of attention in recent decades.

3. Proposed Methodology

This proposal objective is to make it possible for those who are blind to know all of the information available on the internet. They will have a thorough online profile employing content that's absolute by interfacing voice-to-text and text-to-voice conversion technologies. Google API (Application Programming Interface) is used for speech recognition and speech synthesis methodology. Python's internet accessibility module could be developed using Interface. It sorts and output the user-provided Test to Speech and Speech to Text by algorithms.

In this paper, we follow three modes namely, website, youtube and wikipedia. In the website mode, the voice phrase is analysed in the Google API and then prompts the speaker to choose number from the list provided. It is used to convert voice and text and vice-versa to summarize and outputs the preferred topic. In the youtube mode, the API directly moves to youtube module and asks user to select the required video through voice. In the wikipedia mode, the voice topic is scrapped and the content will be summarized in the output area. The wikipedia mode works ideally with website, yet the content will be particularized.

4. Data Sets and Algorithm

Search-ability is voice recognition's main benefit. The development of methods and technology that allows computer recognition and text transcription of spoken language is called speech recognition. This process integrates computational linguistics and computer science. It utilises field-specific analysis and knowledge of languages, computer engineering, and computer science. Speech synthesis is the counterpart procedure.

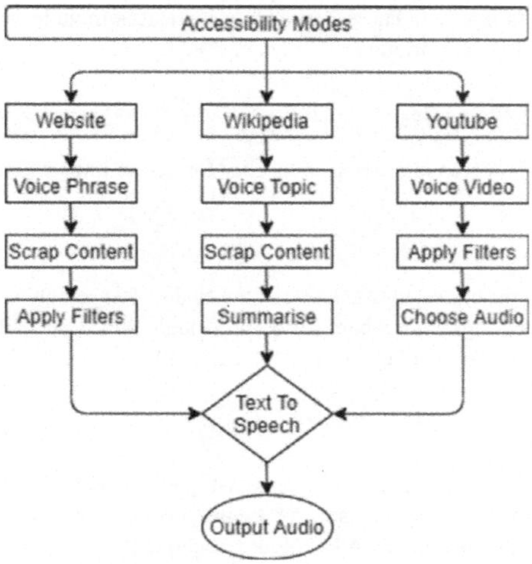

Fig. 29.1 The internet accessibility proposal

4.1 Speech Recognition

The Hidden Markov Model and Dynamic Time Wrapping Algorithm of the Decoder are used in speech recognition. The algorithm uses to transform audio into text and the voice captured from the signal will be processed using estimator with phone probability and decodes to the corresponding text being surfed by internet.

4.2 Speech Synthesis

The technique of converting typewritten text (typographical representation) into spoken language is called speech synthesis. Speech synthesis must sound as natural as feasible and shouldn't use a restricted vocabulary.

As its name implies, the paradigm probes the acoustic waves by "Occupying", "Remuneration Effort" by applying different signals to various regions of the signal, and the transcript "Articulate" one character at a time. The phrase gathered and iterated from the internet will be analysed using phonetic and hence, the speech is produced with the frequency.

5. Module Required

***Summary statistics*:** The idea is to make visually impaired people access the contents of websites available on the internet and also access famous platforms like Wikipedia and youtube through voice recognition are presented in the Table 29.1.

Table 29.1 Modules specification

Sl. No	Framework	Usage
1	Requests	To set up the host computer as a client and send the server a request
2	Beautiful Soup	Web Scraping
3	Python - VLC	To play the youtube video as an audio
4	Pafy	To access the audio from youtube videos
5	Wavio	To save the recorded audio file
6	Pyttsx3	To perform Text – to – Speech Conversion
7	Speech Recognition	To perform Speech – to – Text Conversion
8	Wikipedia	To obtain results from Wikipedia
9	Sound Device	To hear audio from the microphone
10	Youtube Search Python	To search youtube videos and retrieve audio output
11	Keyboard/Touch	Hardware interrupts to stop the voice

The modules are configured to process internet. These application interfaces include voice to measure speed and frequency wave as well as the direction between two-way inputs.

6. Experimentation and Results

This paper has three essential modes to access internet accessibility for vision impairment aid, Website Mode, Youtube Mode, and Wikipedia Mode.

6.1 Website Mode

When using Website Mode, the user is prompted for the name of a website. The search results will include a list of all websites connected to that term, which are then assigned numbers using the Google API. User will be prompted to tell a number to go to that website, read out the material. The keyword entered by the user will be recognized and loaded into a Google search, and the ideal results will be shown.

6.2 Wikipedia Mode

Simply by utilizing voice search, the user can access countless Wikipedia pages. The user is required to specify the page's location. The customized search result will be heard by surfing the internet contents according to the priority. Some wide range of Wikipedia pages can be accessed by the user by merely voicing a search phrase.

6.3 Youtube Mode

The user is instructed to conduct a voice search for a YouTube video in this mode. The results are played as an audio by just stating the number that corresponds to the video. The word is voice-searched on YouTube. The connected video's audio stream will play when a user reads out a number.

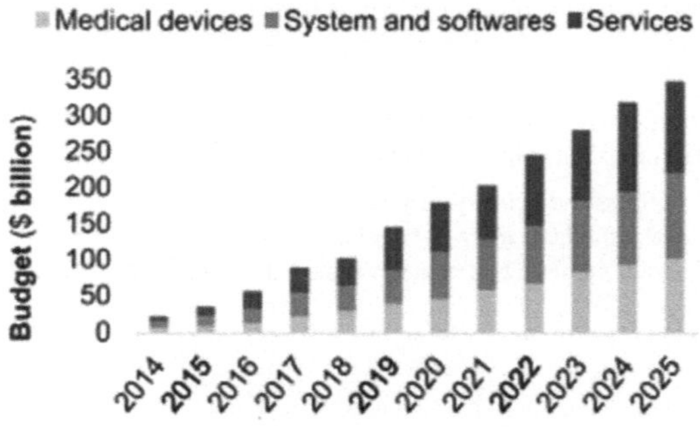

Fig. 29.2 IOT in health care market

The above depicted picture shows that, the use of Internet of Things in diagnostic instruments will proliferate easily and effectively by the disabled. The year depicts systems and software importance with medical devices as well.

7. Conclusion

The study deals with the important issue of providing folks with visual impairments with access to online content. By putting it on a windows computer or a mobile device that supports Google, it can be utilised more conveniently in the digital world. The article's novel feature is the ability of Google Voice Assistant or OK Google to deliver condensed search results in response to a search query. The main page provides access to the whole website, and any of the alternatives can be used to access the most recent information. The most advanced method to achieve this objective is to embed and include internet browsing into e - textiles, such as an attire or handset.

REFERENCES

1. Zhikai Wei, Xuhui Hu, Hong Zeng, and Aiguo Song (2022) Stereo Pilot: A Wearable Target Location System for Blind and Visually Impaired using Spatial Audio Rendering: Transactions on Neural Systems and Rehabilitation Engineering, 30, IEEE
2. Yu Shi, Junwei Liao, and Yong Xu (2022) Automatic Speech Recognition Post-Processing for Readability: Task, Dataset and a Two-Stage Pre-Trained Approach, 10, IEEE
3. Sakriani Sakti, Sashi Novitasari, and Satoshi Nakamura (2022) A Machine Speech Chain Approach for Dynamically Adaptive Lombard TTS in Static and Dynamic Noise Environments, 30, IEEE
4. Jia Liu, Dapeng Chen, Lei Tian, Aiguo Song, and Xuhui Hu (2021) Research on the Method of Displaying the Contour Features of Image to the Visually Impaired on the Touch Screen: Transactions on Neural Systems and Rehabilitation Engineering, 29, IEEE
5. Muna Alrazgan, Sadeen Alharbi, Alanoud Alrashed, Raghad Almojel, Turkiayh Alnomasi, Rimah Alharbi et al. (2021). Automatic Speech Recognition: Systematic Literature Review, 9, IEEE
6. Shah Nazir, Sulaiman Khan, and Habib Ullah Khan (2021) Analysis of Navigation Assistants for Blind and Visually Impaired People: A Systematic Review, 9, IEEE
7. Shinnosuke Takamichi, Takaaki Saeki, and Hiroshi Saruwatari (2021) Incremental Text-to-Speech Synthesis Using Pseudo Lookahead with Large Pretrained Language Model, 28, IEEE
8. Kamran Khowaja, Aboubakr Aqle, and Dena Al-Thani (2020) Preliminary Evaluation of Interactive Search Engine Interface for Visually Impaired Users, Vol. 8, IEEE
9. Joachim Fainberg, Peter Bell, Ondrej Klejch, Steve Renals, Jinyu Li, and Pawel Swietojanski (2020) Adaptation Algorithms For Neural Network-Based Speech Recognition: An Overview, 2, IEEE
10. Ismail Shahin, Ali Bou Nassif, Imtinan Attili1, Khaled Shaalan , and Mohammad Azzeh (2019) Speech Recognition using Deep Neural Networks: A Systematic Review, 7, IEEE

Note: All the figures and table in this chapter were made by the author.

Disruptive Technologies for Sustainable Development – Dr. G. Nagappan et al. (eds)
© 2024 Taylor & Francis Group, London, ISBN 978-1-032-55000-8

Deep Learning for Traffic Sign Recognition in Autonomous Vehicles

30

Karthika R. N.[1]
Associate Professor,
Saveetha Engineering College, Chennai, Tamil Nadu

Mythili P.[2], **Kousalya R.**[3], **Varshini A. P.**[4]
Final Year Student,
Saveetha Engineering College, Chennai, Tamil Nadu

Abstract: Before, savvy things like brilliant vehicles have assumed a major part. To foster savvy vehicles for street security, it is important to know and know the indications of the street. The calculation is the old procurement of LeNet-5 CNN. The method was presented utilizing Gabor's presentation, trailed by the standard revolution and later the reconciliation of parts. The best strategy to use here is the Adams technique. Shade and Saturation The worth of shading mirrors the attributes of speed discovery and low light. The street recognizable proof procedure is being tried using the assistance vehicle identification in German. The innovation gave around almost 100%.

Keywords: Smart vehicles, LeNet-5 Conv.NN, Saturation, traffic sign detection, Identification driving safety, Savvy vehicles

1. Introduction

A key step in ensuring car respectability and wellness is exact traffic stock. Typically, this labor is done bodily. Roadway markers are photographed with a portable digicam, recognizing evidence hints to determine whether they were created outside the person's boundary. Reduce the number of specialists and improve wellbeing by quickly identifying damaged or absent imprints. The process of changing hands, recognizing street signs, and programming position is an important step in computerizing this routine. The problem of acknowledging street signs has long been a point of contention for the PC population group, and estimates have been made to improve agreement and recognition. However, these setups are only intended for a select few classes, especially for street signage related to private cars and advanced driver aid systems (ADAS). Many of them only focus on street sign acknowledgment (TSR), ignoring the more important problem of traffic-sign identification (TSD), which is necessary when a legitimate traffic sign is needed. Other limits which show the TSD location only display the area where the notice plaque is typically necessary when ADAS and car requirements. Several of the groups represented on these measures vary in appearance and minute details, and can be recognized by handmade objects. Models often include roundabout or three-sided images. To quickly reach the city markers, simply follow the road signs. To let us know whenever a traffic sign indicator is getting near, get a street sign recognizing traffic signals Recognize street signage using a real camera.

[1]karthikarn@saveetha.ac.in, [2]mythuparandhaman@gmail.com, [3]kousi291709@gmail.com, [4]varshiniprabhu118@gmail.com

DOI: 10.1201/9781003428473-30

2. Literature Review

2.1 Multi-feature Fusion and the ELM Classifier are Used to Recognise Traffic Signs

Transportation signals and groups play an important role when traveling. Numerous methodologies have recently been proposed to address this issue, but in order to continue meeting functional requirements, the computation's presentation must actually be improved, in light of the Convolutional Neural Network and Machine Vector Machine. The assessment was based on ongoing data and video taken from a typical machine. The evaluation's findings show that our strategy often outperforms the complex method by choosing and positioning the limits of the light signs. The results of the evaluation demonstrate that by selecting and placing the boundaries of the light signs, our approach is typically 98.6% superior to the sophisticated method.

2.2 Data Unbiased Identification of Traffic Signs

The layout is finished using the Extreme Learning Machine algorithm. The Belgian vehicle and the German automobile standards were used to assess how well the framework was implemented. The results of the review demonstrate that everything exhibits a high level of integrity, and the combination of the three components is sensible for suitable use, quickly recognisable, and well correlated. Street markers and markings are used in our method. It uses Bayesian and DtB algorithms to understand traffic signs.

2.3 Road Traffic Sign Recognition System

The free users from the burden of information gathering and evaluating evidence by allowing them to use informational files previously created from various areas to help characterize the ideal objective. This research demonstrates that information exchanged between internal and external inquiries can increase the reliability of street sign identification over formats for acknowledging the region signage from top to bottom.

3. Existing System

Yuan and other people. Although the processing time was excessive, the authentic standard was high. The other method involved organizing marking and quantifiable accents around the review map, but it was difficult.

3.1 Disadvantages of Existing System

- The computations' magnitude should be increased.
- Handling Preparing a variety of preparation components is necessary.

4. Proposed Work

The ability to read street signs is an important component of a private automobile. The true challenge, though, is learning to recognise the traffic signs correctly and gradually in everyday images. In this article, we provide a summary of the creation and implementation of street signs and awareness through an arrangement of tasks that are planned using ongoing details. The framework, which may be used to transport individuals, households, and other vehicles, has been developed in light of this approach in order to maintain driving and reduce the number of people involved in accidents. Self-driving vehicles are often used.

4.1 Advantages of Proposed System

- Interruptions have the effect of lowering the severity of such accidents.
- Increasing the safety of drivers when they are on the road.

5. Modules

- Image pre-processing
- Segmentation module
- Traffic sign-detection
- Traffic sign-recognition

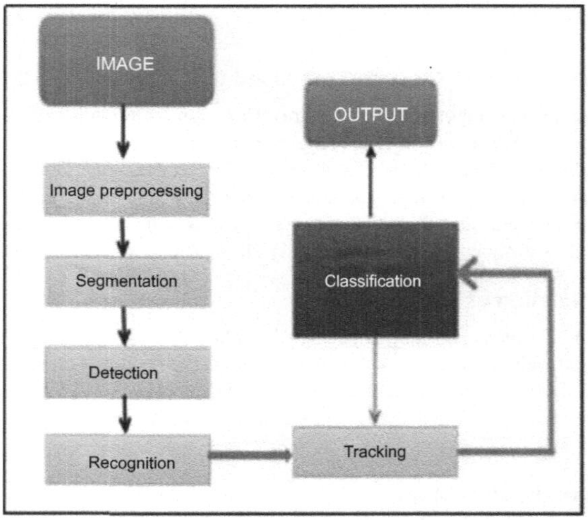

Fig. 30.1 System architecture of proposed work

5.1 Image Pre-processing Module

A common technique for transforming pictures before incorporating them in sketches is to use photos. This includes adjustments, displays, and additional not just tiny reflections. To save time or reduce risk, it may occasionally be a good idea to remove the data from the image. The picture's direction should be changed.

5.2 Segmentation Module

Distributing is a technique for transferring images that is accessible in a block of pixels that lowers the resolution of the image and visual learning. Disallowance is a technique which turns an image into a two-part sad excuse for a picture by using marginal methods.

5.3 Traffic Sign Detection Module

Like driving and driving, street signs and signage are an important component of the expert level. This blatantly helps the driver or framework distinguish and recognise traffic signs with accuracy. This module covers the subject of identifying and recognising the many street signs that are essential to the display of street signs. We asked approaches to understand and distinguish investigating using the RCNN (Region Convolutional Neural Organization) shroud because there are so many minute but significant items. This facility offers a top-notch association that searches for many classes to watch as clear and distinct photographs.

5.4 Traffic Sign-recognition Module

Improvement Learning is an area of machine learning that includes convolutional neural networks. Papers called Street Signs are used to make streets passable for motorists phrases like front, speed, young people, etc. The original ADAS component is this one. It uses the outlines of traffic signs to identify them.

5.5 Model for Convolution and Subsampling Layers for CNN

LeNet was used to recognise handwritten digits, and it easily outperformed all previous techniques.Conv.NN is also powerless to stop fingerprinting work. This section deconstructs existing LeNet-5 problems and changes the programme objective to include Conv.NN benefits. Particularly, the LeNet-5 benefits from recognition and translation. Yet, it might be difficult to predict the exact amount of preparation needed for the afternoon and the network layout.

6. Results

The overall accuracy is 97.92%. This amounted to around 98.84% of the population. But, with such definitions, we may modify the strategy to obtain more specific information on the most effective approach to use this capability. The casing can inform the driver when to call with the use of a recording device.

Fig. 30.2 Traffic signs

Fig. 30.3 Traffic sign detection

7. Conclusion

Overall, uncontrolled preparation leads to better performance and more knowledge. Choose top-to-bottom review skills if you have a lot of data. Also, you have received inside and out certificate preparation as well as support preparation. Now you are aware of the characteristics, functions, and cutoff points of nerve strands. At long last, you've given several development languages, IDEs, and phases some thought in terms of growing your AI machine. The next step is to start studying and practicing each AI technique.

REFERENCES

1. Aziz S, Mohamed E, Youssef F (2018) Traffic sign recognition based on multi-feature fusion and ELM classifier. Proc Comput Sci 127: 146–153.
2. Hatolkar, Y., Agarwal, P., & Patil, S. (2018). A Survey on Road Traffic Sign Recognition System using Convolution Neural Network.
3. Huang, Z., Yu, Y., Gu, J., & Liu, H. (2017). An efficient method for traffic sign recognition based on an extreme learning machine. IEEE transactions on cybernetics, 47(4), 920–933.
4. Ying Sun, Pingshu Ge, and Dequan Liu (2019). Traffic sign detection and recognition based on convolutional neural networks. In 2019 Chinese Automation Congress (CAC), 2851–2854.
5. Ghassen Hammouda, Sellami Dorra, and Hammouda Atef,(2020) Pattern recognition based on compound complex shape-invariant radon transform. The Visual Computer, 1432–2315.
6. S. Ren, K. He, R. Girshick, J. Sun, (2017), Faster R-CNN: Towards Real-Time Object Detection with Region Proposal Networks", IEEE Trans. on pattern analysis and machine intelligence, vol.39 (6), 1137–1149.
7. M. B. Mahatme and S. Kuwelkar (2017), "Detection and Recognition of Traffic Signs based on RGB to RED Conversion," Proceedings of the IEEE 2017 International Conference on Computing Methodologies and Communication.
8. Fistrek, T.; Loncaric, S.(2011), "Traffic sign detection and recognition using neural networks and histogram based selection of segmentation method", Proceedings ELMAR.
9. D. Soendoro and I. Supriana (2010), "Traffic sign recognition with Color-based Method, shape-arc estimation and SVM," in Int. Conf. Elect. Eng. and Informatics, Bandung, July, 1–6.
10. Jang, C., Kim, H., Park, E., Kim, H. (2016). Data debiased traffic sign recognition using MSERs and CNN. In 2016 International Conference on Electronics, Information, and Communications (ICEIC), Da Nang, Vietnam, 1–4.

Note: All the figures in this chapter were made by the author.

Disruptive Technologies for Sustainable Development – Dr. G. Nagappan et al. (eds)

A Novel Algorithm for Detecting and Classifying Human Brain Stroke Using Deep Learning Algorithm

31

S. Gandhimathi@Usha[1], V. G. Janani[2],
Kaviya Prabha M. P.[3], Subasri R.[4], Soundarya G.[5]
Assistant Professor, Velammal College of Engineering & Technology (Autonomous), Madurai

Abstract: The Krill herd Optimization (KHO) algorithm is used in this paper to suggest an unique method for the identification of brain strokes. The suggested technique takes information from brain MRIs using magnetic resonance imaging (MRI) to extract features, and then utilises the KHO algorithm to pick the best subset of features for classification. A population-based meta-heuristic optimization tool called the KHO algorithm imitates the swarming behaviour of krill. The technique is intended to improve the precision of classifying brain stroke from MRI pictures. Using a dataset of brain MRI pictures, the suggested method is tested, and the findings demonstrate that it performs better than conventional classification techniques. The accuracy of the suggested method is 94.56%, which is greater than that of existing methods like Support Vector Machine (SVM).

Keywords: Krill herd optimization, Machine learning, ResNet algorithm

1. Introduction

A common and deadly medical illness known as a brain stroke can cause mortality, cognitive decline, and disability. When a blood artery in the brain becomes blocked or when there is bleeding in the brain, the result is a stroke. In order to reduce brain damage and enhance patient outcomes, prompt diagnosis and treatment are essential. By identifying anomalies in brain structure and blood flow, medical imaging techniques that diagnosis of stroke usually uses imaging techniques like computed tomography (CT) and magnetic resonance imaging (MRI).Due to the complexity and heterogeneity of stroke, it is still difficult to accurately and promptly detect stroke from medical pictures.

A novel heuristic optimization algorithm called "Krill herd optimization" (KHO) is based on the swarming behaviour of krill. KHO has been applied to numerous classification and regression issues, however it has primarily been employed for optimization issues. KHO can be used to optimise a fitness function that takes into account both the classification model's accuracy and the number of features it uses. KHO has demonstrated promise in resolving numerous optimization and classification issues and has application in medical image analysis for brain stroke detection. Data pre-processing, feature selection, KHO optimization, and model evaluation are all part of the suggested methodology for employing KHO to detect brain strokes. Data pre-processing, which includes picture normalisation, filtering, and segmentation to extract pertinent regions of interest from the medical images, is the first stage in the methodology. The following step is feature selection, which involves choosing the pre-processed data's most pertinent features to improve the effectiveness of the classification model. By optimising a fitness function that balances the precision of the classification model and the amount of features employed, KHO can be used for feature selection. The use of KHO for brain stroke detection may have a number of advantages. Secondly, by enhancing the feature selection and parameter tuning processes, KHO can assist in enhancing the accuracy of stroke identification from medical imaging.

[1]gmi@vcet.ac.in, [2]vgj@vcet.ac.in

DOI: 10.1201/9781003428473-31

2. Literature Review

Machine learning (ML) techniques are rarely used in studies to forecast strokes. Significant contributions to research are discussed in this section. In a recent study, the accuracy of predicting the death of stroke patients using the multilayer perceptron (MLP) algorithm-trained artificial neural networks (ANN) technique was 80.7%. It is applied to the classification-based analysis of a number of earlier investigations. Study a number of prior studies from the perspective of classification and explore the categorization of stroke using machine learning methodologies. Their study focused on the k-nearest neighbour and decision tree algorithms (KNN). The decision tree approach was discovered to perform better than the KNN algorithm of the stroke victims is still unknown A recent study asserts that the aetiology of stroke patients is still unknown despite the availability of several ischemic stroke diagnostic techniques. The study concluded that categorising strokes according to their phenotypic characteristics was important, but it also addressed how imprecise and unreliable this classification was using Statistical Region and Contrast Limited Adaptive Histogram Equalization (CLAHE) to increase contrast It was feasible to recognise stroke from a CT scan by combining to segment the image. For the first level of classification, histogram features are used to count the histogram value, whereas wavelet-based features are utilised for the second level of classification to count the result of the histogram value from the first level of classification. The suggested approach successfully classifies stroke with 90% accuracy. This study uses a neural network technique called Extreme Learning Machine to categorise brain haemorrhage from CT scan images. Before an image is segmented using a threshold to obtain an object, it is processed using the Grayscale and Scaling procedure.

3. Data and Variables

Datasets are groups of data. Kaggle's data repository has been used to gather the study's data. Data from a head MRI scan is obtained from Kaggle's data archive. The training and testing datasets each consist of ten digital photos. The information was gathered from brains with and without strokes. The stages in which the photographs were gathered are indicated by the labels and classifications on the folders. Displays an example of head MRI scan images.

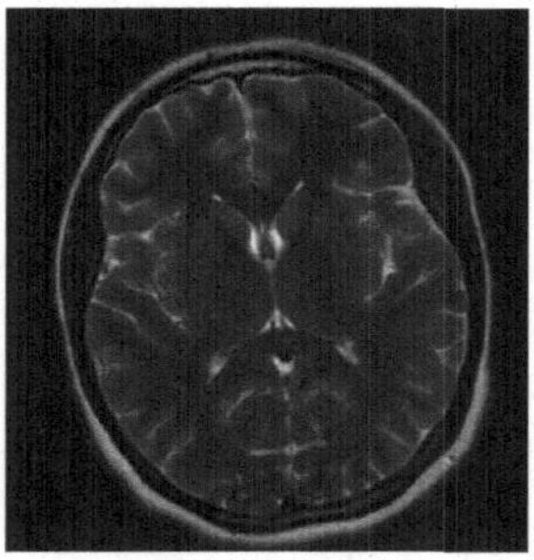

Fig. 31.1 Normal image

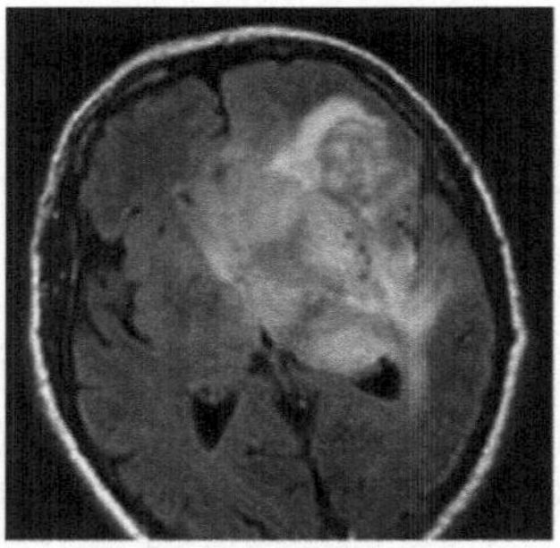

Fig. 31.2 Stroke image

3.1 Pre-Processing

Stock return is a commonly used metric to assess a stock's profitability that has a significant impact on investors' sentiment. In order to identify brain stroke disease and categorise its phases from MRI Scan images, image preprocessing is a crucial step.

Cropping: Cropping is a pre-processing step in brain stroke detection that involves selecting a region of interest (ROI)to isolate the pertinent region for stroke Cropping Filtering Resizing Normalisation detection from the original brain imaging data.

Resizing: Resizing is another pre-processing step commonly used in brain stroke detection, which involves changing the size and resolution of the brain imaging data to a uniform size that is suitable for further analysis and modelling.

Normalisation: Normalisation is a pre-processing step commonly used in brain stroke detection to transform the brain imaging data into a standardised scale or distribution that is suitable for further analysis and modelling.

Gray scaling: Grayscale in pre-processing typically refers to converting an image or video from a colour format to a grayscale format as a part of pre-processing steps

3.2 Segmentation

Segmentation is a pre-processing step commonly used in brain stroke detection to separate the brain structures of interest from the surrounding tissues and to identify the regions of abnormality. Segmentation can help improve the accuracy and efficiency of stroke detection algorithms by reducing the complexity and variability of the imaging data. In this process..

3.3 Thresholding based Segmentation

This method involves setting a threshold value to distinguish the targeted brain regions from the surrounding tissues and background. The threshold value can be determined based on the intensity distribution of the imaging data, or by using other information such as prior knowledge or template-based segmentation.

4. Methodology and Model Specifications

Our suggested strategy is based on the deep learning algorithm Krill Herd Optimization (KHO). Figure displays a flowchart of the suggested approach for classifying brain strokes. Figure 31.3 methodology diagram

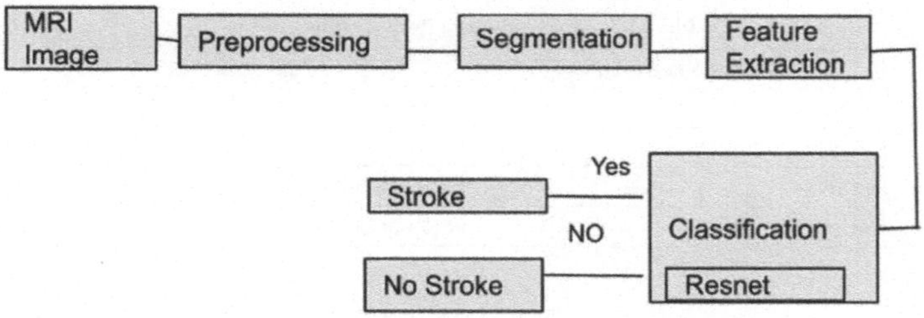

Fig. 31.3 Methodology of proposed model

4.1 Feature Extraction

Feature extraction is a process of selecting and representing relevant information or features from raw data. In terms of computer vision and machine learning, feature extraction is often used to prepare data for further analysis or modelling, such as classification or clustering. The process of feature extraction involves several steps, which may vary depending on the type of data and the specific application. Typically, the first step involves pre-processing the data to remove noise, normalise the data, and enhance important features.

4.2 Classification

The architecture of a convolutional neural network (CNN) for brain stroke disease detection and classification stages from MRI Scan images typically consists of several layers including convolutional, pooling and fully connected, input, Convolutional, Pooling, Dropout and Output layers

5. Empirical Results

Table 31.1 Statistic report

Layer (type)	Output shape	Parameter
input_2(inputlayer)	(none, 65,65,3)	0
conv2d_1(conv2d)	(none, 65,65,3)	84
resnet50(model)	Multiple	23587712
gobal_average_pooling2d_1	(none, 2048)	0
batch_normalization_1(batch)	(none, 2048)	8192
dropout_1(dropout)	(none, 2048)	0

Layer (type)	Output shape	Parameter
dense_1(dense)	(none, 256)	524544
batch_normalization_2(batch)	(none, 256)	1024
dropout_2(dropout)	(none, 256)	0
root(dense)	(none, 0)	514

Total parameters: 24 , 122, 070, Trainable parameters: 24, 064, 342, Non-trainable parameters: 57, 728.

ResNet50 is the latest family of ResNet-based object recognition models from Ultralytics, offering state-of-the-art performance. ResNet50 provides already trained classification models, doing the same video classification inference using ResNet50 model. This is the largest classification model offered by the achieve.

Correlation analysis

Performance Metrics:we assess classifiers with performance metrics: recall, accuracy,precision, X1-Score. Here, OZ: true positive, TO: false positive, ON: True negative, FO: False negative are,

Table 31.2 Performance metrics and equation

Performance Metrics	Equation
Precision	OZ/OZ+TO
Recall	OZ/OZ+FO
F1 score	2*Precision*Recall/Precision+Recall
Accuracy	(OZ+ON)/(OZ+ON+TO+FO)

6. Conclusion

This research utilises a multistage classification-based technique to create ResNet50 architecture which is a convolutional neural network which allows end-to-end training and testing and is capable of accurately detecting and recognizing many targets in images. The model plotted an accuracy of 98.79% for the given MRI scan image dataset after the application of classification parameters. Even if the findings of the classifier appear to be good right now, the paper's technique and the classifier's performance both have room for improvement.

Table 31.3 Performance metrics of different architecture

Architecture	Accuracy
GoogleNet	0.7942
Inceptionv3	0.8142
MobileNetv2	0.8736
ResNet50	0.9879

References

1. A. M. Hasan, H. A. Jalab, F. Meziane, H. Kahtan, and A. S. Al-Ahmad(2019), "Combining deep and handcrafted image features for MRI brain scan classification," IEEE Access, vol. 7, pp. 79959–79967.
2. Tateisi Y, Tsujii J (2004) Part-of-speech annotation of biology research abstracts. In: Proceedings of 4th international conference on language resource and evaluation (LREC2004), pp. 1267–1270.
3. H. M. Ali, M. S. Kaiser, and M. Mahmud(2019),"Application of convolutional neural networks in segmenting brain regions from MRI data," in Brain Informatics. Cham, Switzerland: Springer, pp. 136–146.
4. V. M. Aradhya, M. Mahmud, D. Guru, B. Agarwal,and M. S. Kaiser(2021), "One-shot cluster-based approach for the detection of COVID-19 from chest X-ray images,"Cogn. Comput., vol. 13, no. 4, pp. 873–881.

5. M. B. T. Noor, N. Z. Zenia, M. S. Kaiser, S. A. Mamun, and M. Mahmud(2020),"Application of deep learning in detecting neurological disorders from magnetic resonance images: A survey on the detection of Alzheimer's disease, Parkinson's disease and schizophrenia," Brain Informat., vol. 7, no. 1, pp. 1–21.
6. M. Mahmud, M. S. Kaiser, A. Hussain, and S.Vassanelli(2018), "Applications of deep learning and reinforcement learning to biological data," IEEE Trans.Neural Netw. Learn. Syst., vol. 29, no. 6, pp. 2063–2079.
7. A. M. Hasan, H. A. Jalab, F. Meziane, H. Kahtan,and A. S. Al-Ahmad(2019), "Combining deep and handcrafted image features for MRI brain scan classification," IEEE Access, vol. 7, pp. 79959–79967.

Note: All the figures and tables in this chapter were made by the author.

Disruptive Technologies for Sustainable Development – Dr. G. Nagappan et al. (eds)
© 2024 Taylor & Francis Group, London, ISBN 978-1-032-55000-8

Retinal Image Classification Using Neural Network

32

Kaladevi R.[1]
Associate Professor Department of Computer Science and Engineering,
Saveetha Engineering College

Dinesh T.[2], **Kishore R.**[3]
Master of Engineering Department of Computer Science and Engineering,
Saveetha Engineering College

Abstract: Retinal image processing is a crucial diagnostic tool in ophthalmology as it helps identify various eye conditions. However, using a Convolutional Neural Network (CNN) regarding the designation of the retinal illnesses from FUNDUS images can be challenging, especially when the input image has noise. The most common type of noise in medical images is salt and pepper noise. This noise is also known as the impulsive noise. Impulsive noise affects image processing significantly. To solve this issue, FUNDUS CNV, DME, DRUSEN, and NORMAL images were tested. The noise was removed using an advanced filter technology, which resulted in high precision compared to the conventional filter method. The proposed algorithm has three steps: firstly, adding noise to the test image, secondly, applying a new filter to remove the noise. and finally, using ALEXNET and LENET network to classify the disease.

Keywords: Retinal, deep learning, TensorFlow, Keras, CNN

1. Introduction

Diabetic retinopathy is a major factor in blindness, primarily affecting those having diabetes. Early detection of DR with the use of computer-aided diagnostic (CAD) technologies can avert vision loss. Retinal fundus images must be of high quality, and appropriate image enhancement techniques must be used to improve the efficacy of these systems. Fundus photography entails using a fundus camera to capture an image of the retina by concentrating fundus after dilation of eye. With dyes or colored filters, you can observe the camera's intricate microscope and flash. The fundus camera's optical setup employs the monocular indirect ophthalmoscopy concept to create clear images of the retinal arteries and optic nerve. Depending on how much noise is present in the image, several enhancement techniques like histogram equalization and filtering (HE), morphological or transformations can be used to enhance the segmentation of clinical features, the optic disc, and blood vessels in retinal images. Resizing the image is part of the image resolution technique, which lowers memory usage and expedites the classification process.

In chapter 2, we have the literature survey, followed by the proposed system in chapter 3, which also explains the algorithms used in the system. Chapter 4 has the methodology and model specification from the analysis followed by the result in chapter 5 and the conclusion in chapter 6.

[1]kaladevi@saveetha.ac.in, [2]droiddinesh.tamil@gmail.com, [3]kishorewolfeik@gmail.com

DOI: 10.1201/9781003428473-32

2. Literature Survey

Researchers are investigating the possibilities of deep learning in a variety of medical imaging applications as the discipline expands quickly. Girard and Cheriet (2017) proposed a unique method that combined deep learning and graph propagation methods to classify artery/vein abnormalities from fundus images. Their solution beat the most recent efficiency on the DRIVE database, which was 91.7%, with an excellent precision of 93.3%. Another study by Gleryz and Ulusoy (2018) claimed that identifying and treating conditions like retinopathy of prematurity, hypertension, and diabetes might be aided by segmenting retinal arteries and extracting parameters like tortuosity, breadth, and length. They suggested a convolutional neural network for automatically segmenting retinal arteries and outperformed earlier studies on the DRIVE dataset. Sambaturu and colleagues (2017) used a deep learning technique, CNN to detect, exudates and bleeding in fundus images for the investigation of diabetic retinopathy, achieving a 90% retention rate. According to a different study by Calimeri, Marzullo, Stamile, and Terracina (2016), the exact location of the Optical Disc can be determined using a supervised approach in ocular fundus image data. The models were able to detect the optic disc on the various databases with an excellent level of accuracy by using previous data from a pre-trained CNN, that was already capable to differentiate between faces in a picture. In this area, Triwijoyo et al. (2017) also carried out research These findings suggest that deep learning in medical imaging has the potential to improve illness detection and therapy. Researchers can extract meaningful information from medical images using modern techniques such as CNNs and region-based algorithms, which can aid in illness detection and monitoring. More sophisticated algorithms and strategies to solve even more complicated medical imaging jobs are anticipated to be developed as the field advances.

This study aimed to use retinal pictures to diagnose eye problems such as glaucoma and hypertension. The researchers hoped that by analysing vascular irregularities in retinal images, medical practitioners would be able to give early diagnosis and therapy for a few illnesses such as stroke, cerebral injury, carotid atherosclerosis, both cerebral amyloid angiopathy and vascular disease. The public STARE colour picture collection, which contains 15 courses, and CNNs (convolutional neural networks) were used by the researchers to detect retinal images. According to the statistics, the precision of the CNN simulation was 80.93%.

3. Proposed System

Figure 1 depicts the broad strokes of our strategy for sickness classification. The retinal OCT scan of patient is collected and it is converted into a grayscale image for feature extraction. Using vision algorithms, the region of retinal OCT scan image had been segmented and scaled. The resized images were sent into a basic CNN for prediction. The types of illnesses are detected by a machine. Finally, the system predicts the patient has a specific retinal disease or not.

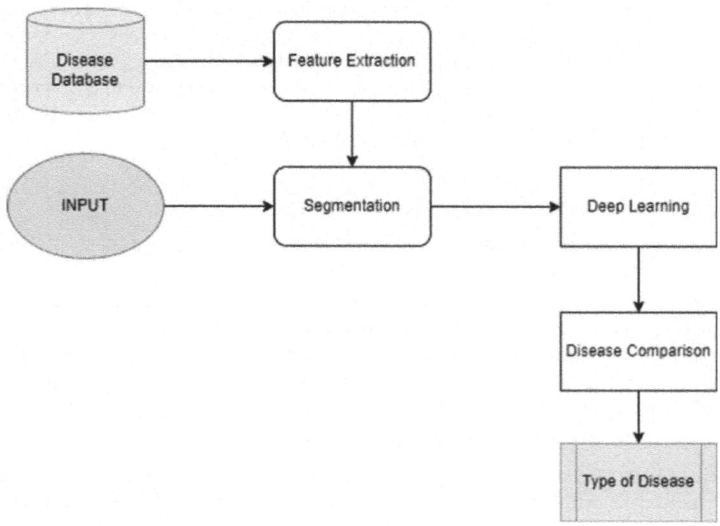

Fig. 32.1 Workflow

3.1 Segmentation

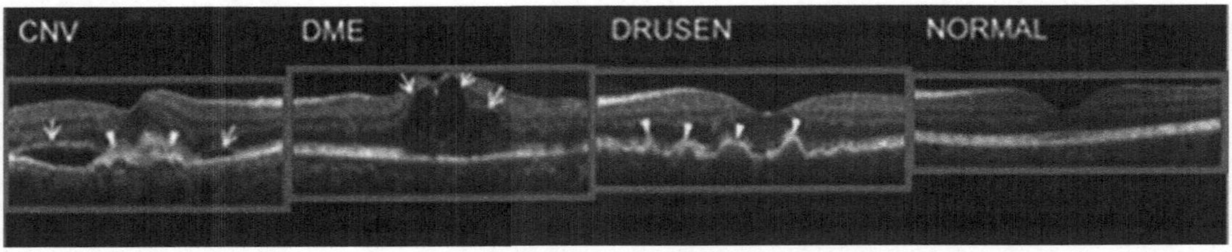

Fig. 32.2 Region of interests

Figure 32.2 depicts a separate Area of Attention for each type of disease where focus is directed in order to identify retinal abnormalities. The Region of Interest in the OCT images was separated out using some fundamental image processing techniques. The photographs' initial processing will help to improve data processing speed, decrease data loss, and increase accuracy. Christian Szegedy (2016) used the Python library OpenCV for the segmentation process. The dataset's photos had white borders around a sizable portion of them. As a result, dark pixels were used in place of the bright ones in the bordering areas. Thresholding created a binary image with white pixels elsewhere and dark pixels in the retinal areas. The image distortion was reduced via a widening procedure accompanied by an expansion, but the thresholding process left some noise behind. In order to prevent unintended cropping of the retinal region, the The region of interest was expanded by 60 pixels in each direction. The area of interest was eventually cut off using the indices discovered in the previous phase.

3.2 Convolution Neural Network

The pre-processed retinal OCT is passed to convolutional neural network layers for prediction.

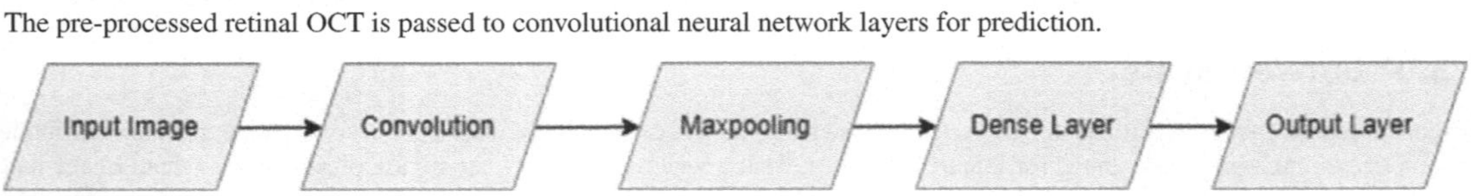

Fig. 32.3 Architecture of the convolutional neural network

Figure 32.3 depicts the CNN fundamental layers used for prediction. Initially, images are scaled to 128 ×128 pixels and read as one channel of Gray-scale pictures before being fed into the CNN. The first layer of the CNN is a two-dimensional convolution procedure utilising thirty-two filters with 3 X 3 kernels with stride sliding window size 1. A flattening operation yields a one-dimensional vector with 127008 components and a max-pooled features are passed to two fully connected layer 256 and 4 the neurons, create the neural network's output, or the target class of any of the four possible disorders. If a person's class or neuron has the highest probability, they may have the related disease or abnormality.

4. Result

Precision, recall, and f1-score, which are validating classification metrics, were used to further validate the accuracy metrics that the neural network's training had produced.

The worst case of grouping in this occasion is DME, which has been wrongly classified as CNV twenty times and Normal condition nine times out of the 242 images in that class. As a result, the recall score is low. The results from the other courses, on the other hand, are highly convincing. The category-by-class accuracy values of our classifier were also compared to the findings of Kermany et al. (2020). Despite being mathematically considerably simpler, our system may predict retinal disorders with nearly equal overall accuracy on the test set (95:66%) as the final study (96:1%).

A more detailed assessment of the several measures, class by class

We used the open-source Optical Coherence Tomography Images for Classification dataset to train our network. Kermany et al. (2020report and analyse the certified OCT image dataset. An independent patient testing set and an optical coherence tomography image training set are produced.

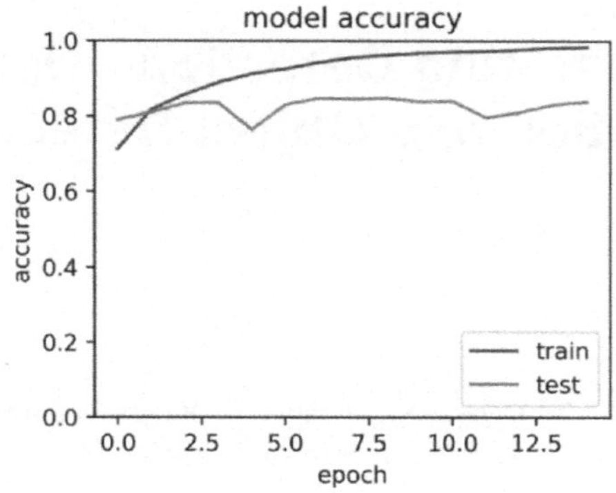

Fig. 32.4 Model accuracy

5. Conclusion

It takes a lot of time to manually classify and identify retinal disorders from retinal OCT scans. Our suggested strategy has a 95.66% accuracy rate for identifying retinal disorders. In comparison to previous methods, the network is computationally effective and light. This reveals some details about how retinal disease might be anticipated, like the CNN classification system's main benefit of automatically classifying images. As a result, our approach may have significant practical consequences and be able to assist medical professionals without the use of costly hardware, saving both money and time.

Reference

1. F. Girard and F. Cheriet, (2017) Artery/vein classification in fundus images using CNN and likelihood score propagation, IEEE Global Conference QC, Canada pp. 720–724.
2. M. Ş. Guleryuz and I. Ulusoy, (2018) "Retinal vessel segmentation using convolutional neural networks," 2018 26th Signal Processing and Communications Applications Conference (SIU), Izmir, Turkey, 1–4.
3. B. Sambaturu *et al.*, "A novel deep learning-based method for retinal lesion detection," 2017 International Conference on Advances in Computing, Communications and Informatics (ICACCI), Udupi, India, 33–37.
4. F. Calimeri, A. Marzullo, C. Stamile and G. Terracina, (2016) "Optic Disc Detection Using Fine Tuned Convolutional Neural Networks," 12th International Conference on Signal-Image Technology & Internet-Based Systems (SITIS), Naples, Italy, pp. 69–75.
5. B. K. Triwijoyo *et al.*, (2017) "Retina disease classification based on colour fundus images using convolutional neural networks", International Conference on Innovative and Creative Information Technology (ICITech), Salatiga, Indonesia 1-4.
6. J. Hu, Y. Chen, J. Zhong, R. Ju and Z. Yi, (2019) "Automated Analysis for Retinopathy of Prematurity by Deep Neural Networks," in IEEE Transactions on Medical Imaging, 38, 269–279.
7. Q. Meng, Y. Hashimoto and S. Satoh, (2020) "How to Extract More Information with Less Burden: Fundus Image Classification and Retinal Disease Localization with Ophthalmologist Intervention," IEEE 17th International Symposium on Biomedical Imaging (ISBI), Iowa City, IA, USA, 1373–1377.
8. V. Das, S. Dandapat and P. K. Bora, (2017) "Automated Classification of Retinal OCT Images Using a Deep Multi-Scale Fusion CNN," in IEEE Sensors Journal, vol. 21, no. 20. pp. 23256–23265.
9. C. Szegedy, V. Vanhoucke, S. Ioffe, J. Shlens and Z. Wojna, (2016), Rethinking the Inception Architecture for Computer Vision," IEEE Conference on Computer Vision and Pattern Recognition (CVPR), Las Vegas, NV, USA, 2818–2826.
10. Daniel S. Kermany,(2020) "Identifying Medical Diagnoses and Treatable Diseases by Image-Based Deep Learning," National centre for biotechnology information pp. 205–225.

Note: All the figures in this chapter were made by the author.

Disruptive Technologies for Sustainable Development – Dr. G. Nagappan et al. (eds)

Real-Time Littering Detection—Using Pose Estimation and Object Detection

33

Kaladevi R.[1], Ganesh K.[2], Kishore P.[3], Vijaya Raghavan V.[4]
Computer Science and Engineering, Saveetha Engineering College, Chennai, India

Abstract: Public littering is a huge problem in Developing nations currently. Even though the world governments have implemented various schemes and awareness programs to clean our surroundings, there are not many ways to prevent people from littering. We see every other street littered with garbage and there is no proper system in place to monitor and hold people accountable for littering. Hence this research aims to solve this problem by developing a real time Litter detection system which uses YOLO, Faster-RCNN and Human Action Recognition methods to identify if a person litters in real time using live footage from IP cameras and captures the face of the litterer if visible and flags them so we can identify if they repeat their actions in the future. This data can be used by the authorities to hold the people who litter accountable for their actions, and it can also help us identify which places require a cleanup.

Keywords: YOLO, Faster-RCNN, Face detection, Human action recognition

1. Introduction

Public littering has always been a big problem in developing countries and there still aren't any better measures taken to prevent people from littering, and people keep littering which leads to accumulation of garbage on the streets. In this research "**real-time littering detection**", we propose a solution to bring about an idea of monitoring the places where people tend to litter garbage by setting up ip cameras. This system runs YOLO, Faster-RCNN trained model on the video in real-time behind the camera to detect whether littering takes place or not and save the video, which is once again processed to detect and recognize the face of the perpetrator (Ren et al. 2015).

There aren't many existing models to predict littering and even in the existing models there are a wide range of issues like, low accuracies, lack of features to hold the litterer accountable and lack of proper data. In systems which have face detection and recognition the implemented detection methods are not applicable for real world scenarios. The systems which have implemented the Human Action Recognition do not consider camera angles and occlusion of objects and faces.

2. Literature Review

A model in which littering action is detected using background subtraction algorithm and human action recognition algorithm has been deployed here. Pre trained face recognition model from openCV which is used for face detection. The drawback is that dataset is really small, and the face detection system is not trained on data similar to real world scenarios, hence leading to lower accuracies. They also don't consider real world camera angles to detect dumping action (Korchani et al. 2020).

The Authors proposes a method in which the dumping action is detected by change in relation between person's joints and garbage held by them using background subtraction algorithm and human joint estimation. The drawback of this model is that

[1]kaladevi@saveetha.ac.in, [2]ghkn2001@gmail.com, [3]pkichu1997@gmail.com, [4]vj123chethanavenky@gmail.com

DOI: 10.1201/9781003428473-33

there is no face detection done, hence the litterer could not be held accountable, and the model cannot detect the dumping action if the garbage is occluded (K. Yun et al. 2019).

A model which uses a UAV to detect and classify garbage using various machine learning algorithms on real-time was deployed, but the model could only be trained on images from an aerial angle which does not correspond to real world scenarios and the use of an UAV is not feasible. It also uses machine learning algorithms rather than deep learning which makes the model less robust (A. Chung et al. 2018).

The Authors have developed a model that has an integrated EfficientDet and efficientNet architectures that are used to perform object detection and object classification. In this paper, the model was used to detect and classify garbage from existing image datasets. The classifier is trained in a weak supervised manner to make use of the unlabelled images. The version EfficientDet-2 is used in localizing the litter whereas EfficientNet-B2 is used to classify the detected waste into one of the seven categories. The drawback of this model is that it only segregates the type of garbage and doesn't provide and actual use in real world scenarios for garbage cleanup also the EfficientDet is slow and not applicable for real-time use (Majchrowska S et al. 2021).

The revolting idea have presented which uses a garbage pickup robot to detect the garbage accurately and autonomously by means of using a deep neural network for garbage recognition on grass areas. A unique navigation strategy is proposed to guide the robot's movement alongside the ground segmentation technique. Results of the garbage recognition accuracy is high, but the only drawback is that the model is limited to grassy areas (J. Bai et al. 2018).

A model which can track multiple people in the video at the same time and can also continue to monitor them for a long time is very useful in crowded area where only face is the only visible part of a person is proposed in one of the articles. The system there can detect a person by tracking their movements by combining a visual tracker and an online face verification strategy. The major advantage of the model is that they have managed to obtain a huge 50% longer tracks than the other state of the art deep learning tracks with no additional computation cost. This system only tracks the pedestrians if their face is visible (Barquero et al. 2020).

3. Proposed System

This model is built so that it can detect and recognize littering in real-time along with detecting the face of the litterer and save the clip for further processing of the data to perform face recognition. This model uses data collected from CCTV cameras capturing civilians littering and other public cameras capturing pedestrians. It uses a pretrained model trained on the CityScapes dataset so that it can perform Pedestrian tracking. The clip is timestamped in the video and the location of the action is saved along with the clip, this clip is later fed into a face recognition model to hold the litterer accountable.

3.1 Dumping Action Recognition

This module is divided into several parts since it involves a crucial real-world scenario. The first step is to verify if there is a pedestrian present in the view of the camera and if there is a pedestrian present we track them. The next step is to check whether they are carrying any garbage and it is one of the tough segments, because the shape of the object is occluded since the person is holding the garbage. We use YOLOv5 pre-trained on the COCO dataset to detect these objects carried by the pedestrians. The next step is to track the object that is being carried to detect if it is being dumped. Finally, the dumping action is detected by calculating the distance between the handheld object and joint closest to the object, this way if the distance increases rapidly we can identify that littering has occurred.

3.2 Human Pose Estimation

Human pose estimation enables to plot the joint positions on humans. It is done using MoveNet model which helps in returning 17/14 keypoints of a human body and draws those keypoints on a person to give us the result like the Fig. 33.1.

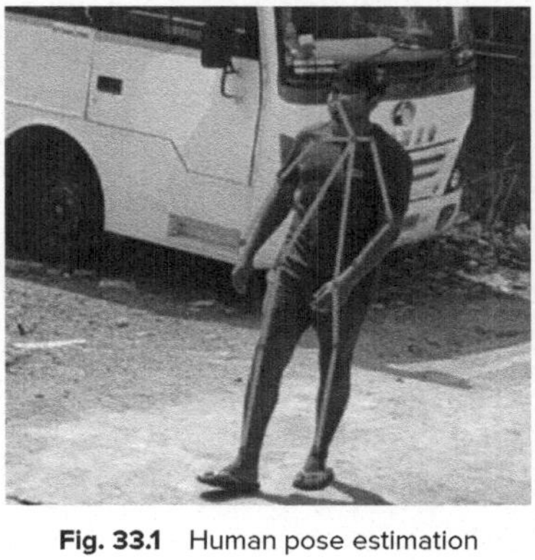

Fig. 33.1 Human pose estimation

3.3 Object Detection

Object Detection is done in real-time along with the other detection such as pedestrian and face. The method that was approached to perform the object detection in this paper was using "YOLO" algorithm. Data are annotated before fed into the model to then produce the results.

Fig. 33.2 Object detection

3.4 Faster-RCNN for Pedestrian Tracking

Faster R-CNN belonging to the family of Region Proposal Networks, this family also includes R-CNN and Fast R-CNN. These networks work by proposing regions using a selective search algorithm, after which features are then extracted from these regions which are in turn used to detect objects in these regions. R-CNN which was the first network introduced works great by is not as fast as it is successors, since we have to track pedestrians in real-time with high accuracy, we chose Faster R-CNN which is pretrained on a dataset. The Faster R-CNN replaces the slow selective search algorithm with a fast Region Proposal Network which will help reduce the computations needed and hence making the algorithm much faster. In our work we have used the ResNet-50 network followed by RPN which is available on the Detectron2 model zoo. The faster R-CNN are further trained on the CityScapes dataset.

3.5 Complete System Architecture

The Dumping Action Recognition is divided into several parts since it involves a crucial real-world scenario as seen in Fig. 33.3. The first step is to track pedestrians in the view of the camera Pedestrian detection is done to detect the people in motion in

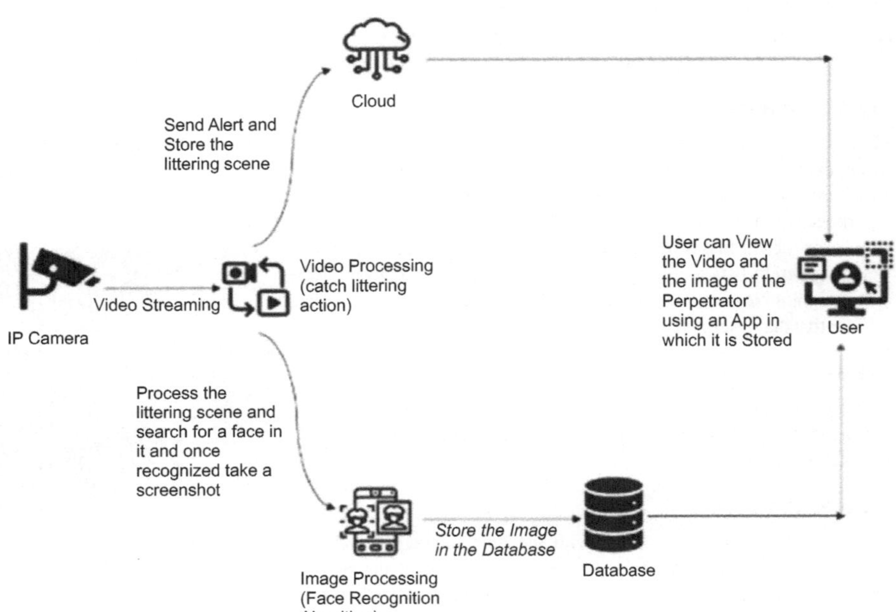

Fig. 33.3 System architecture

camera frame. A pretrained model using Faster-RCNN trained on the CityPersons datasets is used in the process of "Pedestrian Detection", Face detection is done alongside the pedestrian tracking to find the face of the person

4. Result

In this phase we have implemented the three foremost stages of dumping action recognition which are pedestrian tracking, pose estimation and object detection. Only if we track and detect the pedestrian, object and joint estimation of the persons then we could detect whether someone littered or not. Hence, we have implemented pedestrian tracking along with object detection and pose estimation for this research.

The final output after the above mentioned detection processes would give us exactly like Fig. 33.4

Fig. 33.4 Final output

5. Conclusion

Using various deep learning techniques, we perform pedestrian tracking, human pose estimation and object detection using transfer learning followed by dumping action detection which is implemented using human action recognition techniques further we strengthen our research by identifying the perpetrators using face detection to hold them accountable. The proposed system is currently limited to only detecting images. Face recognition, Dumping Action Recognition, storing and displaying functions are yet to be implemented.

References

1. Ren, Shaoqing, et al. (2015). "Faster R-Cnn: Towards Real-Time Object Detection with Region Proposal Networks." Advances In Neural Information Processing Systems 28.
2. Korchani, Basma and Kaouthar Sethom. (2020). "Real-Time Littering Detection for Smart City Using Deep Learning Algorithm." 2020 International Conference on Communications, Signal Processing, And Their Applications (Iccspa): 1–5.
3. K. Yun, Y. Kwon, S. Oh, J. Moon, And J. Park. (2019). "Vision-Based Garbage Dumping Action Detection for Real-World Surveillance Platform," Etri Journal, Vol. 41, No. 4. Wiley, Pp. 494–505.
4. A. Chung, D. Y. Kim, E. Kwok, M. Ryan, E. Tan And R. Gamadia. (2018). "Cloud Computed Machine Learning Based Real-Time Litter Detection Using Micro-Uav Surveillance," 2018 Ieee Mit Undergraduate Research Technology Conference (Urtc), Pp. 1–4, Doi: 10.1109/Urtc45901.2018.9244800.
5. Majchrowska S, Mikołajczyk A, Ferlin M, Klawikowska Z, Plantykow Ma, Kwasigroch A, Majek K. (2022). Deep Learning-Based Waste Detection in Natural and Urban Environments. Waste Manag. 2022 Feb 1; 138:274-284. Doi: 10.1016/J.Wasman.2021.12.001. Epub 2021 Dec 14. Pmid: 34920243.
6. J. Bai, S. Lian, Z. Liu, K. Wang and D. Liu. (2018). "Deep Learning Based Robot for Automatically Picking Up Garbage on The Grass," In Ieee Transactions On Consumer Electronics, Vol. 64, No. 3, Pp. 382-389, Aug. 2018, Doi: 10.1109/Tce.2018.2859629.
7. Barquero, Germán, Carles Fernández, And Isabelle Hupont. (2020). "Long-Term Face Tracking for Crowded Video- Surveillance Scenarios." 2020 Ieee International Joint Conference on Biometrics (Ijcb).

8. W. Yimyam, T. Pinthong, N. Chumuang and M. Ketcham. (2018). "Face Detection Criminals Through Cctv Cameras," 2018 14th International Conference on Signal-Image Technology & Internet-Based Systems (Sitis), 2018, Pp. 351-357, Doi: 10.1109/Sitis.2018.00061.
9. M. T. Bhatti, M. G. Khan, M. Aslam, and M. J. Fiaz. (2021) "Weapon Detection in Real-Time Cctv Videos Using Deep Learning," In Ieee Access, Vol. 9, Pp. 34366–34382, 2021, Doi: 10.1109/Access.2021.3059170.
10. D. Stadler and J. Beyerer. (2021). "Improving Multiple Pedestrian Tracking by Track Management and Occlusion Handling," 2021 Ieee/Cvf Conference on Computer Vision And Pattern Recognition (Cvpr), 2021, Pp. 10953–10962, Doi: 10.1109/Cvpr46437.2021.0108.

Note: All the figures in this chapter were made by the author.

Disruptive Technologies for Sustainable Development – Dr. G. Nagappan et al. (eds)
© 2024 Taylor & Francis Group, London, ISBN 978-1-032-55000-8

Skin Cancer Prediction Using Convolutional Neural Network—Cloud Based Web App

34

N. V. Ravindhar[1]
Assistant Professor, Department of Computer Science and Engineering,
Saveetha Engineering College, Chennai, Tamil Nadu

Janani V.
Department of Computer Science and Engineering,
Saveetha Engineering College, Chennai, Tamil Nadu

Wael M. Alenazy
Associate Professor, Computer Systems,
Kind Saud University, Saudi Arabia

Abstract: Due to the ozone layer depletion and nowadays pollution level people are easily affected by skin cancer . Even though there are a lot of ways to identify the skin cancer people are failing to use in the early stages , the main reason behind that is lack of technology which should be used to detect whether it is skin cancer or not. Most of the skin cancer is curable if it was detected at the early stage, recovery rate is nearly 90 percent. Skin cancers originate on the surface and are typically visible, in contrast to malignancies that develop inside the body. If they own a technology that uses their image to forecast a level of performance based on their current state. For solving this problem we came up with the solution ,which will give the prediction level of their skin cancer at early stage with the help of their current image which has processed. A convolutional neural network and two machine learning classifiers trained a datasets with various feature extraction describing the borders, cell alignment, width, texture and color of the skin pigment. These methods are then combined to improve their performances using various feature engineering processes. The CNN model aims to identify skin cancer in its earliest stages in patients and treat them effectively in order to lower death and impact rates. With cloud deployment, this model will be accessible to all regular people, and a chatbot built using the RASA framework will be beneficial to everyone.

Keywords: Convolutional neural newok, Flask, The cloud, Applications and classifications

1. Introduction

The most prevalent form of cancer is skin cancer. When detected early, it is one of the malignancies that is most treatable. The melanocytes are the site of melanoma, a rare and extremely aggressive kind of skin cancer. They are the cells that create the melanin, the pigment in your body. However, most skin cancers are non-melanoma, which means they do not involve the melanocytes. Basal cell and squamous cell skin cancers are the two most prevalent of these. If detected early, they are almost always curable. Yet melanoma can develop fast into a disease that is more challenging to treat if it is not found and treated early. UV rays is the main cause of skin cancer. UV radiation from the sun and tanning beds can harm your skin's DNA if you don't protect it. Cancer results from the DNA's inability to appropriately regulate the proliferation of skin cells. Your likelihood of receiving it may be increased by a variety of factors. One of the three types of skin cells reproduces erratically, which leads to

[1]ravindhar@saveetha.ac.in, [2]Alenazym@gmail.com

DOI: 10.1201/9781003428473-34

the development of skin cancer. They can spread or metastasis as they continue to multiply and divide. This indicates that they spread through your lymphatic system to other areas of your body. Due to extensive lifetime sun exposure among an ageing population, skin cancer incidence has increased dramatically over the world over the past 20 years, reaching epidemic levels. Skin cancer makes a major contribution to the entire burden of cutaneous disorders in the aged, affecting fertility, mortality, and healthcare expenses significantly.

2. Literature Review

In India and many other nations, skin cancer is one of the most prevalent types of the disease, although compared to Indians, westerners are more likely to contract it. Dermoscopy picture classification is significantly improved by artificial intelligence (2019)especially for vascular lesions. a medical diagnostic. Deep learning (DL) models have unlocked new potential in medical image processing, excelling in tasks like tissue segmentation and classification as well as better clinical prediction. At 83 percent (3-D) against 69.6 percent and 65.2 percent (2-D) accuracy, the classification performance results significantly improved tissue classification accuracy (2-D).For the specific clinical issue, CNNs with different topologies were also created using the same dataset(2018).Many methods for classifying skin lesions focus on the unique characteristics of lesions, including their colour, texture, form, and combination of descriptors. By extracting a variety of colour and texture data and using them to train a classifier, researchers presented a thorough and local strategy to classifying skin lesions(2019).Skin lesion segmentation based on PSO employing a cascade swarm intelligence (SI) approach. It is used to optimise deep CNN hyper-parameters and typical FCM clustering cluster centroids(2018).On top of the reinforced foundation of CNN and hybrid FCM models, two ensemble segments— ensemble deep networks(2020) and hybrid clustering models—are constructed. Segment is composed of three fundamental CNNs, or clustering models,(2018). Section is made up of three basic CNNs or clustering models, or the results of the hair removal algorithm. the impact of different illumination levels on image pixel values(2018). Synaptic learning is carried out unsupervised and layer by layer in each convolutional layer. In convolutional layers, input synaptic weights of neurons are altered(2013)by combining a winner-takes-all (WTA) process with a condensed STDP rule. The DNN algorithms covered in this section all go through a training phase in which data is supplied into the system and its weights or behaviour is modified. Additionally, they demand validation and testing stages when the actual outcome is contrasted with the network's prediction. The framework(2018) suggests a discriminatory network, such as the CNN dubbed Inception V4. A training data set of 17,200 dermoscopy images was produced using random sample generation techniques in the horizontal, vertical, and horizontal vertical directions. the samples that were produced, with vertical flipping on the lower left, horizontal flipping on the lower right, and both horizontal and vertical flipping. The pixels in the original image range from 540x722 to 4499x6748 due to the original image's scale.

3. Implementation and System Architecture

Skin cancers The extensive collection of multisource dermatoscopic images of pigmented lesions is found at MNIST:HAM10000(KAGGLE). With the use of a dataset, we developed a deep convolutional neural network using a training set of 100,000 photos and a test set of 10,000 images, each measuring 28 x 28 in width and height. The model is then processed in a flask web app before being posted to Heroku.

3.1 Sub Architecture

To process the resources, the user submits them. These features are simple to use while still being able to accurately and uniquely describe the real data set. The reduction of the data makes it easier for the machine to develop the model and accelerates the learning and generalization phases of the process. The model is deployed as a (.Pickle) file .it is then embed into flask framework which comprises of HTML , CSS bootstrap as its frontend elements then the entire functional elements are deployed into AWS EC2or Heroku instance .As the user perspective the user uploads the photo the image isthen processed by the convolutional neural network the result is shown in the webUI Based on the results the chat bot suggestions will happen.

3.2 Modules

There are mainly three important modules used in our project. They are:

- Registration module
- Prediction module
- Chat bot Module

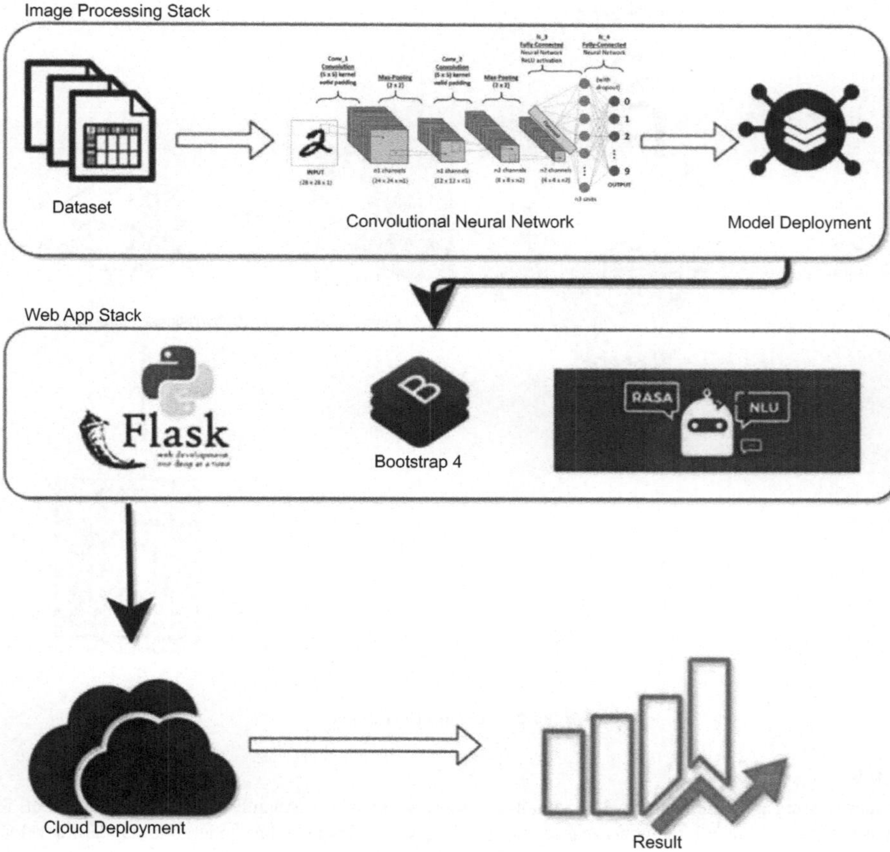

Fig. 34.1 System architecture

Modules Explanation

Registration module

In this module users can the login. New users register their Email, name,username Password, and Confirm Password activity into the registration page. After registration, the next stage is login process, the login page contains user name and password field (Fig. 34.2).

Prediction module

Melanocytes, which are cells found throughout the body, are what create melanoma. One of the main factors contributing to the development of melanomas is skin exposure to UV light. The goal of this project is to automatically segment the boundaries of potential cancerous regions from dermoscopic images. The goal is broken down into three tasks, the first task of this project is to correctly predict the pixel-wise masks from the input image. The second task is more challenging, involving automatically predicting and localizing various attributes of the lesion, such as pigment networks, streaks, milia-like cysts, and more. Finally image classification is performed, given an input image the goal is to classify the lesioninto seven distinct categories such as Melanoma, Basal cell carcinoma, Melanocytic nevus and others.

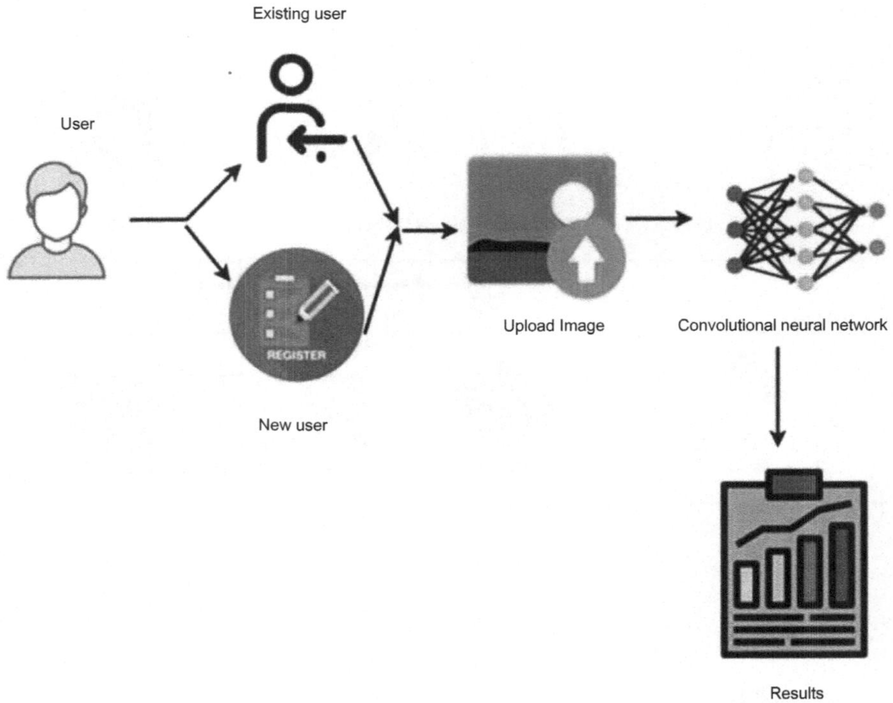

Fig. 34.2 Registration module

Chat bot module

One of the most interesting and promising types of human-machine interaction is frequently referred to as a chat- bot. Technically speaking, a chat-bot is an artificial intelligence-based Question Answering Tool that has undergone standard development.

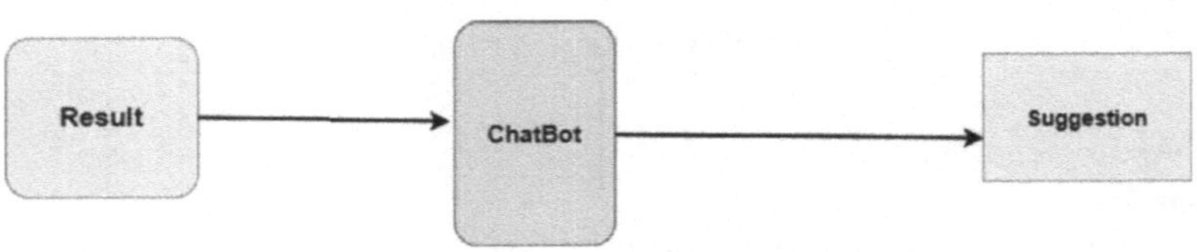

Fig. 34.3 Chat bot module

4. Conclusion

A cancer patient goes through a lot, and we as a group must support them in their time of need. We are taking them a step closer to a better life if we can spare them the time they need to determine and prepare their future by drastically reducing the time it takes to diagnose their form of cancer at an early stage. When you're going through a tough patch emotionally, it's important to stay positive. Identifying the kind of Early detection of cancer ensures that it can be treated, if not cured. This is important for medical institutions all around the world to ensure that cancer patients are successfully treated.

REFERENCES

1. Jianpeng Zhang, Yutong ie, et al, (2019), Attention Residual Learning for skin Lesion Classification 2–5.
2. Seyed Mohammad Alizadeh, Ali Mahloojifar, (2020), Automatic Skin Cancer detection in dermoscopy images by combining convolutional neural networks and texture features, 5–8
3. Servio Lima, Luis Terán, (2018),et al, A Proposal for an Explainable Fuzzy- based Deep Learning System for Skin Cancer Prediction, 6–9.
4. Tryan Aditya Putra ,Syahidah izza Rufaida (2020) ,et al, alEnhanced skin condition prediction through machine learning using dynamic training and testing augmentation, 5–8
5. BENJAMIN ALEXANDER ALBERT, (2020), Deep Learning From Limited Training Data: Novel Segmentation and Ensemble Algorithms Applied to Automatic Melanoma Diagnosis, 2–8.
6. PENG LIU , (Member, IEEE), AND SHUMIN FEI , (Member, IEEE) (2018) et al, Two-Stage Prediction of Comorbid Cancer Patient Survivability Based on Improved Infinite Feature Selection, 4–9.
7. Ahmad Naeem1, Muhammad Shoaib Farooq1, (MEMBER, IEEE), et al, 2019, Malignant Melanoma Classification using deep learning:database, Performance Measurment ,Challenge and Opportunities, 2–7.
8. Peng Tang, Qiaokang Liang , et al, (2020), GP-CNN-DTEL: Global-Part CNN Model with Data-Transformed Ensemble Learning for Skin Lesion Classification, 3–7.
9. Zhen Yu, Xudong Jiang, et al, (2018), Melanoma Recognition in Images via Aggregated Deep Convolutional Features, 2–7.

Note: All the figures in this chapter were made by the author.

Disruptive Technologies for Sustainable Development – Dr. G. Nagappan et al. (eds)
© 2024 Taylor & Francis Group, London, ISBN 978-1-032-55000-8

Design of Student Monitoring System Using Face Detection Technique

35

Nagappan G.[1]
Professor, Department of Computer Science and Engineering,
Saveetha Engineering College, Chennai, Tamil Nadu, India

T. Navya[2], **T. J. Mounika**[3], **S. Tharun**[4]
B.E Scholar, Department of Computer Science and Engineering,
Saveetha Engineering College, Chennai, Tamil Nadu, India

Abstract: The calculation ought to be extremely precise in recognizing frontal area elements of understudies in the homeroom. The number of students enrolled should be immediately visible in the graphical user interface of the system. For student identification, match student features in the database to images in the database with an accuracy of 80 percent. The image's student features should be able to be matched by the algorithm to those in its database. Because of the formula used to make the system work, its accuracy is within 80%. The likelihood of success ought to be high. The graphical user interface of the system makes it simple for users to access the system's features. Overall Attendance of Students based on the monthly report the faculty can monitor about the student presence and the report can be check by the respective faculty members and students.

Keywords: Convolutional neural network algorithm, Face recognition

1. Introduction

A person can be recognized only by his face. It is a rapidly developing field of study and is used in many different fields. There are many proposed methods of attendance management. False signatures from students are rampant these days and pen and paper or attendance sheets for signature are easily circulated your attendance is recorded when you actually come to class, as facial recognition is not disabled as each person has unique characteristics that cannot be copied or altered. This method deals with the fundamental problem of 3D face monitoring related to human-computer communication. A video stream captured by a cheap camera guides the user's vision. Our efforts are focused on full 3D real-time face tracking with six degrees of freedom (DOF) in the face of changing facial expressions, a wide range of pose variables and partial occlusions. To achieve this, system makes far fewer assumptions than the majority of advanced methods. The proposed method uses a basic 3D face model to monitor users' 3D head movement in real time and generate accurate and detailed face models. The system must first generate information to create it. In a face detection system, various processes are initiated when the image clarity is appropriate. Functions are performed using Questions in Python. Information is used as input. Each display in the system is encoded with a precise structure. All the strategies and methods required to detect the visibility of a person from the image provided to the dataset will be included in a separate file. A Python program executes the specified file and the system has the ability to modify or roughly rotate the image to get the desired result. The results of face detection system can be improved by existing algorithm and OpenCV modules.

[1]hod.cse@saveetha.ac.in, [2]navyachowdary.thellapati@gmail.com, [3]tjmounika01@gmail.com, [4]Tharuns9492@gmail.com

DOI: 10.1201/9781003428473-35

2. Literature Review

An "Intelligent Attendance Management System" developed by "Kanij Bhatti, Lareeb Mughal and Faheem Khohor" (2014) The primary objective of this work is to enable the system to automatically track movements before applying facial recognition and recognize and identify the presented image. Selected students indicate their presence and update their presence with appropriate time and subject ID. Here, deep learning is used to design the system. To detect faces in images, a graph-guided gradient approach is used, and deep learning is used to analyze and evaluate the facial features of the students. Our algorithm is capable of quickly recognizing the faces of many students Shavian Kudale, Pooja Humbe, Apoorva Kamshetty (2017) when he enters the classroom, the system records his attendance by capturing an image using the Principal Component Analysis algorithm, which helps to address issues such as time wastage and poor attendance. The computer system immediately documents each student's attendance. A student handbook was compiled; It contains name, photograph and registration number of each student. Each student in each subject has a record on the log report and a PDF summary of the student's attendance is generated. A.P. Visalakshi (2018) With the OpenCV tool, this is achieved. Native graphing technology is used to select the image. There are cameras in the classroom where students sit. In video recording, the camera continuously monitors the students. That is "Aparna Trivedi" K (2022) The proposed method involves directly processing the measured face using the local binary samples histogram algorithm, compressing the detected face to the required size, and detecting human features from the camera using the Viola-Jones technique. The relevant data will be added to the presence in the SQLite database as soon as the approval is complete. Numerous businesses profit greatly from this project. Because it takes less time and has fewer human errors, this makes it more effective.

3. Research Basics

3.1 Convolutional Neural Network Algorithm

For the most part, speech and image recognition are where convolutional neural networks are put to use. They accomplish this by using an internal convolutional layer to reduce the high dimensionality of images without altering the quality of the images themselves. If this framework must employ a fully convolutional neural network to deal with the images, it finds that it does not scale well. The color displayed by each pixel in an image can be described using one of three matrices in an RGB notation. Each color's "red," "green," and "blue" components are defined by their respective matrices. Through average pooling, input is divided into rectangular pooling regions and the average values of each region are computed before down-scaling begins. Step is the quantity of pixels which are shift over the info lattice..While developing the convolution neural network, padding is an essential step. This system will give us a small filtered image if the image gets smaller and a neural network with hundreds of layers used. The sub-region is determined in exactly the same way as for maximum grouping, such as sum grouping or average grouping, but this method uses the sum or average rather than the maximum. Data from other levels are compressed into a vector and sent to the strictly bound layer. The score is converted to the number of combinations required by the grid

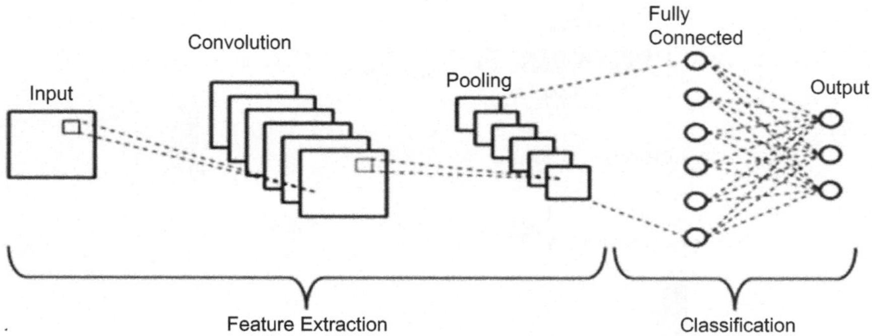

Fig. 35.1 Convolution neural network algorithm

4. Proposed Approach

As the need for face recognition grew, a number of algorithms were developed in the early 1990s. Live footage management systems were developed. As a result, research into this technology is ongoing. Commercial products are beginning to appear on the market as a result of the most recent advancements in technologies that track faces and recognize facial features. Even though the business is doing well financially, there are some problems that need more research. There are two primary functions of facial recognition: Verification is the process of comparing an unidentified person's face to that of someone who claims to be the person in the picture. Identification is a one-to-one comparison in which the person's name is verified by comparing the face in the raw photo to a group of photos of known individuals. Face Acknowledgment, then again, can be used in a large number of utilizations, including security, observation, general character confirmation (discretionary enlistment, public ID cards, visas, driving licenses, understudy IDs), law enforcement frameworks, picture data set examinations, savvy cards, media conditions, video ordering, and witness face remaking. In this system the Student login allows the students to log in with their credentials they can see their attendance per course and time of attendance taken on the particular lecture. Faculty login allows the admin to access the database and it contains of student data. The admin can add the student with there sportive mailed and pass word and also it allows to capture the face by pressing the' C' to capture the image that to be stored in the database Faculty and admin can check the no. of students present according to the each course or lecture Faculty Register Admin can add the faculty by adding their login credentials along with the course name and the lecture that faculty is taken for the student .Marking Attendance Faculty allows to open the webcam in front of the students the system can able to detect faces of their glistered students with more than 80% accuracy and the attendance will be marked.

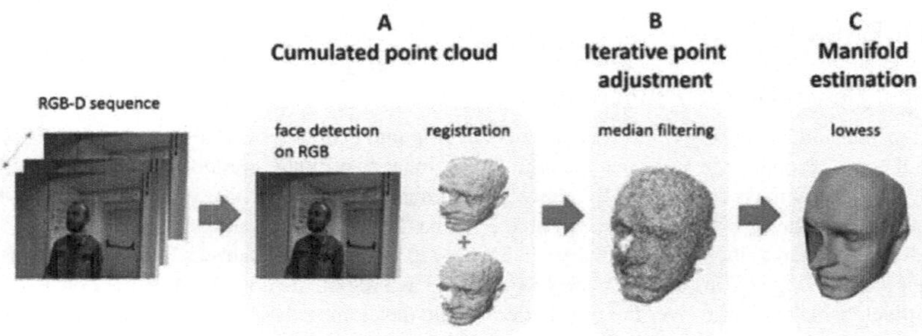

Fig. 35.2 Proposed approach monitoring attendance

4.1 Architecture Diagram

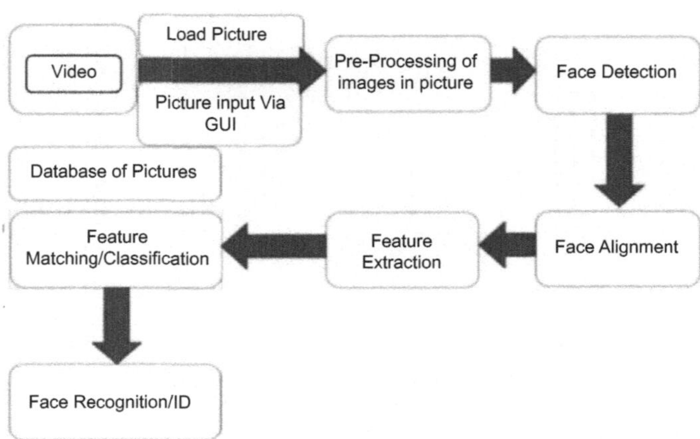

Fig. 35.3 Proposed student monitoring system architecture

The system's adaptability to changing face projections and poses is demonstrated by this result. It was discovered from the face recognition using a deep learning algorithm that the brightness problem is solved during face detection by converting the original image into a HOG representation, which captures the majority of the image's features regardless of the brightness. As a result, an Excel sheet is generated, and a weekly PDF of it is sent to students and professors. This An individual can without much of a stretch work the gadget and it gives great security

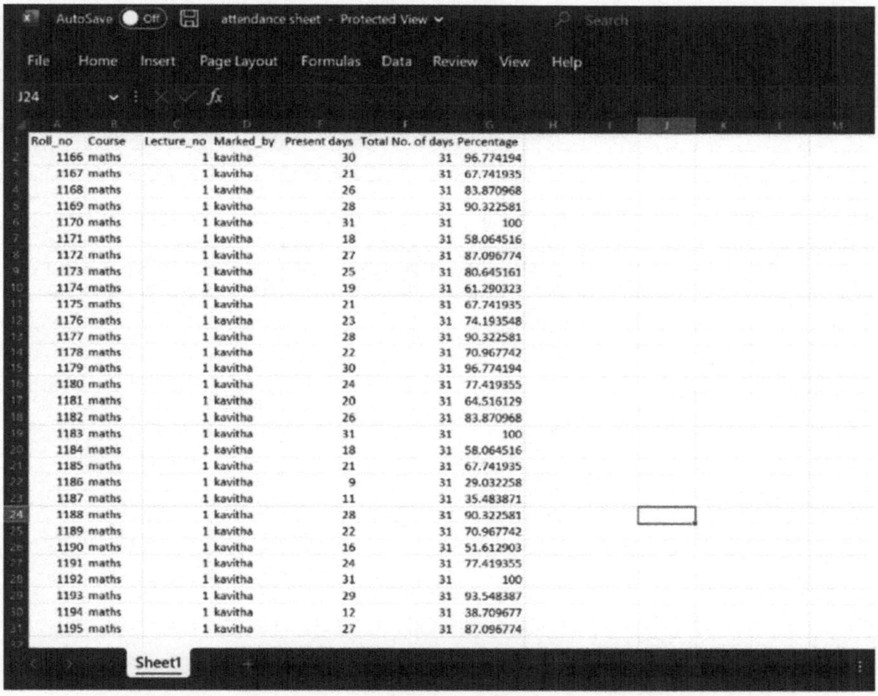

Roll_no	Course	Lecture_no	Marked_by	Present days	Total No. of days	Percentage
1166	maths	1	kavitha	30	31	96.774194
1167	maths	1	kavitha	21	31	67.741935
1168	maths	1	kavitha	26	31	83.870968
1169	maths	1	kavitha	28	31	90.322581
1170	maths	1	kavitha	31	31	100
1171	maths	1	kavitha	18	31	58.064516
1172	maths	1	kavitha	27	31	87.096774
1173	maths	1	kavitha	25	31	80.645161
1174	maths	1	kavitha	19	31	61.290323
1175	maths	1	kavitha	21	31	67.741935
1176	maths	1	kavitha	23	31	74.193548
1177	maths	1	kavitha	28	31	90.322581
1178	maths	1	kavitha	22	31	70.967742
1179	maths	1	kavitha	30	31	96.774194
1180	maths	1	kavitha	24	31	77.419355
1181	maths	1	kavitha	20	31	64.516129
1182	maths	1	kavitha	26	31	83.870968
1183	maths	1	kavitha	31	31	100
1184	maths	1	kavitha	18	31	58.064516
1185	maths	1	kavitha	21	31	67.741935
1186	maths	1	kavitha	9	31	29.032258
1187	maths	1	kavitha	11	31	35.483871
1188	maths	1	kavitha	28	31	90.322581
1189	maths	1	kavitha	22	31	70.967742
1190	maths	1	kavitha	16	31	51.612903
1191	maths	1	kavitha	24	31	77.419355
1192	maths	1	kavitha	31	31	100
1193	maths	1	kavitha	29	31	93.548387
1194	maths	1	kavitha	12	31	38.709677
1195	maths	1	kavitha	27	31	87.096774

Fig. 35.4 As a result of attendance Sheet in PDF format

5. Conclusion

This automatic face recognition attendance system implementation aims to eliminate the errors that often occur with traditional attendance systems. This study uses deep learning to perform facial recognition, which demonstrates flexibility towards user identification with a high accuracy of 98.3%. The results provide an illustration and specific description of the system's ability to adapt to changes in position and face projection. The lighting puzzle during face recognition is solved thanks to face recognition using deep learning, and the original image now takes the form of a HOG representation that captures the majority of image features regardless of image brightness. Local facial features must be taken into account for the initial processing of the facial contouring method. After these procedures, it generates 128 measurements of the recorded face which is encoded and the best face recognition is achieved by extracting the person's identity from the coding process. As a result, an Excel document is created which is sent in PDF format to students and teachers on a monthly basis. One person can easily operate this device and it provides excellent protection.

References

1. V. Blanz and T. Vetter (2003), Face recognition based on fitting a 3dmorphable model. IEEE Transactions on Pattern Analysis and Machine Intelligence, 25(9): 1063–1074.

2. Kaneez Bhatti, Laraib Mughal, Faheem Khuhawar (2018), "Smart Attendance Management System using Face Recognition", EAI.EU, 5(17):159713, DOI:10.4108/eai.13-7-2018.159713.
3. Pooja Humbe, Shivani Kudale, Apurva Kamshetty(2017), "Automatic attendance using Face Recognition", International Journal on Recent and Innovation Trends in Computing and Communication, Volume : 5, Issue : 12, ISSN: 2321–8169, 24–27.
4. X. Lu, and A. Jain(2008), "Deformation Modeling for Robust 3D Face Matching." Transactions on Pattern Analysis and Machine Intelligence, Vol. 30, No. 8, pp. 1347–1356.
5. S. Ramanathan, A. Kassim, Y. Venkatesh, and W. Wah(2006), "Human Facial Expression Recognition Using 3D Morphable Model." ICIP, pp 661–664.
6. P. Visalakshi, Sushant Ashish(2018), "Attendance system using multi face detection", International journal of Pure and Applied Mathematics, Volume 118 No. 22.
7. Aparna Trivedi , Chandan Mani Tripathi, Dr. Yusuf Perwej, Ashish Kumar Srivastava, Neha Kulshrestha(2002), "Automated attendance system based on facial recognition", International Journal of Scientific Research in Science and Technology Volume 9(Issue 1): Pages 261–268 DOI:10.32628/IJSRST229147.
8. M. Gaparovi and D. Gajski (2016), Realtime quasi dense two-frames depth map for autonomous guided vehicles. In Intelligent Vehicles Symposium (IV), IEEE, pp. 497–503.
9. Mrs. Madhuram .M, B. Prithvi Kumar, Mr. Lakshman Sridhar, Mr. Nishanth Prem, Mr. Venkatesh Prasad(2018), Face Recognition and Detection using OpenCV, International Research Journal of Engineering and Technology, volume no:05(9001:2008), pp: 474–477.
10. F. Schroff, D. Kalenichenko ,and J. Philbin(2015).Face Net: A Unified Embedding for Face Recognition and Clustering, IEEE conference on computer vision and pattern recognition, DOI: 10.1109/CVPR.2015.7298682 Corpus ID: 206592766, pp: 815-823.

Note: All the figures in this chapter were made by the author.

Disruptive Technologies for Sustainable Development – Dr. G. Nagappan et al. (eds)
© 2024 Taylor & Francis Group, London, ISBN 978-1-032-55000-8

Prioritize Task Scheduling in Fog Computing Using Machine Learning

36

Jayapriya R. M.[1]
Master of Engineering, Department of Computer Science and Engineering,
Saveetha Engineering College (Autonomous), Chennai

P. Preethi[2]
Assistant Professor, Department of Computer Science and Engineering,
Saveetha Engineering College (Autonomous), Chennai

Abstract: The Internet of Things (IoT) concept, first introduced by a member of the RFID development community, has recently become more applicable to the real world due to the propagation of portable devices, cloud computing, embedded and ubiquitous communication, and data analytics. Developing IoT applications with strict constraints on latency and data handling has created numerous challenges for cloud-centric systems for smart cities. The rapid speed of technology development could have a big impact on how we do business. The many benefits that cloud services offer have increased their popularity. In this study, we examine fog computing, a relatively recent idea in computer architecture. Fog computing has been suggested as a workable method to support these unique applications because of the ever-increasing capacity of data and devices connected to the Internet of Things (IoT). In the fog computing platform, which is defined by a horizontal system-level design, devices are close to end users, and Internet of Things devices are employed for storage, networking, and processing. In order to address this, the paper suggests a machine learning algorithm of the Reinforcement Learning (RL) framework that makes the assumption that the agent generally satisfies the necessary latencies and preventing overloading of fog nodes to decrease latency, security requirements should be met. According to the analytical outcomes, the proposed approach performs satisfactorily, with a more uniformly distributed load and fewer efficiency violations in delay and security.

Keywords: IOT, Cloud computing, Fog computing, Machine learning, Reinforcement learning

1. Introduction

The amount of people using the Internet will reach 5.3 billion in 2023, or over 71 percent of the world's population, thanks to the development of 5G networks that enable ubiquitous broadband internet at any place and at any moment. The majority of them make use of wireless gadgets like smartphones, tablets, smartwatches, home metres, and wearables, which result in a proliferation of portable IoT data that necessitates real-time and secured services. As a result of these services' significant differences from earlier cloud-based data services, IoT data service demands are being met by fog nodes at the edges that have abundant computing, bandwidth capabilities, and rich storage. The main goal is to efficiently offload internet of things tasks to fog nodes while still meeting the QoS constraints using the least amount of fog resources. A decentralized block chain-based approach is put forth with the goal of minimizing system latency while maximizing user data security. This study suggests setting up a RL agents on one of the edge devices in the fog nodes to realize the aforementioned concepts. The agent divides an IoT job into many fragments depending on the safety requirement and task size before scheduling it to any fog node or cloud. To

[1]jaya.priya17jp@gmail.com, [2]ppreethi@saveetha.ac.in

DOI: 10.1201/9781003428473-36

prevent overwhelming just the nearest fog nodes, the agent keeps track of how long each cloud node needs in order to complete the task piece at that period step. As a result, we develop a Q-learning method that directs the agents to allocate arriving internet of things tasks between fog nodes and the cloud while balancing real-time, bandwidth needs, and security, while ensuring that specific fog nodes don't become overloaded.

2. Literature Survey

To attain effective task offload for real-world applications while taking into account communication and computational costs, Chiti et al. introduced a distributed algorithm in to effectively pick fog nodes centered on matching principle. Using the algorithm, the maximum time needed to accomplish the task is minimized. The resource-aware task offloading system introduced by Tran Dang et al. reduces the average delay by selecting the best fog node to offload tasks based on an optimum policy. Yousef pour et al. were able to reduce the latency of service from offloading tasks to fog nodes. An imitation learning-based approach to update vehicle networks was developed by Wang et al. Then, he solved the problem of organizing UAVs from many service providers into a common space so that the profit of the owner as well as the benefit of every end user is maximized. They have created a technique to deploy mobile peripheral nodes to drones based on imitation learning.

In delegating the work to fog nodes, some publications has addressed the issue of energy efficiency. An energy-minimizing algorithm was suggested by the author Zhang et al. that chooses a fog node for task offloading while taking into account the energy consumption, fog node priority, fog node use, and historical data regarding the typical energy. Jalali et al. contrasted the centralized cloud data centers with the nano data centers (nDCs) employed in fog computing. The authors suggested that the energy efficiency of nano servers is more heavily influenced than the number of hops by the types of applications, access networks, and idle and active times.

3. Proposed Methodology

3.1 System Design

Centralized architectures are not appropriate for implementing IoT smart city scenarios since existing storage systems and transmitting all of the information gathered by smart gadgets to the cloud layer both need a substantial amount of network capacity. As a result, the Fog Computing paradigm enables local data analytics and processing processes, it significantly reduces the amount of data being sent to the cloud. In the meantime, we must balance the workloads distributed across the edge nodes to prevent some of them from becoming overloaded and affecting the performance of the system as a whole. We identify IoT tasks as per their safety need, real-time requirement, and processing time, and partition IoT jobs into small parts centered on their confidentiality levels in order to plan the sensitive IoT tasks correctly and also managing the traffic load. Three edge devices (IoT devices) send data to a cloud server with fog computing in our real-time prototype modeling. Before scheduling IoT tasks, machine learning is used at the fog node. In light of this, In order to reduce average latency, we provide a RL machine learning model where the agent tries to meet the necessary delay and necessities for security without causing congestion at any fog nodes. Figure 36.1 displays the overall proposed methodology.

Features of the proposed system design

- Multi-Edge Device Support.
- Storing, shorting, and analyzing all incoming data.
- Auto task scheduling.
- Dashboard for visualization parameters like incoming data's, network status, data processing comparison, data latency and storage.

(a) Reinforcement model for task scheduling

The goal is to reduce the average task offloading delay in order to stabilize the stream of traffic loads on fog nodes and fulfill numerous QoS necessities, particularly the privacy necessity.

State: Centered on its obtain available sources ES_i, security level ER_i, reaction time EQ_i, and cloud EQ_b response time, fog node i status during time slot l is determined. The following equation represents the fog node state i, which consists of a few sub states, at time slot l:

$$\Delta^i(l) = (ES_i, ER_i, EQ_i) \tag{1}$$

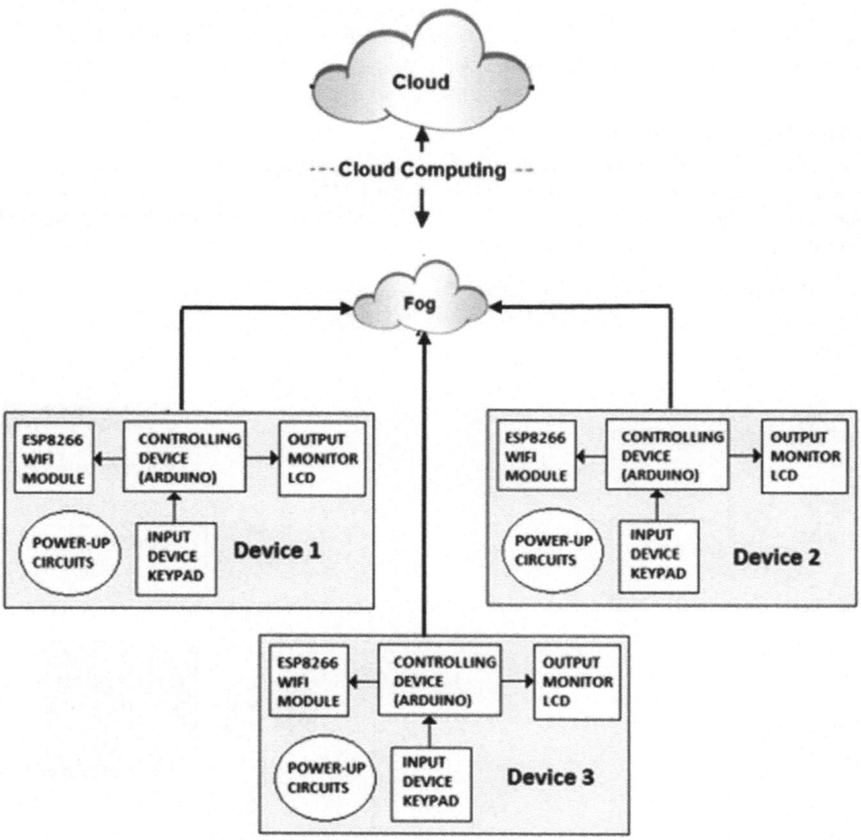

Fig. 36.1 Overall proposed system architecture

The fog nodes state space and cloud response time are described in equation (2).

$$R(l) = [\Delta^1(l), \Delta^2(l), \Delta^3(l), \dots \Delta^m(l), EQ_b] \quad (2)$$

Action: The agent's available actions at each possible state are all contained in the system action space.

The total number of possibilities oriented on nodes of fog and IoT is defined as follows:

$$M = \binom{m+1}{TE_j} = \frac{m+1!}{TE_j!\left((m+1)-TE_j\right)!} \quad (3)$$

where $m + 1$ is the total fog nodes. TE_j is the total fragments in IoT task. In light of this, the action space can be illustrated as follows:

$$B_j(l) = [h_1(l), h_2(l), h_3(l), \dots, h_M(l)] \quad (4)$$

State Transition: The system's state will change from r to r' when the agent performs an action $b \epsilon B_j(l)$ in state R at time slot l. The system's transition probability matrix is stated as follows:

$$Q^b_{r,r'} = Qs\left[r(l+1) = r' \mid r(l) = r, b(l) = a\right] \quad (5)$$

Reward: While a Reinforcement Learning Agent learns from the reward provided by the surroundings. The fog node's safety and reaction time i shows how well it can carry out an IoT operation with security requirements, represented by $RW_s^b(l)$, and calculate the agent reward on for each time step l as follows:

$$RW_s^b(l) = RR_i(l) + RS_i(l) \tag{6}$$

(b) Task scheduling based on Q-Learning

The system status space $R(l)$, the action space $B(l)$, the resource availability states R, the set of nodes associated with fog F, the set of internet of things tasks to perform T, and the cloud C are all inputs. The algorithm's goal is to choose the agent's optimal course of action so that the greatest cloud or fog node, the IoT task has been chosen for offloading. Our algorithm's output in this regard is the ideal Q-table's value and policies.

3.2 Hardware Design

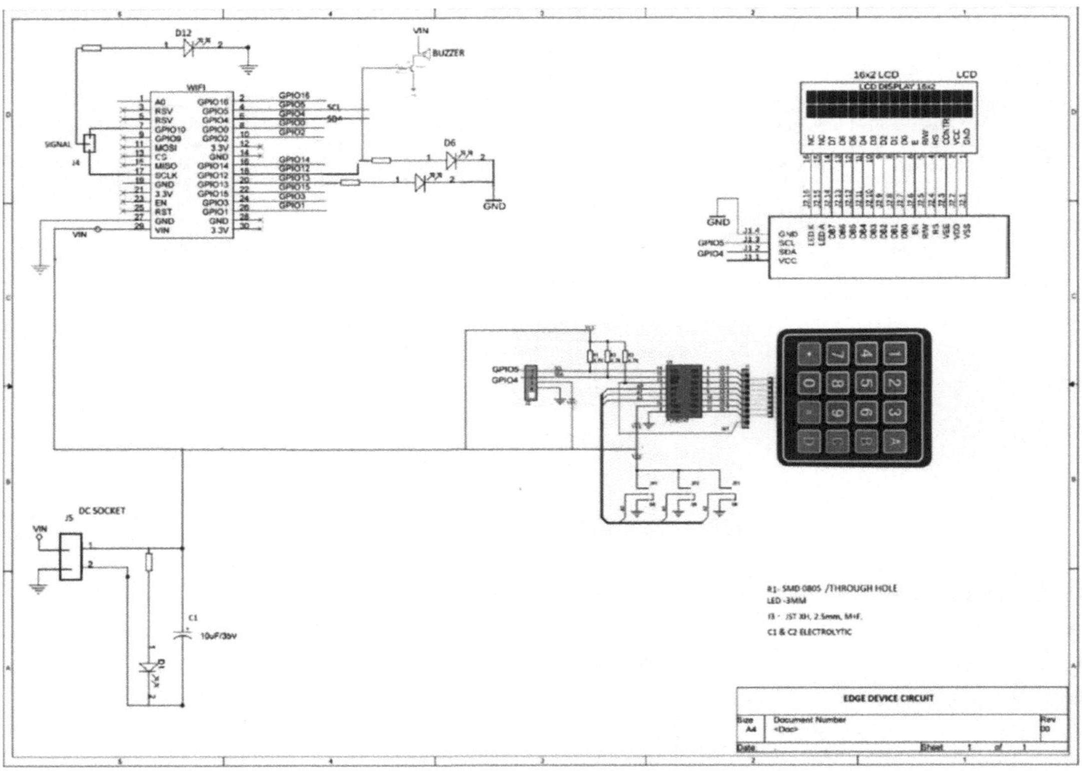

Fig. 36.2 Hardware circuit diagram

4. Experimental Analysis

A RL algorithm evaluation experiment was conducted on an Intel® Core i5 computer with 12 GB of Memory. Real time integrated hardware nodes shown in Fig. 36.3. Generated Dataset and Data analytics result on 3 Edges result shown in Fig. 36.4. Simulation result on Message Reception shown and the graphical representation of total data processed result are shown in Fig. 36.5.

Fig. 36.3 Real time integrated hardware nodes

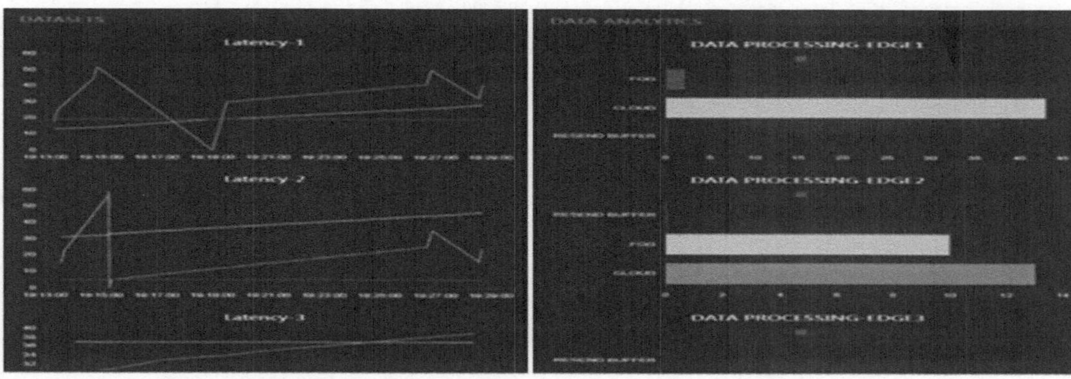

Fig. 36.4 Generated dataset and data analytics result on 3 edges

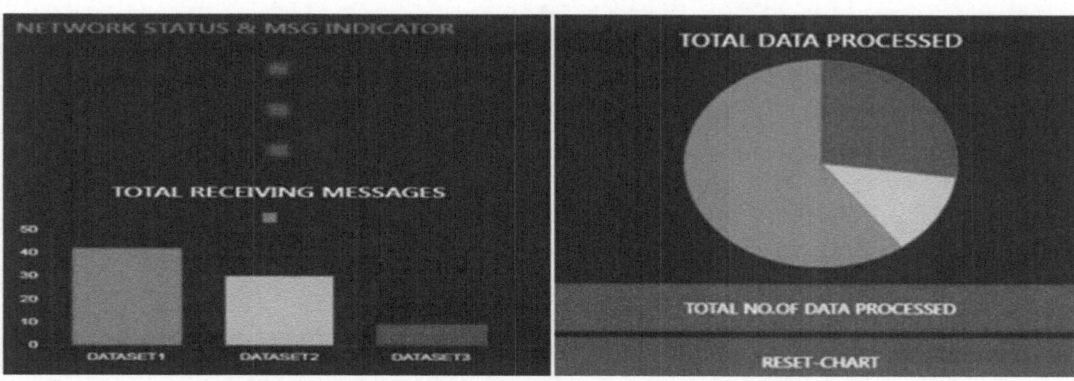

Fig. 36.5 Simulation result on message reception and number of total data processed result

5. Conclusion

This study suggests a Q-learning-based Internet of Things task allocation method that compels the agents to schedule incoming IoT jobs to fog entities with minimal traffic loads and adequate security. First the agent divided an IoT job into smaller components based on its size and requirements for security, then scheduled the components to fog nodes with required security credits. The agent attempted to maintain the network's workload during this process by taking into account the fog nodes response times for specific IoT tasks, which decreased the overall latency and also maintaining security requirements.

References

1. Manisha Verma, Neelam Bhardwaj, and Arun Kumar Yadav. (2016). Real time efficient scheduling algorithm for load balancing in fog computing environment, International Journal of Information Technology and Computer Science. 8(4): 1–10.
2. Tian Wang, Md Zakirul Alam Bhuiyan, Hui Tian, Yiqiao Cai, Yong Hong Chen and Bineng Zhong. (2017). Trajectory privacy preservation based on a fog structure for cloud location services, IEEE Access. 5: 7692–7701.
3. Goiuri Peralta, Markel Iglesias Urkia, Marc Barcelo, Raul Gomez, Adrian Moran, and Josu Bilbao. (2017). Fog computing based efficient IoT scheme for the Industry 4.0, in Proceedings of the 2017 IEEE International Workshop of Electronics, ECMSM. (pp. 1–6).
4. Francesco Chiti, Romano Fantacci, and Benedetta Picano. (2018). A matching theory framework for tasks offloading in fog computing for IoT systems, IEEE Internet of Things Journal. 5(6): 5089–5096.
5. Zhaolong Ning, Shouming Sun, Xioajie Wang. (2021). Blockchain-enabled intelligent transportation systems: a distributed crowd sensing framework, IEEE Transactions on Mobile Computing, 21(12): 4201–4217.
6. Flavio G Bonomi, Rodolfo A Milito, Jiang Zhu, and Sateesh Addepalli. (2012). Fog computing and its role in the internet of things, in MCC'12- Proceedings of the 1st ACM Mobile Cloud Computing Workshop. (pp. 13–15).
7. Zhaolong Ning, Peiran Dong and Miaowen Wen. (2021). 5G-enabled UAV-tocommunity offloading: joint trajectory design and task scheduling, IEEE Journal on Selected Areas in Communications. 39(11): 3306 –3320.

Note: All the figures in this chapter were made by the author.

Disruptive Technologies for Sustainable Development – Dr. G. Nagappan et al. (eds)
© 2024 Taylor & Francis Group, London, ISBN 978-1-032-55000-8

Analysis and Detection of Gas Leakage in Industries

37

A. Manonmani[1]
Assistant professor (SG), Department of EIE,
Saveetha Engineering College, Chennai

Praveen Kumar R.[2], Karthik Raja K.[3], Naveen M. D.[4]
UG scholar Department of EIE,
Saveetha Engineering College, Chennai.

Abstract: Liquefied petroleum gas (LPG) leakage in work place, domestic and also, container gas, stored gases which could exhibit ideal characteristic is used. Because of that, a buzzer is used to vibrate an alarm which operates as an alarm unit. The beep provides an auditory signal to show the existence of LPG level. ISO Butane, LPG, propane, and even pollution can all be detected using the instruments. The devices' ability to blend sensitivity and reaction speed is a benefit. The sensor's output changes to an active low (logic-0) state if it detects gas leaks from a building or a home. In this undertaking, an Arduino UNO is utilized; the Arduino detects gas leaks but ignores weak impulses. The alarm and liquid crystal display (LCD) are turned on by the Arduino UNO. In order to warn the user that there is a risk, it also activates the Global System for Mobile Communication (GSM) modem. After that, it keeps sending message warnings to the phone number that is explicitly stated in the provided programme of the source code. Hazardous gas leaks (LPG) in a home or workplace, as well as container gases and kept gas with optimal characteristics, are all factors. As a result, an alert that serves as a notification device vibrates using a siren. The beep provides an auditory signal to show the existence of LPG level. Iso Butane, LPG, propane, and even pollution can all be detected using the instruments. The devices' ability to blend sensitivity and reaction speed is a benefit. The sensor's output changes to an active low (logic-0) state if it detects gas leaks from a building or a home. In this undertaking, an Arduino UNO is utilized; the Arduino detects gas leaks but ignores weak impulses. The alarm and liquid crystal display (LCD) are turned on by the Arduino UNO. Additionally, the Global System for Mobile Communication device is activated. (GSM) After that, it keeps sending message warnings to the phone numbers listed in the source code programme that is provided to warn people of the risk.

Keywords: LCD, LPG, GSM, Arduino UNO

1. Introduction

The term "gas leakage" referred to an unintentional escape of natural gases or other gaseous products into any place where they shouldn't be present from a conduit or other confinement. The health and well-being of civilization are both at risk from gas leaks. In a structure or other enclosed area, even a tiny escape can steadily grow into a deadly or fatal gas accumulation. The ozone loss and global warming potential of refrigerant gas and natural gas make their leakage into the atmosphere particularly detrimental. Fugitive emissions are another name for the leakage of gases connected to industrial activities and the machinery

[1]manoraj1983@gmail.com

DOI: 10.1201/9781003428473-37

they use. Fugitive gas pollution are caused by the capture and use of natural gases that escape from fossil fuels. The Gas leakage detectors are used in following industries mentioned below:

1.1 Semiconductor Manufacturing Industries

For the production of semiconductors, gases like carbon monoxide (CO), phosphine (PH3), and methane (CH4) are essential. These gases are used in a variety of locations, including wafer furnaces, wafer dryers, and gas cases, throughout semiconductor production sites. To create the chemical interactions that result in superconducting devices like Transistors and Diodes, processes like doping, etching, casting, and photo-lithography are used. These vapours are undoubtedly a crucial quality control measure for the production of semiconductor devices, but they are extremely poisonous and can cause injuries when ingested. Additionally, these businesses employ acids like HCl for comparable purposes; however, the vapours from these acids can irritate inhalers by affecting their ability to breathe. Gas leaks that could cause a serious disaster can be quickly detected using Internet of Things (IoT)-based gas monitoring systems that are simple to incorporate into an organisation's current systems and equipment. There are quick steps that can be taken to not only stop the spread of toxins over a large area, but also to find the leak's source and fix it.

1.2 Wastewater Treatment Plants

To safeguard employees from any accidents, it is crucial to use gas monitoring systems, particularly for H2S gases. These solutions' related equipment can further result in lives being spared by assisting in the detection of such pollutants.

1.3 Oil and Gas

The downstream, middle, and upstream sectors of the oil and gas industry deal with extremely flammable vapors that can blow up when subjected to high temps, having a detrimental impact. There is always a high likelihood that such vapors will catch fire, whether they are being used for onshore processing or offshore extraction of petroleum into later goods. Supporting environmental standards and laws for the release of waste fumes in plants presents another task for these energy businesses. In this instance, factories with a tone can regulate the discharge of any hazardous gases into the environment with the aid of gas tracking systems. Additionally, gas monitoring systems that are placed in particular locations by energy companies can aid in the discovery of gas leaks, enabling the officials to move quickly to fix the sources and avoid serious fatalities.

1.4 Chemical Industry

Hazardous poisonous gas leaks that can seriously irritate the eyes, esophagus, and nostrils are caused by certain factory-made chemicals. When ingested, some of them can even cause convulsions and vertigo, which can lead to coma and even mortality. Non-toxic gases have the ability to replace oxygen (O2) in a confined space, suffocating occupants and making respiration difficult. The big metal receptacles are used to hold and move these vapours. There is always a possibility that these receptacles could discharge the fumes while being moved from one location to another, even though the danger is minimal. Therefore, while the receptacles are permanent, movable gases detectors can be used with them, and moved detectors can be used in factories to find dangerous gas leaks.

1.5 Mining

Toxic and combustible gases like H2S and methane can be found in mines using monitoring devices for gas in the mining industry. However, these fumes, also referred to as asphyxiants, remove air from the tunnels, which may have an impact on the health of the miners. The absence of air can have an adverse effect on employees' mental health, lowering their output and even leading to lasting mental incapacity. Gas tracking systems can be used to make sure that employees are working in an atmosphere that is abundant in oxygen and that poses no health risks.

2. Literature Review

R Kaur, M Mandot, et al. (2021) presented IoT based smart industrial Automation for gas leakage system it is a comparative analysis ofany technologies which allows a device to connect to the internet is referred as the"Internet of Things". K Umapathi, K Pranathi, LR Sarma, M Nisanth et al. (2021) presented a Bluetooth controlled Electronic Industrial gas pipe controller: Bluetooth is used for exchange of data using UHF radio waves. It is a short range wireless communication technology. In this designing controlling of pipelines using smart phone via Bluetooth technology. The project is built on android application.

K Mehta, A Sharma, L Bhalla et al. (2020) presented college automation security management by using messaging service instructions, the motive is to automate and control lights of University to be assurance of specified mobile number. Future smart household products using the Internet of Things were introduced by Pattlola Srinivas, M Swamy Dasz, YL Malathi Latha, et al. in 2021. The Internet of Things is a tangible network of items and devices that gather, store, and analyse data. The approach that is suggested is the ideal way to use some smart home applications, including contact between devices, smart home activities, and management with mobile apps and Arduino. Hamidi, EAZ Design and implementation of a Bluetooth-based smart outlet prototype for automatic homes using an Arduino UNO. One of the simple daily tasks at home is house automation. Switch action may be shortened through automation of the halt contact mechanism. The HC-05's Bluetooth chip set serves as a wireless communication means. As a consequence, messages can only be sent up to 17 meters away from a wall and 20 meters away from other objects. 400 metre per second time delay.

2.1 Block Diagram

Figure 37.1 represents the block diagram of gas leakage detector.

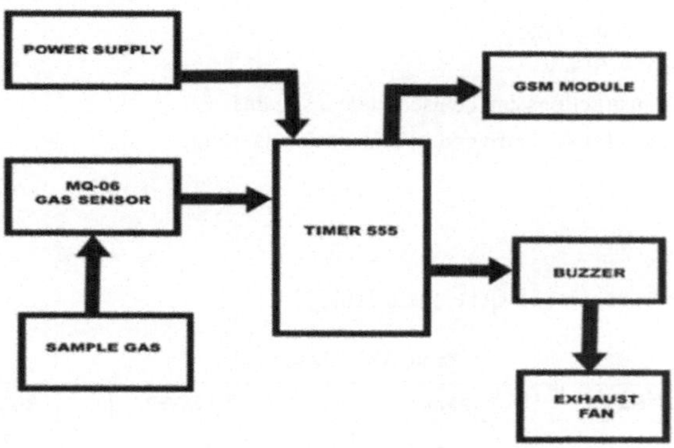

Fig. 37.1 Block diagram of gas leakage detector

2.2 Industrial Model Expansion

The simple mode of mini project in Gas leakage detector can be extended and developed in advanced manner of adding few advanced components like GSM module and using Arduino it can be used in the mode of I.o.T. Further expansion of this project can be done using the following components:

- Arduino board
- GSM module
- LCD screen
- Stepper motor

GSM Module

Model recommended: SIM800 GSM

For the global market, SIM800, a quad-band GSM/GPRS module, was developed. It operates on the GSM 850 MHz, EGSM 900 MHz, DCS 1800 MHz, and PCS 1900 MHz frequency channels. The SIM800 provides GPRS class 12/10 multi-slot functionality and is compatible with the GPRS CS-1, CS-2, CS-3, and CS-4 security standards. (optional). The SIM800's small 24243mm form is able to meet the vast majority of users' space requirements for M2M, smart phones, PDAs, and other mobile applications. The SIM800's 68 SMT ports serve as the only physical link between the module and the customer's board. The SIM800 was designed with low-power consumption in mind, drawing just 1.2mA while dormant. Data transfer is greatly aided by the integration of TCP/IP protocol and sophisticated TCP/IPAT instructions by SIM800.

Features of GSM module

- Supports up to 552 keyboards. A completely working UART interface with the ability to be set up as two separate serial lines.
- Firmware updates and troubleshooting are both possible with USB ports.
- An audio channel with radio output and microphone input.
- General-purpose inputs and outputs that are programmable.
- Interface for SIM cards.
- Assistance with Bluetooth features.
- Support for one PWM.
- SD card port, PCM, and SPI. A single function can only be used simultaneously.
- Power supply: 3.4 to 4.4 volts
- The average battery usage while in slumber state is 1.2mA.
- Class 12 GPRS multi-slot frequency range
- SIM device support: 1.8V, 3V – 19 – V
- The serial interface: Yes, AT instructions can be used in data streams.
- The USB interface: can be used for software update and troubleshooting.
- GSM MODULE : SIM800

3. Result

Tabulation of the conducted experiment is shown below in Table 37.1

Table 37.1 Result table

NAME OF GAS	PPM	SAFE / DANGER	BUZZER
Oxygen	58	SAFE	OFF
Carbon dioxide	212	DANGER	ON
Methane	1900	DANGER	ON
Methyl iso cynade	2200	DANGER	ON

4. Conclusion

This paper proposes and describes the design of a sensor-based automated gas escape detection with alert and management system. The complicated gases and basic gases monitoring system gadget is low-cost, low-power, lightweight, movable, secure, simple to use, effective, and flexible. The need to ensure labor safety is seen as a major driver of the industry in the future years. We can infer from his project success that his LPG gas escape spotting in the project system was incredible. Convenient for industrial and home use. In dangerous situations, this system can save lives. A warning is displayed by the GSM module. Sensor nodes detect gases such as CO2, oxygen, and propane. Get estimated transmission range and power consumption. Simple steps and areas of the Arduino UNO micro-controller used to build the sensor.

References

1. Platform, Date Added to IEEE Xplore: 27 December 2018 INSPEC Accession Number: 18381682DOI: 10.1109/ICSENS.2018.8589853
2. Automated unified system for LPG using load sensor, 2017 International Conference on Power and Embedded Drive Control (ICPEDC)
3. Smart gas cylinder: Leakage alert and automatic booking, 2017 International Conference on Energy, Communication, Data Analytics and Soft Computing (ICECDS)2018INSPEC Accession Number: 17859626DOI: 10.1109/ICECDS.2017.8389616,Publisher: IEEE.
4. UAV environment monitoring system, Published in: (2017) International Conference on Electrical, Electronics and System Engineering (ICEESE)Date of Conference: 9-10 Nov. 2017,Date Added to IEEE Xplore: 22 February 2018,INSPEC Accession Number: 17598984,DOI: 10.1109/ICEESE.2017.8298395

5. Micro controller based low cost gas leakage detector with SMS alert, Published in: 2018 Emerging Trends in Electronic Devices and Computational Techniques (EDCT),Date of Conference: 8–9 March 2018, IEEE.
6. Novel Sensing Approach for LPG Leakage Detection: Part I—Operating Mechanism and Preliminary Results, IEEE Sensors Journal (Volume: 16 , Issue: 4 , Feb.15, 2016 ,Page(s): 996–1003.
7. Smart gas cylinder: Leakage alert and automatic booking, 2017 International Conference on Energy, Communication, Data Analytics and Soft Computing (ICECDS), IEEE.
8. Smart Gas Level Monitoring, Booking & Gas Leakage Detector over IoT,2017 IEEE 7th International Advance Computing Conference (IACC),Date of Conference: 5-7 Jan. 2017,Date Added to IEEE Xplore: 13 July 2017,Electronic ISSN: 2473–3571.

Note: The figure and table in this chapter were made by the author.

Disruptive Technologies for Sustainable Development – Dr. G. Nagappan et al. (eds)
 ISBN 978-1-032-55000-8

Review of the Literature Concerning the Application of Sentimental Analysis in Business Reputation

38

G.Dharanidevi*, R. Dhanush Aravind, R. Pradeep, K. Gokulakrishnan
Department of Computer Science and Engineering Rajalakshmi Engineering College, Thandalam, Chennai, Tamil Nādu, India

Abstract: Today's consumers depend increasingly on online recommendations from "people they know" as opposed to a company's own content, and a number of social media platforms have evolved into useful sources of recommendations and opinions on products and businesses. It is crucial for businesses to keep an eye on how their brands are mentioned for this reason. It is a big challenge and difficult to manage a company's internet reputation. The sharing of modern information and opinion relies heavily on digital media. The influence of social media has grown, and now people's opinions are what determine if something is worthwhile or not. In order to maintain and gauge their reputation, businesses use a variety of social media platforms, but the results aren't always reliable. Thus, the main objective of this article is to determine whether a company's reputation can be assessed using sentiment analysis and to achieve the result with better accuracy.

Keyword: Sentiment analysis, Block chain, Smart contracts, Reputation

1. Introduction

A novel approach to text analysis called sentiment analysis seeks to ascertain the viewpoint and objectivity of reviewers. With the rising recognition of new businesses, including start-ups, etc, the internet is awash in reviews, comments, and ratings from people who can express their opinions and assess various things. Thus, finding unbiased reviews of particular products is simple. The collective opinion of the internet community on a particular item is referred to as its online reputation.

Company Reputation refers to the public's perception of a particular company, whether it has a good or negative reputation. It is the confluence of a customer's thoughts and feelings about a brand. A company's reputation must be an intangible asset. Sentiment analysis gives information on a subject's feelings, whether expressed in written or spoken language. Sentiment analysis is a technique which is already used by companies in order to find their reputation value, but it doesn't give enough closure. One-dimensional scales are frequently used to gauge reputation in order to determine whether it is favourable ("good") or negative. ("Poor"). This approach is the easiest way to comprehend reputation, but it doesn't take into account the reasons why one company's reputation may be better or worse than that of its competitors.

Our system aims at finding the reputation of a company by performing sentimental analysis and topic modelling, by performing data cleaning, exploratory data analysis and keyword extraction, by doing this we can get a more accurate value and companies can get a clear picture about their social standing. The programming is done in Python. For web scraping, we used requests and Beautiful Soup. The Python libraries used are NLTK, TextBlob, Gensim and Spacy. For data analysis and visualization, we used Numpy, Pandas, matplotlib and seaborn.

*Corresponding Author: dharanidevi.g@rajalakshmi.edu.in

DOI: 10.1201/9781003428473-38

2. Related Work

Because of the ubiquity of spam attacks, user reputation in online rating systems is extremely significant. As a result, assessing user image in online rating systems is required. As a result of the iterative reputation-allocation procedure, several group-based ranking (IGR) approaches has been integrated to the system. Ratings from users with more favorable reviews are given greater weightage when calculating the group sizes and categorizes user ratings into distinct rating categories based on their historical readings. However, it is still unclear how continuous and discrete-valued ratings affect user ratings, and they cannot be applied to ratings with higher resolution scales. (Gao, J., & Zhou, T, 2017).

Here, a hybrid sentiment analysis was carried out by the researchers. This is achieved by using a dataset that includes only good and negative tweets. The amount of retweets and favorite tweets are included. The majority of the sentiment analysis done was polarity-only. A hybrid strategy that blends textual and non-textual elements was used to improve precision. However, some of the spam tweets were not deleted during the procedure, therefore it is not accurate enough. There is a chance that some word polarity in the data label will go unnoticed (Khatoon, S, 2017).

Given that the majority of the established mobile phone brands will go out of business due to a lack of innovation in their offerings, Indian consumers were at a loss as to which mobile phone to purchase. Consequently, the researchers suggested a system in which they identify the most well-known brand in India. They collected 7330 tweets for their project, which required ranking six famous smart phone manufacturers based on brand recognition. Textblob, Tweepy, and APIs were utilized as open source libraries for visualization. As the year changes, the ranks of smart phones will also vary because this methodology is not real-time (Azzeh, M., Hijjawi, M., & Altamimi, A. M, 2018).

This paper explains that companies aren't able to know the public's opinion on their product. As a result it's not good for the company's image. They created the "SENTIMINER" web application to analyze, build, and provide public attitudes with geographical limits. This system display tweets across the worldwide (Ahmad, Sharik, et al. 2018).

The researchers discuss how brands need to have a good exposure in order to become profitable and how the absence of such interaction between customer and the company is detrimental for them. The data is cleaned, preprocessed, and analyzed in steps. Machine learning is used for classification. The text is classified as positive or negative based on the attitude of expression. Utilizing neural networks, the labelled data is compared to a big corpus/lexicon. The disadvantage is that it concentrates entirely on the ultimate sentiment. This is inefficient with multiple emotions in the comment. (Păvăloaia, Vasile-Daniel, et al. 2019).

This paper shows how the increasing quantity of publications makes keep track of information challenging. Based on large-scale, the researchers experimented by drawing on bibliographic data from 5885 papers published between 2000 and 2016, and includes co-citation and bibliographic coupling with different kinds of network visualization. It provide a systematic analysis to reviewing literature but the raw data of bibliometric analysis must be gradually reduced in order to achieve high levels of aggregation. (Veh, A., Göbel, M., & Vogel, R, 2019).

By utilizing a naive bayes algorithm, the authors suggest a framework for sentiment analysis. The classification phase comes first. This distinguishes between favorable and unfavorable evaluations. The fusing and grouping phase is the next stage. In the third stage, custom reputation values are created independently for the good and negative groups. Final Reputation generation is the last stage, where the reputation values of the positive and negative sets are combined using weighted arithmetic mean to determine the final reputation value of a dataset. There were no restrictions in the system as it was (Gupta, A., Priyani, S., & Balakrishnan, R, 2020).

For online corporate reputation, convolutional neural network (CNN), is used to detect and categorize the reputation in various dimensions from genuine user online comments. The magnitude of the data collected was insufficient to represent the internet consumer and stakeholder populations.. Some of the attributes chosen were also not captured by the instrument (Rantanen, Anette, et al. 2020).

The researchers created a classification framework which uses machine learning model by leveraging 19,991 social media comments regarding two banks and applying a NN to automatically identify the remarks based on manually annotated training data.. However the drawback was the accuracy is less than 60 percent which is not feasible to accurately analyze (Alqwadri, A., Azzeh, M., &Almasalha, F, 2021).

In order to evaluate reputation in social media, an unique reputation analysis method based on Bidirectional Encoder representation from transformers was suggested in this study. The algorithm for logistic regression was employed. The disadvantage of BERT application, is the reputation polarity to categorize texts, has not yet undergone validation, and the prediction was thus not precise enough. Using the RepLab 2013 dataset, the proposed technique was assessed for the reputational polarity challenge (Rahman, Mohammad Wali Ur, et al. 2022)

Here, a technique for evaluating online reputation is created, and 100 companies' news and articles are examined. The sentiment analysis procedure is run on the news story, which is used as a dataset. However, the algorithm generates noisy results that require additional filtering, breaking the automation chain and making it challenging to meet the objective of avoiding information overload. Additionally, it falsifies firms' online reputations, which mostly harms their online reputation (Syllaidopoulos, I., Skraparlis, A., & Ntalianis, K., 2022).

3. System Design

The project aims at enhancing the accuracy of the dataset by working on the text columns and performing sentiment analysis. Sentiment of words can vary based on where it is in a sentence. The TextBlob module labels the text based on sentiment of words in a sentence and each word in a corpus is labeled based on polarity and subjectivity.

- Polarity specify how much sentiment words are positive or negative and it take value -1 is extremely negative and +1 is really positive.
- Subjectivity specifies how much subjective or opinionated a word is. The number 0 is true. +1 is purely subjective.

This is the proposed architecture for the project. The processes can be divided into two modules. The first module consists of data cleaning and preprocessing. Preprocessing data removes inconsistent, noisy, and incomplete data. Any data mining functionality requires preprocessing of the data. To eliminate inconsistencies during the assignment of polarity, special characters like., []()/' should be eliminated. All punctuation is then eliminated, with the exception of apexes because they are a component of grammatical structures like the genitive.

The text is then entirely changed to lowercase, and any excess spaces are eliminated. Some columns in the original dataset include numerous features, which violates the clean data principle. Fortunately, some of the missing data can be discovered in other text columns. We can use the text extracted from these columns to construct additional features. The second module consists of processing the now cleaned dataset by using machine learning techniques. Keywords from the text data are extracted and topic modeling is performed.

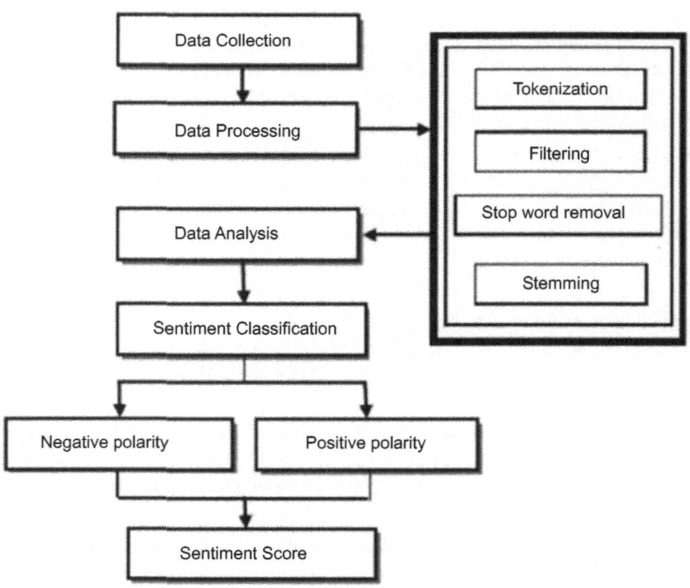

Fig. 38.1 Data cleaning and preprocessing

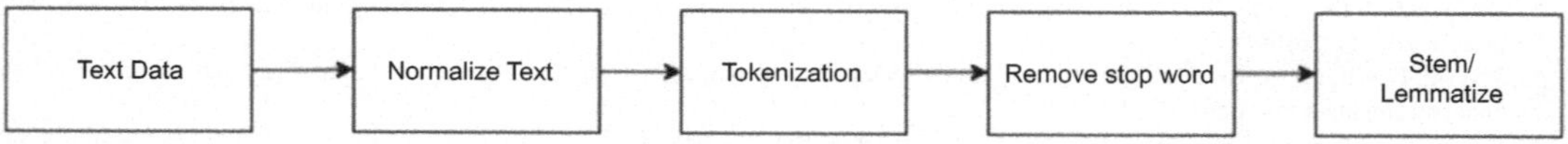

Fig. 38.2 Lemmatization

The above diagram showcases the flow of the first module. The raw data is cleaned and preprocessed and undergoes lemmatization before moving on to the next step.

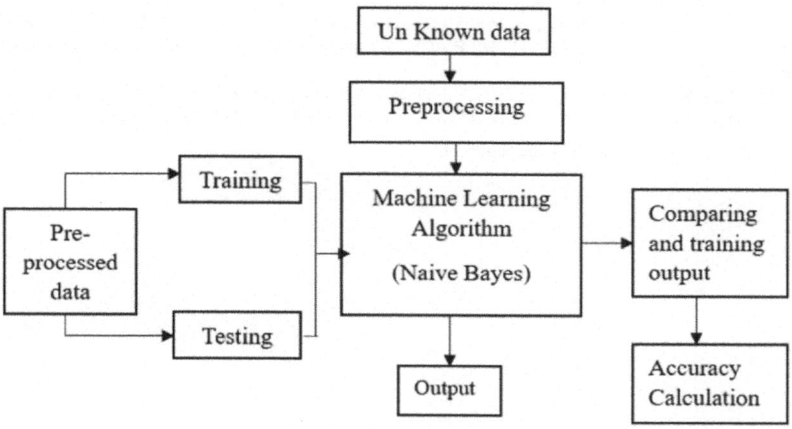

Fig. 38.3 Analysis of polarity values

The above diagram showcases the flow of the second module. The data is sent to the machine learning process where sentiment analysis is performed. Keywords are taken and the polarity is determined

4. Conclusion

This research work was carried out to assess the research on corporate reputation and business studies to deepen the scholarly understanding of the topic. There have been several proposals put forward that focus on calculating the reputation of a company. Each proposal contains techniques to address the problems encountered when implementing this concept, such as NLP and the Naive Bayes technique, etc., and involves the use of existing databases.

But in this paper, web scraping is used to extract the data, and the sentiment analysis tool is used to find the accuracy of the analysis. This method is especially beneficial when large amounts of literature make it difficult for researchers to keep up with an ever-increasing number of articles. This is undeniably true of corporate reputation. We extracted a large number of research maps that provide orientation for both established researchers and newcomers to the corporation field, with a focus on both previous and current trends, and supplemented by different visualizations. In future, system could do a similar analysis on this company's competitor and see how they compare and what they can learn from each other.

References

1. Gao, J., & Zhou, T. (2017). Evaluating user reputation in online rating systems via an iterative group-based ranking method. Physica A: Statistical Mechanics and its Applications, 473, 546–560.
2. Khatoon, S. (2017). Real-time twitter data analysis of Saudi telecom companies for enhanced customer relationship management. International Journal of Computer Science and Network Security, 17(2), 141–147.
3. Azzeh, M., Hijjawi, M., &Altamimi, A. M. (2018). Online reputation model using multiple quality factors. International Journal on Advanced Science, Engineering and Information Technology, 8(6), 2612–2619.
4. Ahmad, S., Tyagi, N., Chandra, U., &Maaz, M. (2018). Sentiment Analysis Evaluating Net Brand Reputation of Mobile Phones Using Polarity. (PDGC) (pp. 358–363). IEEE.

5. Păvăloaia, V. D., Teodor, E. M., Fotache, D., & Danileţ, M. (2019). Opinion mining on social media data: sentiment analysis of user preferences. 11(16), 4459.
6. Veh, A., Göbel, M., & Vogel, R. (2019). Corporate reputation in management research: a review of the literature and assessment of the concept. Business research, 12, 315–353.
7. Gupta, A., Priyani, S., & Balakrishnan, R (2020). A Naïve Bayes Reputation Generating Model Based on Sentiment Analysis and Opinion Fusion. J. Mech. Cont.& Math. Sci., Special Issue, No.- 6, January (2020) pp. 78–94
8. Rantanen, A., Salminen, J., Ginter, F., & Jansen, B. J. (2020). Classifying online corporate reputation with machine learning: a study in the banking domain. Internet Research, 30(1), 45–66.
9. Alqwadri, A., Azzeh, M., & Almasalha, F. (2021). Application of machine learning for online reputation systems. International Journal of Automation and Computing, 18(3), 492–502.
10. Rahman, M. W. U., Shao, S., Satam, P., Hariri, S., Padilla, C., Taylor, Z., & Nevarez, C. (2022). A BERT-based Deep Learning Approach for Reputation Analysis in Social Media. In 2022 IEEE/ACS 19th International Conference on Computer Systems and Applications (AICCSA) (pp. 1–8). IEEE.
11. Syllaidopoulos, I., Skraparlis, A., &Ntalianis, K. (2022). Evaluating Corporate Online Reputation through Sentiment Analysis of News Articles: Threats, Maliciousness and Real Opinions. International Journal of Cultural Heritage, 7.

Note: All the figures in this chapter were made by the author.

Disruptive Technologies for Sustainable Development – Dr. G. Nagappan et al. (eds)

Green Energy Charging Station

39

A. Manonmani[1]
Associate Professor, Department of Electronics and Instrumentation,
Saveetha Engineering College, Chennai, Tamil Nadu, India
A. Sarigasri[2], **G. Brindha**[3]
Department of Electronics and Instrumentation,
Saveetha Engineering College, Chennai, Tamil Nadu, India

Abstract: This work is based on the solar and wind charging mechanism (SWCM) that can be used to refuel Electric Vehicles (EVs). Solar photovoltaic (PV) modules and a wind generator make up the renewable charging station. The use of fossil fuels to generate power is drastically cut down by the SWCM, which in turn leads to a significant decrease in Carbon-di-oxide and Carbon monoxide related emissions. Analytical modelling for wind energy generation has been performed, and the single diode model has been used to represent other renewable sources including wind and solar. The suggested SWCM's simulation model was built using MATLAB-Simulink. Distinct parameters of wind turbine have been investigated at two different loading levels (1 kW and 3 kW), and the (current-voltage) I-V and PV properties of the solar panel have been analysed at a range of irradiance levels. Two one-way direct current (DC) to direct current (DC) converters link the PV modules and wind turbine to the EV charging infrastructure, while six two-way DC-DC converters link the ten charging stations. The suggested system uses a three-phase bidirectional DC-AC (alternating current) converter to connect to the grid and equalise the load. Results obtained validate the viability of the suggested renewable charging system for EVs, leading to a cleaner, more sustainable environment.

Keywords: Charging station, Solar-PV system, Wind system, SWCM

1. Introduction

The purpose of this work is to make it easier for renewable energy sources to be adopted at the same time as electric car charging loads. The distribution firms in several jurisdictions already provide discounted charging prices for electric vehicles (EVs), but this cannot be sustained permanently. For this reason, Roof Top Photovoltaic systems (RTPV) (not including BESS) are available option as a cheap energy source because their generating cost is now very competitive with the grid pricing. Project results indicate that solar RTPV charging stations for EVs are feasible from a financial standpoint and are ready for widespread implementation (When compared to the retail power rate for EV charging, the levelized cost of RTPV now is quite low). On the other hand, the cost may rise if solar power production and consumption don't sync up (as could happen when charging electric vehicles). The installation of a net-metering system at charging stations is the answer to this problem. Public EV charging stations that install solar RTPV systems would benefit from net-metering because it would increase usage and decrease maintenance costs. Among the most abundant features of Earth's biosphere are the renewable resources it provides. The viability of a resource can be gauged in large part by how well it scores on life-cycle assessments. Renewable energy resources are yet another kind of non-depletable materials. Energy derived from the sun, the earth's geothermal system, and the

[1]manonmani@saveetha.ac.in, [2]sariga072002@gmail.com, [3]brindha.eie.sec@gmail.com

DOI: 10.1201/9781003428473-39

wind are all examples of renewable energy. "Renewable energy" describes the generation of power from resources that can be replaced at a rate that keeps up with their consumption. Energy sources like the sun, wind, biomass, downpour, tides, waves, and geothermal intensity are completely included. Power age, high temp water/space warming, transportation energizes, and off-framework energy administrations are just a few examples of where renewable energy has the potential to supplement or entirely replace fossil fuels.

2. Literature Survey

The impact of electric cars on distribution system expansion plans was analysed by L. Li, D. Yang, H. Pandžić and N. Tomin in 2022, taking into consideration renewable energy, storage, and charging stations. In this context, Energy Storage Systems (ESS) have taken on a new function thanks to the proliferation of electric vehicles (EV) and alternative energy sources (RES). Business models for solar-powered charging stations have been developed for electric car infrastructure by K. Parida and A. Kumar Bohre in 2022. There is an immediate need to build EVCSs to meet the charging demand as the number of EVs on the road continues to rise rapidly. Optimal scheduling of electric vehicle charging stations is addressed in a number of works, and research has been done on two-stage robust energy management of a hybrid charging station linked with the photovoltaic system by Liu, L. Wang, W. Wang, H. Li, M. Liu and X. Xu in 2021. In the study, a stochastic model is converted into a deterministic process by employing an interval optimization strategy for modelling price uncertainty. Off-structure electrical vehicle blaming station hybridized for renewables, containing battery cooling system and different energy accumulating units, created by M. Amir, Zaheeruddin and A. Haque in 2021. E-huge mobility's impact on cutting down on fossil fuel use and carbon emissions is spurring the rapid growth of the electric vehicle market. It would put a significant strain on the electricity industry if millions of electric vehicles had to be charged from the grid at once Nanostructured TiO2 and Sn-based anodes for lithium-molecule batteries have been created by Q. Darn, D. Wu and B. Boulet in 2021. As there are as of now no contending options with the adaptability of lithium-particle batteries for fueling versatile and compact gadgets; what's more, for buffering the fluctuating reserve of sporadic energy sources like breeze and sun powered, it is anticipated that lithium-ion batteries will continue to dominate the market for at least another decade. The recent boom in the development of hybrid electric vehicle technology is largely attributable to the negative effects of the depletion of fossil fuels on global warming. This document synthesises the most up-to-date research on electric vehicle applications for energy storage systems and power conversion topologies (EVs).

3. Methodology

Here, a battery is charged utilizing sustainable power sources like sunlight based chargers and wind turbines, and then it is used to power electric automobiles (EV). For the same reason, we can also refer to this as a wind and solar power recharging station. There are several options available with the green energy charging gadget. Dual-Source Energy Production: Solar + Wind. Properly Oriented Wind Turbines for Maximum Power Capture. In reality, this serves as the prototype for a green energy charging station that can be utilised to power a big fleet of electric cars. This small-scale prototype can be expanded into a full-scale station that uses renewable energy to meet growing global demand. Electricity costs in the city can be lowered by 28% by installing green energy charging stations. Costs for charging an EV are lower than for a traditional vehicle, and EV owners can even make money by selling additional services back to the power company. A capacitor (100 mF 10 V) is used to provide a steady flow of power from two solar panels connected in series, much as how a wind turbine's blades are joined together. This output is wired to an LED to indicate that the battery is being charged, and the other two wires are linked directly to a 9V rechargeable battery through a diode that prevents current from flowing in the opposite direction from the battery to the LED, solar panel, and windmill shown in Fig. 39.1.

Here, we've already connected the solar panels, wind turbines, LED, and battery as discussed in the previous section. We'll finish up by connecting the battery's two terminals to the 7805 IC and a capacitor (470 mF, 25 V) to convert the 12V input to 5V output. This serves as a prototype for charging stations for electric vehicles. With this 5V output, the charging-readiness indicator light can be activated. The IC's input is wired to the battery, and the on/off switch controls when the circuit is active. Nine volts are produced by connecting two 4.5V batteries in series.

4. Results and Discussion

Green energy charging stations are similar to gas stations in that they may be found in a number of different parts of a city, and their models simply represent the end result—In this case, a glowing light—as an indicator of a fully charged battery. Powering

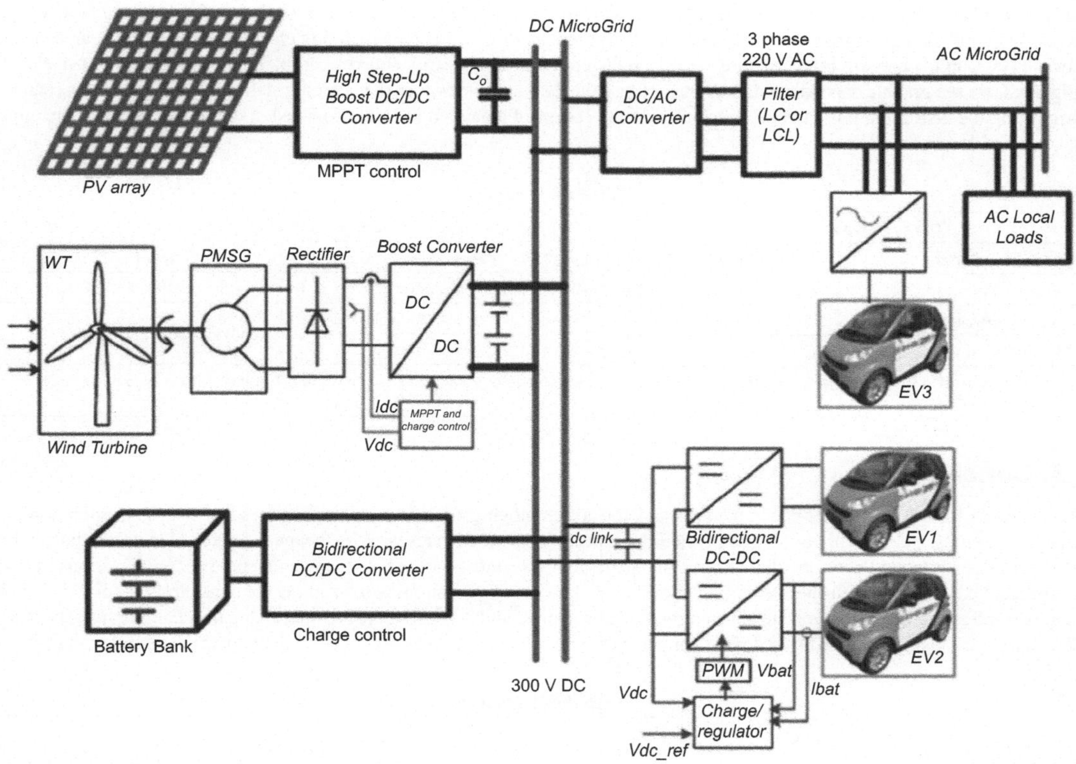

Fig. 39.1 Proposed architecture

up electric cars (EV). An integral part of the sustainability spectrum to keep in mind as we make the shift to clean electrified transportation is the necessity of enabling clean energy conditions to move towards a green-to-green worldview from the very beginning of any plan for the introduction of electric vehicles. For the "green electro mobility" concept to be successfully implemented, it is essential that electric vehicle charging stations (EV-CS) be fueled by power age frameworks that utilize sustainable power sources. Because of this work, we will have a superior comprehension of the natural effect of green, standalone energy systems from the perspective of their CO2 emissions, and we will be able to estimate the ecological advantages of carrying out such frameworks in the power supply of EV-CS according to the viewpoint of kept away from CO2 emanations in contrast with the regular power supply lattice. The outcomes exhibit the viability of renewable energy sources as alternatives for grid-tied EV charging infrastructure.

Fig. 39.2 EV charging stations powered by renewable energy

1: Wind 2: Solar panel Unlike standard charging stations, this one can reliably produce electricity at all hours of the day and night, thanks to the presence of wind. This is in contrast to solar-powered generators, which can only charge electric vehicles during the day. The average monthly number of times an electric car can be charged at public stations drops when outside temperatures are low. There should be a rise in the frequency with which people charge their vehicles using the local power system. This is because electric vehicle owners have to wait longer to charge their batteries when the weather is cold. After dark, they actively plug in their electric vehicle to their home's electrical system to charge. For latitudes north of 50, winter is in full swing. Degrees north of latitude, which presents difficulties for generating electricity. When the peak demand for a city's public charging infrastructure can be predicted, renewable energy sources can be employed more extensively. In Fig. 39.2, we see graphic dependencies that characterise the daily consistency of the charging station's operating factor change. Analyzing the distribution of charging sessions over the course of a day led us to the conclusion that the majority of charging takes place during the day.

Table 39.1 Outputs

Our prototype output		Actual project's output	
Panel	**Wind**	**Panel**	**Wind**
1 Panel = 3V	1 Turbine = 1.5 V	1(32 cell solar panel) = 14.72 V	Rooftop wind turbines (400 W to 1 kW)
2 solar panels used	2 Turbines used	To use 16 solar panels + 4.48 V solar cells	Based on Wind Speed it will work
1 Panel * 2 = 6 V	1 Turbine * 2 = 3 V	(16 panels = 235.52) + 4.48 = 240 V	
Stored in 9 V rechargeable battery		As per the energy requirements, Panels and Turbines will be fixed and stored in storage system	

5. Conclusion

The research for this project looked at the pros and cons of employing alternative energy sources as power generators for EV charging stations. Queueing theory-based methods have been widely used in research on the management of metropolitan public transportation systems. Therefore, there is no agreement among scientists about how many and where charging stations should be placed. One method for further modelling of the network of charging stations is to examine the characteristics of the current charging infrastructure. Improvements in charging infrastructure's ability to generate income. Using the charging infrastructure for shorter periods of time. Reduced demand on the city's electrical grid thanks to more efficient charging facilities.

References

1. R. Madhumitha, Priya, and S. Saravanan, (2022), Half breed Environmentally friendly power Based Electric Vehicles Charging Station in Second Global Gathering, Advance Processing and Imaginative Advancements in Designing, Noida, India. ICACITE: 2348–2352.
2. P. Shukl and B. Singh. (2022), Normal DC Transport Accusing Framework for EVs of Sunlight based, Wind, Battery and Energy unit Sources, Worldwide Meeting on Figuring, Power and Correspondence Innovations (GlobConPT), New Delhi, India, IEEE: 1–6.
3. Bowen, J. Engelhardt, T. Gabderakhmanova, M. Marinelli and G. Rohde. (2022), Battery Supported EV Quick Chargers on Bornholm: Charging Examples and Matrix Combination, 57th Global Colleges Power Designing Gathering (UPEC), Istanbul, Turkey.
4. Verma and B. Singh. (2021), AFF-SOGI-DRC Control of Innocuous to the environment power Based Network Wise Faulting Station for EV for Power Quality Improvement, IEEE Exchanges on Industry Applications, vol. 57, no. 1, doi: 588–597.
5. Q. Darn, D. Wu and B. Boulet. (2021), Electric Vehicle Battery as Energy Stockpiling Unit Think about Sustainable Power Vulnerability, IEEE Energy Change Congress and Piece (ECCE), Vancouver, BC, Canada, doi: 668–673.
6. L. Li, D. Yang, H. Pandžić and N. Tomin. (2022), Ideal Dispatch of Territorial Coordinated Energy Framework Thinking about the Combinations of Electric Vehicles under the Double Carbon Objective, Fourth Worldwide Gathering on Power and Energy Innovation (ICPET), Beijing, China, doi: 1036–1045.
7. K. Parida and A. Kumar Bohre. (2022), Ideal Measuring of PV, Wind based Framework Associated Crossover Environmentally friendly power Frameworks for Country Regions in Presence of EVs, Global Gathering on Choice Guide Sciences and Applications (DASA), Chiangrai, Thailand, doi: 1311–1316.
8. Liu, L. Wang, W. Wang, H. Li, M. Liu and X. Xu. (2021), Methodology of Huge Degree Electric Vehicles Soaking Practical power Abandoned Power considering Master Slave Game, IEEE Access, vol. 9, doi: 92473–92482, 2021.
9. M. Amir, Zaheeruddin and A. Haque.(2021), Hardening of EVs Aggregator with Microgrid and Impact of V2G Power on Culmination Rule, IEEE fourth By and large Gathering on Joining up, Power and Correspondence Advances (GUCON), Kuala Lumpur, Malaysia.

Note: All the figures and table in this chapter were made by the author.

Disruptive Technologies for Sustainable Development – Dr. G. Nagappan et al. (eds)
© 2024 Taylor & Francis Group, London, ISBN 978-1-032-55000-8

Prediction of Chronic Kidney Disease Using Machine Learning Classification Techniques and ANN

40

S. Vidhya[1]
Associate Professor Department of Information Technology,
Saveetha Engineering College, Chennai, India

S. Balachandar[2]
M.E Scholar, Department Of Information Technology,
Saveetha Engineering College, Chennai, India

Abstract: Nowadays, everyone is trying to raise their awareness of health, but due to the busyness of work and busy schedules, we must be careful when some symptoms appear. However, CKD is a disease that has no symptoms at all, and in some cases, no disease-specific symptoms. Predicting, detecting and preventing such diseases can be difficult and can lead to permanent health hazards, However, machine learning shows promise in solving this issue and is perfectly suited for analysis and prediction. Using data from 400 records of CKD patients with 14 attributes, determine whether they have CKD and, If so, describe its severity using various machine learning methods as a whole, like SVM and decision trees. We have created a model with the highest level of accuracy possible for Abnormal kidney function or kidney failure that lasts for months or years is referred to as A renal condition that is ongoing. Patients with diabetes and individuals with high blood pressure, and ongoing kidney problems are all known to be at risk for developing kidney issues, are typically screened. Therefore, early diagnosis is essential for treating the condition effectively.

Keywords: Health care, ML-Techniques, Chronic kidney disease, Support vector machine, ANN, KNN

1. Introduction

The importance of the kidney to the human body is well known. It has vital functions like excretion and osmotic regulation. In essence, it can be said that all poisonous and undesirable substances are drawn out of the body and eliminated via the excretory system and kidneys. Persistent Kidney Disease affects about one million people in India each year, which is incrementing every year. There may be a significant renal ailment present when kidney function is rapidly decreased. A moderate, irregular outcome of kidney function over an extended period of time may be CKD. A person produces constant renal disappointment. Patients with blood weight, iron deficiency, childbirth, dietary deficiencies, health and neurological damage, and debilitated resistant reactions may have dangerous levels of water, electrolytes, and waste materials if CKD is not identified and treated in its early stages. It may gather inside the body and develop in stages.

Early detection of CKD is therefore essential, but its symptoms develop gradually, are not disease-specific, and are consequently unpredictable. A lot of individuals have no symptoms at all, but machine learning as a whole can help solve this conundrum and forecast if a patient has a chronic illness. With the use of the machine learning process,this by creating a predictive demonstration using historical CKD understanding data. It can be determined by taking into account factors like age, race,

[1]vidhya@saveetha.ac.in, [2]balachandar6100@gmail.com

DOI: 10.1201/9781003428473-40

sexual orientation, and blood creatinine results. The earlier a disease is discovered, the more likely it is to develop or spread slowly. Five Levels of CKD-induced renal harm are categorized based on the GFR.

Table 40.1 Levels of CKD

Level	Character	GFR (mL/min)
_	A greater risk of acquiring CKD	>=90 major risk factors
1	With a normal or higher GFR, kidney injury	>=90
2	Significant reduction in GFR	60-89
3	Fairly fall in GFR	30-59
4	GFR declines significantly	15-29
5	Renal Disorder	<15 or dialysis

2. Literature Review

According to experimental data gathered utilising quick mining software, Naive Bayes took less time and generated more accurate results than the artificial neural network. Kunwaar et al (2018). worked with artificial neural networks, also referred to as "Naive Bayes", to predict chronic renal failure (CKD).For their investigation, they used 600 clinical records from a renowned Chennai diabetes research institute. The authors tested the dataset using the WEKA tool, utilising the decision tree and Nave Bayes-categorised algorithms. K. A. Padmanabhan and G. Parthibhan (2020) identified chronic renal illness in diabetic individuals using machine learning approaches. They discovered that the decision tree method performed better than the Naive Bayes method by 91%. A datamining tool, was used to implement experiments using these algorithms on these datasets. E. A. Sharma and M. Arora (2017) employed three classification algorithms: To identify CKD in its early Levels J48 was ultimately shown to be the most suitable algorithm for an accurate and prompt diagnosis of chronic renal failure.

2.1 Proposed Methodologies

Data gathering and dataset preparation in Step 1 This will entail gathering medical information artefacts from a variety of sources, including hospitals, patient discharge papers, and the UCI repository. The dataset will then go through pre-processing, which will eliminate any extraneous data and extract key features from the dataset. Building a Deep Learning and Probabilistic Modeling Approach (ANN) for Disease Prediction This stage involves developing a probabilistic modelling and deep learning method based on RNN that can operate successfully on sizable databases related to healthcare. Additionally, it can create a decision tree and handle a sizable amount of information variables without variable deletion, creating and testing datasets.

2.2 Classification Performance

(i) Naive Bayes: Naive Bayes classification technique places a subject I with characteristics set XI in the class c with the greatest probable outcome after using a Bayesian probabilistic model.

(ii) Decision tree : Decision trees are easy to interpret and visualize, making them a popular choice for data analysis and decision-processing. In addition, they are able to handle categorical as well as numerical data, making them versatile and helpful for a wide range of tasks.

(iii) Artificial neural network (ANN):A **ANN** is a processing system made up of a sizable number of simple processors (neurons), which accept inputs and produce outputs by modifying values to correspond with the predefined activation functions. By effectively updating the weight coefficients in each epoch and minimising mistakes, feed-forward back-propagation learning is used to train artificial neural networks.

(iv) K-Nearest Neighbours: The k-Nearest Neighbours classifier identifies the class where almost all of an unlabeled instance's k proximal neighbours originate by measuring the distance between it and each subsequent training instance.

(v) Logistic Regression: A popular supervised learning technique in the medical field is logistic regression. By aggregating a number of independent features, logistic regression may be used to the forecast likelihood of a class output.

(vi) SVM: Support Vector Machines are controlled models of learning with related algorithms of learning used in machine learning to evaluate data used in regression and classification.

3. Result and Discussion

The metrics listed below provide data on the calibre of the findings from this study. We can better understand the classifier's performance using a confusion matrix.

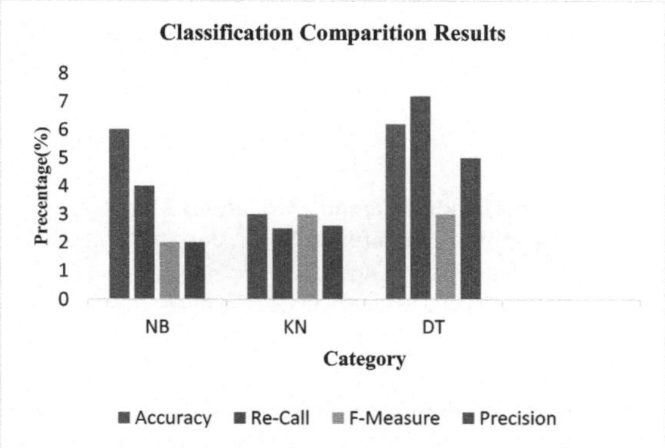

Fig. 40.1 Classification comparition results

4. Conclusion

It is commendable that the medical community is using machine learning techniques to address the challenge of early detection and treatment of Chronic kidney failure (CKD), a condition that can lead to long-term kidney failure. The use of supervised learning planning with probabilistic, tree-based, and ensemble learning-based models such as SVM, DT, LR, ANN, and k-NN is a step towards improving the accuracy of CKD risk prediction models. The fact that CKD is a quiet illness that shows no symptoms makes it imperative to develop accurate and reliable prediction models. The reported performance of the model, which combines the other models in terms of accuracy, precision, recall, F-measure, and AUC, is impressive, with an AUC of 100% and 96.2% accuracy.

5. Future Enhancement

Based on trade insights and information analysis, five mining locations are described. In order to evaluate the objectives, a prepared show is used. In terms of show interpretability, access to specific data, and precision, all three models were able to provide answers to challenging questions. Four out of the five targets could be answered by ANN-MDRP. Three different trees, neural networks twice. Choice tree results are easier to read and understand, despite not being the most compelling demonstration. Within the choice tree, drill-through access to point-by-point persistent profiles is essentially available. Choice trees were outperformed by ANN-MDRP because it was able to identify all significant restorative indicators. It is more difficult to obtain connections between the qualities produced by neural systems.

References

1. K. A. Padmanabhan and G. Parthibhan (2020) "Predicting the Risk of Chronic Kidney Disorder Using Machine Learning Techniques," The Indian Journal of Science and Technology, Vol. 9, 204–214.
2. G. Muhammad and K.A.Otunaiya (2019)" The paper "Performance of Analysis of Data Methodologies in Chronic Kidney Disorder Prediction" Computer Science, Information Technology, Vol. 7(2), 48–53.
3. E. Hodneland et al. (2019In vivo detection of chronic kidney dysfunction using tissue misshapening fields from dynamic MR imaging,"IEEE Trans. BioMed. Eng., vol. 66(6), 1779–1790.
4. A. S. Sabetha, V. Kunwaar, and K. Chandhal, (2018) and A. Banzhal,"Analysis of chronic kidney failure using data mining and classification approaches."6th International Conference on Cloud Systems and Big Data Engineering, Vol., 300–305.
5. E. A. Sharma and M.Aror (2017)"Detecting chronic kidney disease from medical records using Weka,"International Journal of Computer Applications, Vol. 6(4), 1797–2250.

Note: All the figure and table in this chapter were made by the author.

Disruptive Technologies for Sustainable Development – Dr. G. Nagappan et al. (eds)
© 2024 Taylor & Francis Group, London, ISBN 978-1-032-55000-8

Development of an Android Support System for Rainwater Harvesting

41

P. Bakkiyalakshmi[1], Mohamed S.[2], Swathi S.[3], Malaravan R. K.[4], Jagan M.[5]

Department of Agriculture Engineering,
Saveetha Engineering College, Chennai, India

Abstract: Rainwater harvesting is a system of collection and storage of rainwater into reservoirs, lakes, infiltration into the subsurface or storage structures. More than 50% of the population lives in urban area, as a result of which there will be increase in increase in water demand and decrease in the infiltration due to impervious layer. This will in turn decrease in water supply and accelerate groundwater exploitation. As roof top has maximum harvesting potential, it can be utilized for water harvesting to meet the demands of urban population, particularly in non-monsoon period. In this study, we propose to develop an android based application, named "Rain Water Harvesting Assessment Tool" (R-WHAT) for effective rooftop runoff harvesting in urban area. 75% dependable rainfall was used to calculate harvesting potential of chosen rooftop surface in the urban area using curve number method. This application calculates the area of the rooftop using Google Earth Pro by drawing the polygon along the edges of the identified rooftop area. Based on the generated runoff, the storage potential in a land parcel and the required footprint for the storage structure will be determined from R-WHAT. The application also determines the runoff contribution of the land parcel to that catchment area and the potential of rooftop runoff. Therefore, it will be a useful tool for water resources management in urban area where intense urbanization increases flooding potential and reduces infiltration into the subsurface soil.

Keywords: Urbanization, Rain water harvesting, Storage, Android application, IoT

1. Introduction

Water scarcity is the major issue in urban areas as the population increases in urban area occur at an alarming rate and is expected to host over 60% of world population by 2050. The impervious layers in urban areas significantly reduce infiltration and the need for conservation of fresh water available in the form of rainfall becomes essential. Climate change and increasing population pressure are exacerbating these challenges, making it more important than ever to conserve and manage our water resources effectively. Due to irregular rainfall pattern (Palanisamy et al. 2020)the conventional water sources namely rivers, aquifers, and reservoirs are depleted of its capacity and failing to fulfill the water demand by the increasing urban population. As the rainwater is not harvested effectively, about 80% of rainfall gets discharged into sea (Markovic et al. 2014). By maximizing the rainwater collection, the dependence on groundwater and surface water sources, which are often overexploited and at risk of depletion, is reduced. It can also help to reduce stormwater runoff which can cause frequent flooding and other environmental problems.

[1]bakkiyalakshmip@saveetha.ac.in, [2]mohammednagoor70@gmail.com, [3]swathisri3148@gmail.com, [4]malaravankarthi2002@gmail.com, [5]jagan2002in@gmail.com

DOI: 10.1201/9781003428473-41

The benefits of rainwater harvesting are well documented in the literature as it significantly reduces the demand for freshwater resources particularly in regions facing water scarcity (Singh et al. 2022) The rainwater harvesting can reduce the demand for potable water, particularly during periods of water scarcity, and improve the quality of water in urban areas (Nolde ,2007). Rainwater harvesting can also mitigate urban flooding and improve the resilience of urban infrastructure to climate change impacts (Rahman et al. 2023).

Mobile applications have become an integral part of water management, providing with an easy and accessible way to perform various tasks related to urban water at large spatial scale. One such application for rainwater harvesting developed by Sam water foundation includes implementation features, such as calculators and planning tools, to help users design and implement rainwater harvesting systems but is not available for use. To best of our knowledge, the mobile applications for decision making of rainwater harvesting is yet to be made available. Therefore, we propose to develop an android based application to automate the process of determining the available rooftop area, runoff generated from the rooftop area, provide suggestions on the capacity of storage structure required to meet the runoff, footprints required on the ground and percent of contribution from runoff to the parcel area.

Based on the above observations, the objectives of the study area are as follow:

- To develop a mobile application to automate the process of area available for rainwater harvesting
- To summarize the rooftop runoff characteristics and recommend storage structures
- To automate the process of determining the footprints required for the storage structures

2. Data

2.1 Study Area

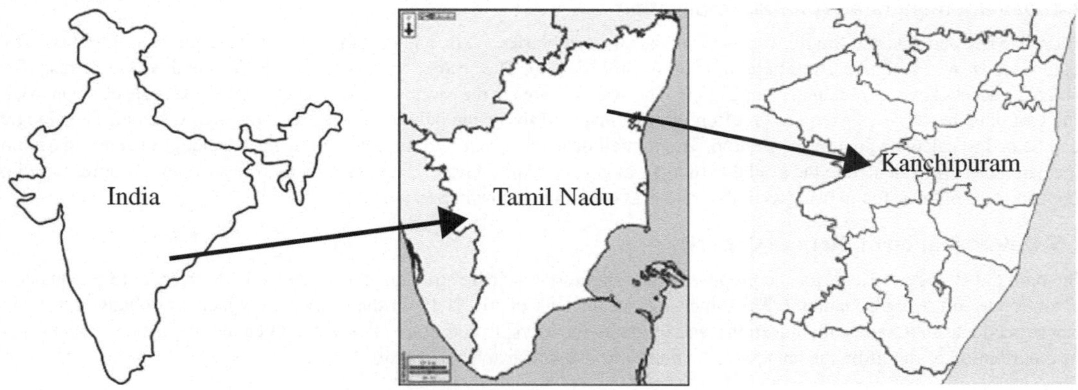

Fig. 41.1 Study area chosen to demonstrate the rainwater harvesting application

Source: (1) https://www.alamy.com/stock-photo/india-map-outline.html?sortBy=relevant, (2) https://d-maps.com/pays.php?num_pay=437&num_pag=1&lang=en, (3) https://commons.m.wikimedia.org/wiki/File:TN_Taluks_Kanchipuram.png

In this study, Kanchipuram was chosen as the study area as it the suburban area that is undergoing urbanization with land use changing from agriculture. Kanchipuram district is located on the northeastern coast of Tamil Nadu, bordered by the Bay of Bengal and Chennai city. The district spans from 11° 00′ to 12° 00′ North latitudes and 77° 28′ to 78° 50′ East longitudes, covering a total area of 1704.79 square kilometers with a coastline of 87.2 kilometers. The district is divided for administrative purposes into two revenue divisions consisting of five taluks and 520 revenue villages, and for development purposes into five blocks with 274 Village Panchayats. Northeast and Southwest monsoon gives at most 54% and 34% to the annual rainfall. The rainfall is almost even throughout the district. The monthly rainfall data of Kanchipuram district was taken for 30 years (1991-2020) to calculate the runoff.

2.2 Rainfall-Runoff Estimation

The rainfall data of 30 years was collected from WRIS website(India-WRIS, 2022)om which the 75% probability occurrence of the rainfall in the study was carried out using Weibull algorithm (Equation 1). In our study area, the 75% rainfall occurred during the year 2022 and hence the maximum monthly rainfall occurred during that year was used to calculate the runoff generated from the rooftop according to Equation 2.

Weibull equation, (Weibul, 1951)

$$\text{Probability} = m \div (n + 1) \tag{1}$$

Where,

m = Rank

n = Number of years

2.3 Calculation of Available Runoff

The available runoff was calculated by

$$\text{Available runoff } (m^3) = \text{Area } (\text{m}^2) \times \text{Rainfall (mm)} \times \text{Runoff coefficient} \tag{2}$$

Where,

Area (m^2) = Area of the catchment (roof top)

Rainfall (mm) = Depth of rainfall

Runoff coefficient = Dimensionless coefficient relating the amount of runoff to the amount of rainfall received (GRAF, 2022).

2.4 Development of an Android Application

Android SDK and Android Studio was used and necessary libraries, APIs, and development environment were installed. The login window was designed using basic tools in Android Studio. The second window in R-WHAT contains Google map, for which the access key was obtained from Google. The available area in the rooftop was found by using the Google earth pro tool. The tool pins the corners of required rooftop through map and forms the polygon from which the area is found. The Google map helped to calculate area of the rooftop. The rainfall data was given into the software using JavaScript in Android Studio The parameters for calculating the available runoff was converted into Javascript syntaxes and given as input in Android studio. The tank capacity was found using available runoff as explained below in Section 2.5.

2.5 Determination of Storage Capacity

The required storage capacity for the proposed water harvesting storage structure was calculated with respect to the quantity of rooftop runoff using Equation 2. The dimensions of the tank (Table 41.1) for the study area, which was obtained from an experienced mason (Oral Communication), was output to the users. In addition to this, the application was designed to display the contribution of runoff by the land parcel in question, to the its catchment runoff.

Table 41.1 Dimensions of the storage tank

Tank Capacty (Litres)	Diameter of the tank (M)	Length of the tank (M)	Footprint for storage structure (M^2)
10000	2.4	2.4	5.76
15000	2.438	4.267	10.4029
20000	2.4	4.8	11.52
25000	2.25	5.52	12.42
30000	2.25	6.32	14.22
35000	2.5	7.2	18
40000	2.34	10.38	24.28
45000	2.34	11.59	27.12
50000	2.34	12.8	29.95

3. Results and Discussion

The application R-WHAT was tested successfully (Figure 3a) to identify the rooftop area and ability to calculate the runoff generated from the calculated rooftop area. The users can either enter the location manually or allow the application to automatically zoom into the current location (Figure 3b). Based on the rainfall input into the application, the runoff was calculated with an information is provided at the bottom, "Note: The average rainfall data is calculated in basis of referring the past 30 years of rainfall in your region". The application also output the storage capacity with an additional information output to the screen based on the calculated rooftop runoff. Based on the storage capacity, the application also successfully provided the footprints required on the ground to install such structures. This is particularly useful in case of commercial building where runoff produced will be much higher than that of residential building.

The results of the study showed that the application developed (Fig. 41.2a) was able to automate the process of delineating the rooftop catchment (Fig. 41.2b) and was able to calculate the amount of runoff.

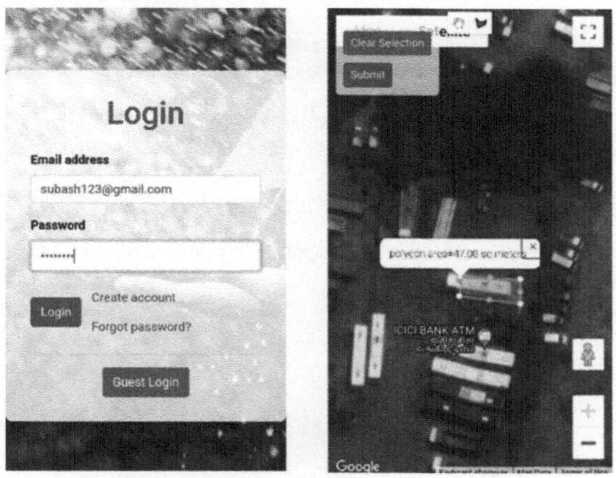

Fig. 41.2 (a) Login figure 2(b) Rooftop catchment

Fig. 41.3 (a) The starting screen of the mobile application R-WHAT (b) The delineation of rooftop catchment

4. Conclusions

In this study, we developed a mobile application R-WHAT based in Android to assist the users in determining the storage capacity required for rainwater harvesting. This application has a user-friendly interface wherein the calculation of rooftop area is automated and outputs the dimensions of the storage structure required and footprints required in the land parcel. With this application, the rainwater harvesting can be effectively planned and implemented by the people, government officials working on urban water management, aquifer recharge and scientific community that works on devising new methods to encourage rainwater harvesting. The application also outputs the runoff contribution from the land parcel, it can be highly useful in flood mitigation measures.

References

1. Nolde (2007). Possibilities of rainwater utilisation in densely populated areas including precipitation runoffs from traffic surfaces. Desalination, 215(1–3), 1–11.
2. Palanisamy, B., Shaurabh, S. and Narasimhan, B., 2020. Analysis of challenges and opportunities for low-impact development techniques in urbanizing catchments of the coastal city of Chennai, India: Case study. Journal of Hydrologic Engineering, 25(10), 05020033.
3. Rahman, A., Yildirim, G., Alim, M. A., Amos, C. C., Khan, M. M. and Shirin, S., (2023), January. Rainwater harvesting systems to promote sustainable water management. In AIP Conference Proceedings (Vol. 2643, No. 1, 020001). AIP Publishing LLC.
4. Singh, R., Akuraju, V., Anantha, K. H., Garg, K. K., Barron, J., Whitbread, A. M., Dev, I. and Dixit, S., (2022). Traditional Rainwater Management (Haveli cultivation) for Building System Level Resilience in a Fragile Ecosystem of Bundelkhand Region, Central India. Front. Sustain. Food Syst. 6: 826722. Doi: 10.3389/fsufs.
5. Weibull, W., (1951). A statistical distribution function of wide applicability. Journal of applied mechanics.
6. Zeleňáková, M., Markovič, G., Kaposztásová, D. and Vranayová, Z., (2014). Rainwater management in compliance with sustainable design of buildings. Procedia Engineering, 89, 1515–1521.

Note: All the figures and table in this chapter were made by the author.

Disruptive Technologies for Sustainable Development – Dr. G. Nagappan et al. (eds)

Deep Learning-Based Facial Emotion Recognition and Behavior Monitoring: An Intelligent System for Human-Computer Interaction

42

Raghavendra Reddy[1], Bindu Bhargava Reddy Chintam[2], Anwar Basha H.[3]
School of Computer Science and Engg, REVA University, Bangalore

Abstract: In this paper, machine learning algorithms are used to implement the proposed approach to identify social distance, drowsiness detection, age-gender detection, and emotion detection. While dealing with social distancing, we need to detect humans faces which are available by using Common Objects in Context (COCO) datasets. Similarly, we predict the age-gender, emotion, and drowsiness altogether using frontal-face detection and eye-detection via haarcascade dataset loaded into Convolutional Neural Network (CNN) to train and test the models on color mapped images. In the proposed model, we are using Linear Discriminant Analysis (LDA), Independent Component Analysis (ICA), Principle Component Analysis (PCA) for age-gender detection and emotion detection, K-Nearest Neighbors (KNN) for Social Distancing, and Support Vector Machine (SVM) for drowsiness detection. The accuracy even depends upon the distance away from the camera. The resultant average accuracy of all the models is 96.3% which is capable to predict various tasks as said above. This complete model is made accessible to users via a standalone software/Desktop GUI. The proposed approach is promising for performing all the tasks and activities more accurately and efficiently.

Keywords: Social distance, Drowsiness detection, Age-gender detection, Emotion detection

1. Introduction

A. M. Mandu (2020), S. Ghosh(2017) and G. Li(2022) stated that, Human facial expressions are a common and straightforward way to communicate feelings and thoughts, particularly in nonverbal communication. Perceptions, emotions, behavioral reactions, and a degree of pleasure or frustration are all part of the emotional state, which is linked to the nervous system. In video surveillance, tracking children, injured, and aged persons, emotion recognition has a lot of potential. To solve these issues, we do need a system that helps to prevent losses and gain features from resources. drowsiness of driver while driving causes lots of accidents in and around the world. As per the Fatality Analysis Reporting System (FARS) survey, 63% of accidents were due to drowning while driving for the past 2 years.

In the proposed work we are performing various tasks Social Distance Detection, Emotion Detection, Drowsiness Detection, and Age-Gender Detection. Social Distance Detection is implemented by using KNN. for Gender Detection, we use SVM. For Age prediction, we use multi-label classification and multi-class classification techniques SVM. For drowsiness detection, we consider eyes that are completely open by excluding all closed and partially closed eyes. It is implemented by using logistic regression.

Paper is structured in the following way: Section II outlines the associated studies in the field of activity detection research. Section III explains the new method's ultimate perspective. Section IV describes the implementation specifics, followed by Section V's conclusion and future enhancement work.

[1]reddycs2004@gmail.com, [2]bhargavchintam@gmail.com, [3]anwar.mtech@gmail.com

DOI: 10.1201/9781003428473-42

2. Literature Review

B. Subarna (2018), Sahoo G. K(2022), C. Anyanwu (2022) and P. Tiwari (2022), proposed a multi-layer NN model with the FERN-2013 dataset loaded and significantly used to capture the face from an image and video track using face frontal structure. Generating results even for background bright conditions. Where they used 2 groups of activities are loaded into the system and increased the accuracy by implementing logistic regression and SVM (T. Luong (2022) and D. Farcas (2019)).

L. Guo (2022), A. Darwin Nesakumar (2022), X. Xu (2020) and R. Deshmukh (2022), proposed that Dark Net is a multi-layered feature accretion & study framework planned to recognize and classify unfocussed driving conduct. Driver distraction has been identified as a significant contributor to motor vehicle crashes and injuries. A sporadic CNN that can deconstruct a driving picture is known as Dark Net. Learn how to correctly identify up to six types of driving events using IMU sensor data.

Table 42.1 Summary of the various papers studied under literature survey

Ref No.	Approach	Advantages	Limitations	Accuracy
B. Subarna et al. [2018]	CNN	Intermitted CNN	Fails capture in low light conditions.	93.2%
Sahoo G. K et al. [2022]	Open-CV and SVM	Works in bright backgrounds.	Fails while wearing glasses.	89.2
C. Anyanwu et al. [2022]	3D-CNN with HMM	Detects both images and videos	Fails to detect multiple faces.	91.5%
P. Tiwari et al. [2022]	R-CNN	Works even in-crowd	Fails in dark background.	95.2%
I. Oliveira et al. [2021]	CNN and LDA	Detects multiple faces	Scope to improve Accuracy.	85.3%

Table 42.1 shows the summary of various existing work, advantages, limitations, and accuracy. From the literature survey, we identified that object prediction at night mode is much difficult. Face identification while wearing sunglasses is too unconditional for prediction. Multiple face prediction is complex in identifying in a single frame. Few models even failed to perform predictions with good accuracies. Limitations towards the datasets even prevent getting good results quickly and precisely.

3. Methodology

Human facial recognition methods were used to develop image processing tasks. There are six fundamental feelings that are uniformly felt. A smile on a person's face denotes joy, and the mouth and eyes have a curled appearance. With rising tipped eyebrows and frowns, the face expresses sorrow with bagginess. Cuddled lashes, lean and strained eyelids highlight the human face's indignation. Disgust is conveyed by drooping ears and a wrinkled nose, while surprise is expressed by widening eyes and a broad jaw. We need to incorporate LDA, KNN, and Naive Bayes so we need to distinguish between different groups.

Meanwhile, the Yolo and COCO identity datasets are used to identify gender. Computers process human face photographs to remove characteristics, which are then classified to determine gender. As there are two genders to identify as they belong to the same class. Hence, we used SVM. Age prediction deals with various features like skin wrinkles, eyes, bread, nose shape, cheeks, Hairstyle, and color. These features are extracted from images and processed to predict the age from the age range table. We use multi-label classification and along with multi-class classification techniques of PCA, Random forests, and Naïve Bayes. Age ranges are fixed as 0-6, 7-12, 12-18, 18-25, 25-32, 32-42, 42-50, 60+ where each category loaded with model images of both male and female.

The drowsiness of drivers has been measured using a variety of methods. Here, we pick the driver who asleep for 3 seconds and alert him with a high-pitched alarming sound. Here, we capture the image of the driver. Later on, we catch up with the eyes of the driver by extracting features of the eyes. Here, we consider eyes that are completely open by excluding all closed and partially closed eyes. Therefore, we found it deals with a single class as we implement binary classification techniques of LDA, SVM, and KNN.

3.1 The Architecture of Proposed System

The complete flow of the system is given in Fig. 42.1 and explained in the algorithm. Image max face area to be captured and crop the face using the haas-cascade frontal face and eye detection model dataset. Background blurring, brightness control,

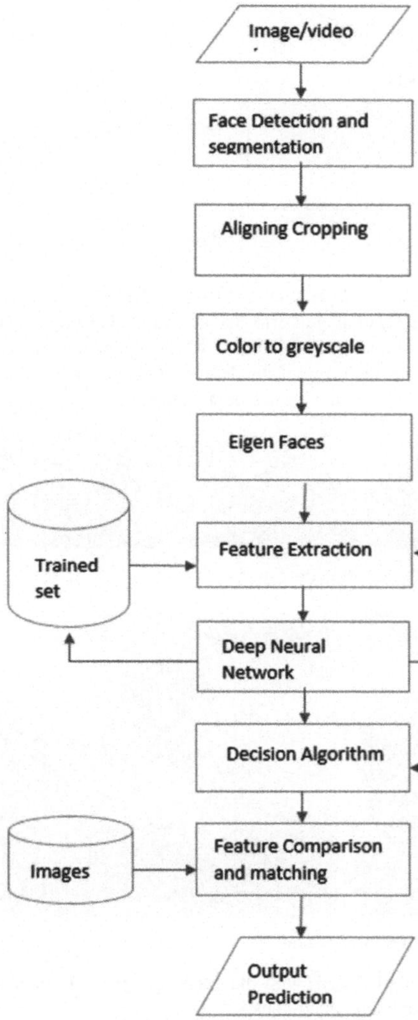

Fig. 42.1 Simplified flow-chart for the entire system

aperture, and sharpness to be monitored based on image collection in real-time. We train the network with GPU for 20 epochs to ensure that the precision converges to the optimal level.

Algorithm: Algorithm for face detection from an image/video/stream
Step 1: Input Face Images, Videos, stream Step 2: Preprocessing the Face Images 2.1 Face Detection and Segmentation 2.2 Aligning & Cropping 2.3 Color to Grayscale 2.4 Eigen Faces 2.5 Capturing Features

Step 3: Feature Extraction from Images
- 3.1 Feature-based /Appearance-based approach
- 3.2 Object detection and template building (Testing)
- 3.3 Matching with trained dataset utilizing CPU (GPU, if available)
- 3.4 improving trained dataset (Training)

Step 4: Classification Methods
- 4.1 Binary Classification or Multi-label classification
- 4.2 Decision Algorithm

Step 5: Final Prediction with an output frame

4. Empirical Results

The proposed work is implemented by using python and its modules such as OpenCV, Keras, Matplotlib, SciPy, TensorFlow, NumPy. Webcam has been used as a part of the hardware component. Performance and prediction even depend upon the GPU performance and CPU speed as well. Figure 42.2-42.3 shows the snapshot of different operations performed by using the proposed system. Here we are taking images, video, or live capturing of the data (by using camera) as an input to our system for performing Social Distancing Detection, Emotion Detection, and Drowsiness Detection.

Fig. 42.2 Emotion detection

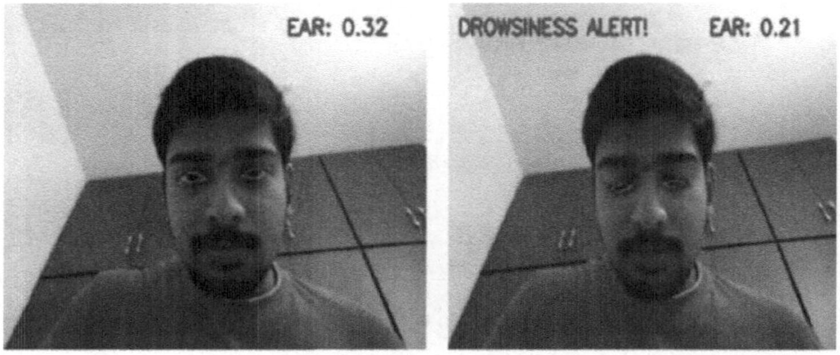

Fig. 42.3 Drowsiness detection

Figure 42.4 show the Accuracy comparison of the proposed system with the existing system. Proposed method helps us to achieve 92.1%, 93.3%, 98.6%, 91.3% as compare to the existing techniques.

5. Conclusion

Finally, we conclude that our model is more effective at multitasking and has a higher rate of accuracy. We are confident that our model can provide information on Social Distancing, Emotion Recognition, Gender/Age Classification, Driver Drowsiness, and Illegal Behaviour with a 96.3 percent accuracy. This model has the potential to inspire large-scale enterprises to incorporate and operate them in a way that maintains stability and robustness. We would like to improve and sort our algorithms to make them more accurate in the upcoming days. We would like to add few addons that could deliberately improve our model on both sides, Trust & Security, improve hardware config's: GPU Enhancement, Algorithm Tuning: Improving Stability and Feature Engineering: More Refined and Accurate.

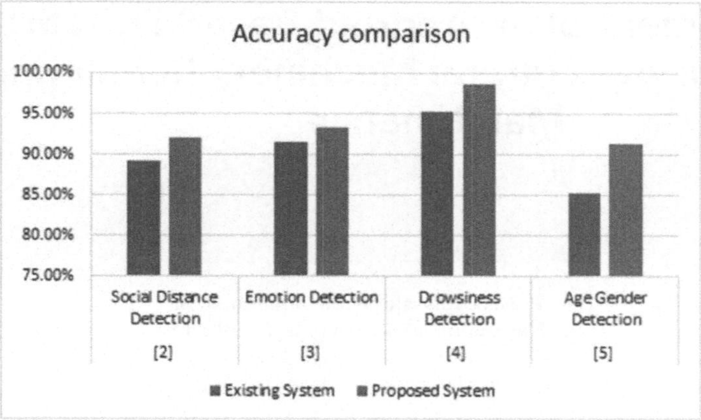

Fig. 42.4 Accuracy comparison of the proposed system with the existing system

REFERENCES

1. B. Subarna and D. M. Viswanathan (2018). Real Time Facial Expression Recognition Based on Deep Convolutional Spatial Neural Networks. International Conference-ICETIETR. (pp. 1–5). IEEE.
2. Sahoo G. K., J. Ponduru, P. Singh and S. K. Das (2022). Deep Leaning-Based Facial Expression Recognition in FER2013 Database: An in-Vehicle Application. International Conference-INDICON. (pp. 1–6). IEEE.
3. C. Anyanwu and C. Hays (2022). A Convolutional Neural Network for Emotion Assessment and Recognition. International Conference-MASS. (pp. 759–763). IEEE.
4. P. Tiwari and S. Veenadhari (2022). An Efficient Classification Technique for Automatic Identification of Emotions Leading to Stress. IEEE 6th Conference -CICT. 1–6.
5. I. Oliveira, J. L. Silva, F. P. Quispe and A. B. Alvarez(2021). A visual monitoring module based on the recognition of facial emotions with CNN. IEEE Conference-EIRCON. 1–6.
6. A. S. M, S. S and S. K. S (2022). A Study on various Applications of Computer Vision for Teaching Learning in Classroom. 6th International Conference on ECAT. 896–900.
7. Z. Lin, X. Yan, Z. Lin, and B. Vucetic (2022). A Novel Exploitative and Explorative GWO-SVM Algorithm for Smart Emotion Recognition. IEEE Internet of Things Journal. 50–56.
8. A. Darwin Nesakumar, R. Priyadharshini, K. Pavithra, L. Sherin, K. N. Pavithra and P. Mugilan (2022). Emotional Based Playback System for Autism. International Conference-IIHC. 1328–1332.
9. L. Guo, Y. Shen and P. Ding (2022). An Algorithm of Emotion Recognition and Valence of Drivers on Multimodal Data. IEEE International Symposium-BMSB. 1–6.
10. A. M. Mandu, C. Vizitiu C. Dinculescu, and L. Petrescu(2020), Emotion Detection System Oriented on Anthropological Face Landmarks Complying with Isolated and Confined Environmental Conditions. International Conference-EHB. 1–6.
11. S. Ghosh (2017). Emotion-aware computing using smartphone. International Conference on COMSNETS. 592–597.
12. G. Li, H. Fu, Y. Sun and Y. Zhang (2022). Intelligent Stable Monitoring Algorithm of Students' Online Learning Behavior Based on User Portrait. Global Reliability and Prognostics and Health Management (PHM-Yantai). 1–8.
13. L. Wang, X. Zhang, Y. Li and X. Guo (2022). Underlying Emotional Mechanisms of Routine m-Health Use in Chronically Ill Patients. IEEE Transactions on Engineering Mgmt. 69(6): 2658–2669.
14. X. Xu, J. Niu, J. Sun, H. Zhang, and Q. Zhu.(2020).Teaching Quality Evaluation System for College Teachers Based on Artificial Intelligence Facial Expression Recognition Technology. IEEE International Conference on AEECA. 997–1001.
15. D. Farcas and A. Marginean. (2019). Emotion Sense: Real-time Emotional Feedback from the Audience. IEEE International Conference on ICCP. 387–393.
16. R. Deshmukh, M. A. Raj, Gupta and K. Singh. (2022). REDE - Detecting human emotions using CNN and RASA. International Conference on ICONAT. pp. 1–6.

Note: All the figures in this chapter were made by the author.

Disruptive Technologies for Sustainable Development – Dr. G. Nagappan et al. (eds)

Development of an Android Based Support System for Agricultural Machinery Repair and Maintenance

43

P. Bakkiyalakshmi*, Bavatharani A. M., Naganandhini S., Resmi S., Shankari S.

Department of Agriculture Engineering,
Saveetha Engineering College, Chennai, India

Abstract: Farm mechanization transformed the agricultural sector by increasing the crop yield, and reducing the dependence of manpower. However, mechanization in India is yet to gain its momentum due to the presence of small and marginal farmers, which comprises 86% of the total farming community. Small and marginal farmers own less than 1-2 ha of lands that are highly fragmented for any farm machineries to be adopted for the cultivation practices. High investment, repair and maintenance cost are also major limitations and hence they, depend on manual labours. In this study, we aim to close the gap between the farmers and repair-service providers through the advent of mobile applications. We have developed an android based application named as "Farm Machinery and Equipment Mechanics (Farm-MEchs)" through which the farmers can find the repair and service providers in nearby locations, who can render the services either on-the-field or off-site. The application also enables crowd-sourcing option wherein the services providers can register themselves. The results of the application showed that the farmers can find the service providers successfully. The application will be a helpful tool for farmers as it facilitates the dependency of manpower and reducing the operational difficulties.

Keywords: Farm machinery, Repair and maintenance, Mobile application, Crowd sourcing

1. Introduction

Agricultural mechanization plays a key role in improving agricultural productivity particularly in developing countries. Common farm machinery includes the combine harvester, rotavator or rotary tiller, plough, tractor, trailer, harrow, leveller, ripper machine and disc harrow. Of these machineries, tractor and rotavator are the machines that often need to be serviced. Repair and maintenance services will be based on the types and degree of operational difficulties in the machinery. According to Mehta et al., (2014), there is a direct correlation between farm power availability and productivity during the past six decades. Despite the fact that sales of transplanters, power weeders, combine harvesters, rotavators, and threshers are increasing in India at respective rates of 50, 50, 28, 20, and 10% CAGR (CAGR, 2013), their usage in India is particularly limited by the highly fragmented agricultural lands because small and marginal farmers with 1- 2ha of land comprises approximately 86% of the India's total farming population as compared to the European Union (14 ha) and the United States (170 ha). Therefore, high cost and energy efficient farm machinery are capital intensive, and repair and maintenance of farm machinery adds to the financial crisis of the small and marginal farmers (Mehta et al., 2014).

Moreover, the availability of mechanics for farm machinery and equipment are not well known by the machine holders. With the advent of mobile technologies, identification of service providers has been made easier. For example, mobile applications have been developed on custom hiring of farm machineries and implements (Oluwasefunmi et al., 2021). A mobile application

*Corresponding Author: bagya1123@gmail.com

DOI: 10.1201/9781003428473-43

named Uzhavan from Government of Tamil Nadu has many features to help the farmers such as crop insurance, stock position of fertilizers, seed stock, custom hiring centres for primary machines in each district, market prices for various crops in regulated markets, weather forecasts (Mathuabirami, 2019). Local farmers and residents of the many States in the nation will be assisted by "FARMS- Farm Machinery Solutions" with the help of Farm Machinery Banks and Custom Hiring Centres (NIC, 2019). These two apps lacked a repair feature; therefore, we created the *Farm-MEchs* app to address this problem. To further the potential of the applications being developed, the practise of gathering information from users by enlisting the assistance of several people, whether paid or unpaid, generally via the internet, is also used (Jang et al., 2018).

Based on the literature review as cited above to our best of knowledge, there is a lacuna in terms of finding the repair and service providers of farm implements and machineries. In this study, we intend to develop a mobile application wherein the service providers of farm machineries and implements are through surveys to help the farmers to avail the service. We also incorporated the concept of crowd sourcing through which the repair and service providers may input their availability. The tool will be highly useful to farmers, small and marginal farmers. For creating mobile applications, most widely used programming language nowadays is Python. It has consistently ranked high on lists of the "Best Programming Languages" for five years (Thaker and Shukla, 2020).

2. Materials and Methods

2.1 Study Area

The area chosen for the study was Tiruvallur district, Tamil Nadu, India (Figure 1). The district consists of three divisions, nine taluks, 14 blocks, and 792 villages, as the cultivation practices and usage of farm machinery was higher with 38% of total geographical area under cultivation. Tiruvallur district typically receives 1152.8 mm of precipitation per year and there is 89383.39 acres of net cultivable across the entire area. The district of Tiruvallur is located in the North Eastern Agro Climatic zone and paddy is the major crop that is sown for three seasons Kharif, Rabi and Zaid. INR 5,00,000 is given to the farmers producer group in the district as a corpus fund to purchase farm machineries (NIC, 2023). The most common types of machinery in Tiruvallur are the tractor, rotavator, and combine harvester.

2.2 Methodology

In this, we discussed about the building of the mobile application and the surveying of the study area.

Surveying

A structured and unstructured survey was conducted in the following villages of the district: Thamaraipakkam, Perambakkam, Poonamalle, Nazrathpet, Putlur, Ikkadu, Irulanjeri and Kaavanur. Structured questions are descriptive, close-ended wherein the service providers can be limited to certain questions or responses whereas unstructured questions are spontaneous and exploratory in nature. A sample of the structured questionnaire prepared for the surveys and the unstructured questions used to interview the service providers in the chosen villages are given in Table 43.1.

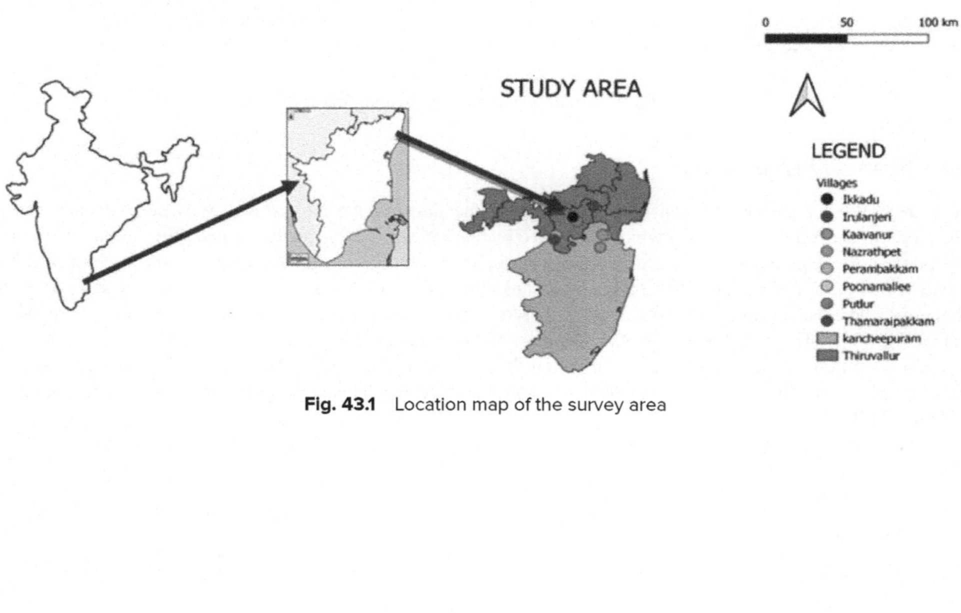

Fig. 43.1 Location map of the survey area

Table 43.1 Sample questionnaire prepared to survey the repair and service providers

S. No	Structured	Unstructured
1.	Whether they are authorized or unauthorized technicians?	If they are unauthorized, do they have their own workshop?
2.	What is the minimum charge for the service?	How much does the charges vary with distance?
3.	What are the types of farm machines handled till now?	Are they aware of new models in farm machinery?
4.	What is the preferable distance?	How does billing process is done for the intended services?
5.	What is their experience in servicing farm machinery?	How do they calculate the estimate for the intended service?

Creation of the mobile application

Python was used to create this Android programme. The Android SDK tools create an APK by compiling code, data, and resource files. The file that Android-powered devices use to install an app is known as an Android package, and it is an archive file with an.apk suffix that includes the contents of an Android app that are necessary at runtime. Additionally, the Android API 30 platform was used to create this application. Sqlite 3 and Kivy was used for the framework and front-end user interface architecture. The workflow of the proposed application is as given below (Fig. 43.2). Upon login, either as a registered user or as a guest, the application provides the options to select the 'Farm Equipment Technicians, Maintenance Guide or Crowd Sourcing'. This app will serve as a platform to gather information from mechanics who are interested in contributing to the application through crowd sourcing.

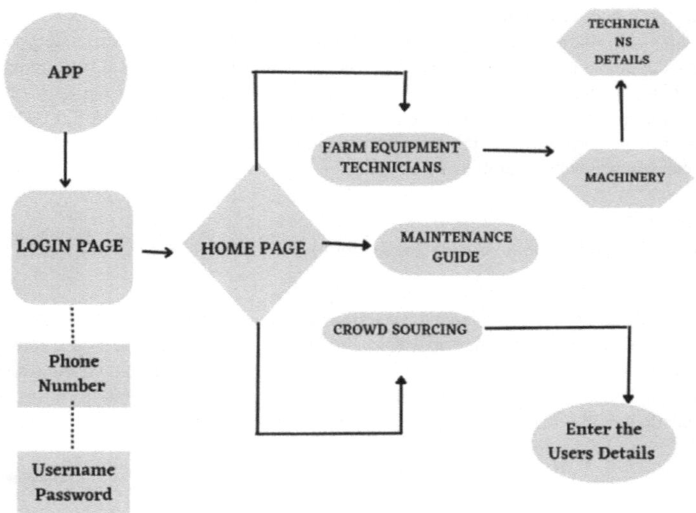

Fig. 43.2 Work flow of the proposed application

3. Result and Discussion

Based on the responses for the survey taken in the eight villages, the input into the application was carried out without producing errors; the output was extracted accurately which was tested by random verification by the farmers (Fig. 43.3a). Also, the crowd sourcing option in the application was tested and extracted successfully. For example, the first menu in *Farm-MEchs* app is labelled Farm Machinery Technicians. When users select this option (Fig. 43.3b), a list of farm machinery icons appears, from which they may select the appropriate farm machinery based on their needs , then it shows the details of the Farm Machinery Technicians (Fig. 43.3c).The second menu in *Farm-MEchs* is called Maintenance Guide, and when a user taps on it, a selection of agricultural operations icons is displayed from which they can choose based on their needs. This menu outlines how to maintain farm equipment. Our third menu is crowd sourcing, based on the users willingness they can be a part of this app by entering their details.

4. Conclusions

The aim of this study was to close the gap between farmers and farm machinery technicians, so that the farmers benefit from lower service cost and increased production potential. The mobile application was developed in Android platform with Python language as a coding language. The application was tested and validated successfully to input and extract required information by . The application will be a useful tool that can facilitate farmers to adopt the farm machinery technologies to increase agricultural production.

Fig. 43.3 Working example of the application

References

1. Degu, A., 2019. Android application memory and energy performance: Systematic literature review. IOSR J. Comput. Eng, 21(3), pp. 20–32.
2. Dewi, D.A., A Review on Agricultural Mobile Apps for Sustainable Agribusiness: before and during Covid-19 Pandemic.
3. Gope, H.L., Uddin, M., Islam, M.A., et al., 2019. Development of an android application for SAU diary in Bangladesh.
4. Jang, E., Barela, M.C., Johnson, M., et al.,2018, April. Crowdsourcing rural network maintenance and repair via network messaging. In Proceedings of the 2018 CHI Conference on Human Factors in Computing Systems (pp. 1–12).
5. Kumar, S.A. and Karthikeyan, C., 2022. 'Uzhavan app'as a conduit to reduce the digital divide by fostering vital agricultural extension services in the state of Tamil Nadu, India. International Journal of Agriculture Innovation, Technology and Globalisation, 3(1), pp. 1–27.
6. Mathuabirami, V., Makokha, J. and Karthikeyan, C., 2019. Constraints faced by extension officers of Coimbatore district, Tamil Nadu in using Uzhavan app. International Journal of Farm Sciences, 9(1), pp. 126–130.
7. Mehta, C.R., Chandel, N.S. and Rajwade, Y.A., 2020. Smart farm mechanization for sustainable Indian agriculture.
8. NIC (National Informatic Centre), 2023. Ministry Of Electronics And Information Technology , Government Of India. Accessed on 15 March 2023.
9. Thaker, N. and Shukla, A., 2020. Python as multi paradigm programming language. International Journal of Computer Applications, 177(31), pp. 38–42.
10. Thaker, N. and Shukla, A., 2020. Python as multi paradigm programming language. International Journal of Computer Applications, 177(31), pp. 38–42.
11. Unhelkar, B. and Murugesan, S., 2010. The enterprise mobile applications development framework. IT professional, 12(3), pp. 33–39.

Note: All the figures and table in this chapter were made by the author.

Disruptive Technologies for Sustainable Development – Dr. G. Nagappan et al. (eds)
© 2024 Taylor & Francis Group, London, ISBN 978-1-032-55000-8

Customer Emotion Analysis on Food Review Images Using Deep Learning: A Review

44

M. Jagadeesh[1]
Assistant Professor(SG), Rajalakshmi Engineering College
Akhilesh Ravikumar[2], **Anjana Deivasigamani**[3], **Dharshini S. S.**[4]
UG Student, Rajalakshmi Engineering College

Abstract: Images communicate more powerful insights than words. The use of images for analysing customer reviews is not a familiar approach due to the abundance of textual data available and the time-intensive nature of image analysis. Customer evaluations posted online are extremely important in improvising a company's reputation. Such knowledge is useful for marketing, product creation, and other endeavours. There are several initiatives centred on supplying the business insights from product reviews using textual data, [sentimental analysis] but the majority of contemporary reviews are multimedia-rich. To find the customer's emotion, image processing of review photographs is highly useful. In the Proposed system, When analysing review images, three key factors are taken into account: first, the image's quality; second, the customer's emotion; and third, the product itself. When all three are considered together, they provide useful information on understanding the customer's emotions. This survey believes that there are only two sides to a review (either positive or negative) and it involves the analysis of review images from three different perspectives and utilises it to accurately predict customer emotion based solely on an image. An application that conveys the customer's emotions of the review image is the final outcome.

Keywords: Convolutional neural networks, Deep learning, Emotion analysis, Image sentiment analysis, Review image classification

1. Introduction

The aim of visual sentiment analysis is understanding how visual images affect emotions. Despite the fact that this field is still quite young, a substantial body of research has been produced with numerous solutions for diverse data sources and issues. This analysis attempts to provide a thorough overview of the subject by reviewing relevant literature. Following a description of the assignment and its important applications, coverage of the problem under many major headings follows. The problem is formalised, taking into consideration different levels of details and factors which may influence how individuals feel by an image. Since so much information is now accessible to the general public, algorithms for sentiment analysis are being researched and developed. To comprehend the mind set of users who produce and share media content on the internet (Biswajeet Pradhan, 2022), these algorithms examine a lot of user data to determine how people feel about certain topics, ideas, and trends. Sentiment analysis is used to ascertain readers' opinions on a subject or the emotional impact an author wants to make on them. The image categorization will enhance the performance of sentiment analysis utilising the popular deep convolutional neural network (Ajay Indian et al. 2022). Since viewpoints affect many human choices in both professional and social contexts,

[1]jagadeesh.m@rajalakshmi.edu.in, [2]akhileshravi.kumar.2019.cse@rajalakshmi.edu.in, [3]anjana.deivasigamani.2019.cse@rajalakshmi.edu.in, [4]dharshini.ss.2019.cse@rajalakshmi.edu.in

DOI: 10.1201/9781003428473-44

sentiment analysis has a number of real-world applications. Businesses, for instance, are curious in how customers perceive their products or services, while buyers look to user evaluations to judge a product before making a purchase. More people and organisations are consulting the public's opinions when making decisions as a result of the expansion of social media (such as blogs, forums, social networks, and reviews).

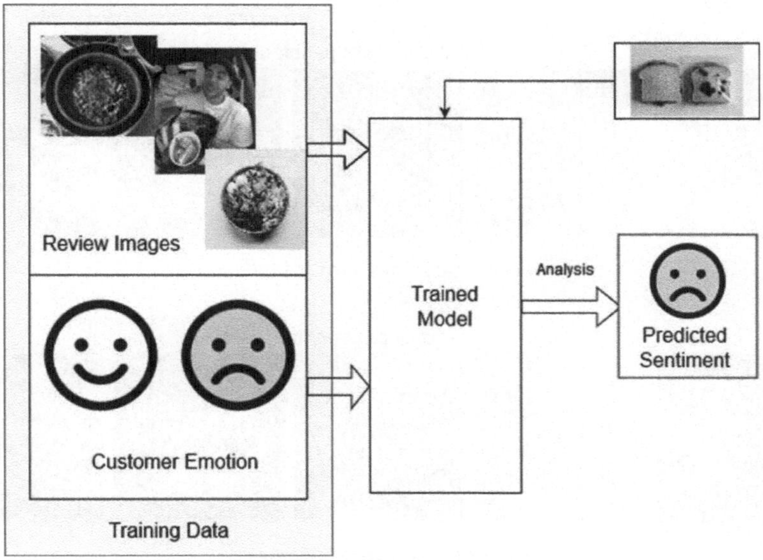

Fig. 44.1 Sentiment Analysis

2. Literature Survey

2.1 Visual Sentiment Analysis

The task of identifying the sentiment or emotion portrayed in an image is known as visual sentiment analysis. It is a branch of sentiment analysis and closely related to computer vision, through the mathematical techniques to manipulate digital images by automatically extracting data from visual content. Visual sentiment analysis seeks to ascertain the emotions that a picture is communicating. There are a variety of techniques that can be used for visual sentiment analysis, including feature extraction, representation learning, and classification. Feature extraction involves identifying and extracting relevant features from the image, such as texts (Thakre et al. 2020), colours, shapes, and textures. Representation learning involves learning a compact and informative representation of the image, such as using a deep neural network. Classification involves using the extracted features or learned representation to classify the image into one of several predefined sentiment categories, such as positive, negative, or neutral. CNNs (convolutional neural networks) and RNNs (recurrent neural networks) have become increasingly popular for visual sentiment analysis in recent years. These techniques are capable of learning complex and rich representations of images, and can be trained on large datasets to achieve high accuracy (Justin Zhan et al. 2015). It has a wide range of applications, such as social media analysis (Joan Santoso et al. 2021), brand monitoring, News analysis (Zhang, Shitao, 2021) and marketing research.

The table lists the top articles on visual emotion analysis in brief. While early systems mostly relied on manually created visual features, modernised techniques utilise the textual metadata and characteristics that may be quickly gleaned from the raw image (i.e., CNN-based representations) (Zhang, Shitao, 2021).

2.2 Datasets Used

We perform research on a set of photos obtained in online restaurant reviews crawled from Yelp in search of an empirical response to what may have also been a philosophical query. Due to the widespread use of smartphones, many online reviewers now frequently add images in their evaluations, which allow them to both textually and visually narrate their experiences and feelings. Although we are working with data from online reviews, the issue with visual sentiment analysis is that preferences are not always present. We characterise the issue as visual sentiment analysis with various contexts for generality's sake, where users and items are the contexts in this case. Contexts may be as particular to each user or as general as the data sources from which the images are derived.

Fig. 44.2 Customer Image Dataset (Truong et al. 2017)

Fig. 44.3 Food Image Dataset (Truong et al. 2017)

2.3 Previous Works

This study (Biswajeet Pradhan, 2022) aimed to investigate explainable AI techniques for forecasting customer emotions in the FDS industry, as well as machine learning or deep learning models. Because 77% of the models were illegible, organisations may disagree on system interpretability and trustworthiness. Although the DL models used in other areas successfully perform with regards to accuracy, they lack the explainability that XAI implementation would bring.

The goal of this research (Ajay Indian et al. 2022) was to enhance picture classification performance applying deep CNN, VGG19. CK+, FER2013, and JAFFE datasets can all be used to simply conduct emotion classification with varied emotions using a VGG-19 architecture. The outcomes indicate that the proposed approach can achieve 99% accuracy. This method may be used to investigate a person's psychological condition.

This paper (Justin Zhan et al. 2015) elaborates on sentiment polarity classification, a crucial problem in sentiment analysis. Each stage of the recommended categorization task is thoroughly explained. The feature delivers an F1 score of 0.73 for classification on review-level using the whole set and more than 0.8 for sentence-level categorization.

The effectiveness of Multi-view Sentiment Analysis was evaluated in (Joan Santoso et al. 2021). They included notions taken from text and picture as a new modality. This model combines three viewpoints using a meta-classifier's ensemble method. By using ensemble learning, predicted probabilities for Image, Text, and Concept were combined from each classifier. A meta-classifier was used to combine Image, Text, and Concept information in order to predict the final emotion. Applying the ensemble method and using Logistic Regression as the meta-classifier resulted in an accuracy of 70%.

They developed a transfer learning classifier in (Zhang, Shitao, 2021) for inter text sentiment categorization. The integration of transfer learning and subspace projection into the dictionary learning framework. The results of the experiment show that the performance of classification in the target domain is improved by the algorithm deployed here.

A novel approach to measuring customer happiness has been presented in this paper (Bouzakraoui et al. 2020) the categorization of numerous facial expressions utilising a variety of classifiers, including Svm, Random Forest, and KNN The aspects of the face are represented by the distances between landmark points. The customer's facial expression is represented by a vector with eighteen values. When their approach was evaluated against the JAFFE database, the accuracy rate was 80.3%.

Paper (Hong-Wei et al. 2015) demonstrates the methods used in emotion recognition. The goal was to categorise the emotions shown by the main human subject in still photos taken from movies. Regarding deep Convolutional Neural Network (CNN) architectures, a transfer learning strategy was used. According to experimental findings, this cascading strategy to fine-tuning outperforms a single stage approach when using the combined datasets. In comparison to the respective accuracy rates of 35.96% and 39.13% of the challenge baseline, their top submission had a test set accuracy of 55.6% and a validation set accuracy of 48.5%, respectively.

This paper (Turgut et al. 2014) presents a highly automated face emotion detection system that does not pre-select distinctive blocks. The accuracy of the face emotion recognition system is about 90%, according to performance statistics, regardless of whether one or more types of facial expressions are included in a photo sequence.

Facial detection is done in various methods like the grey-level co-occurrence matrix (GLCM) texture feature is the foundation for emotion recognition. The extracted GLCM features are trained using different kernels in an SVM. The SVM receives the GLPM's second order statistical texture features as input for classification. The recognition system used here operates without the use of distinctive blocks.

In this study(Lakshmi et al. 2010), YCbCr, HSI, and RGB are applied to the photos in order to recognize faces using skin colour detection techniques.

3. Technologies in Use

Convolutional neural networks (CNNs) are utilised for image and video recognition applications. They are intended to learn spatial feature hierarchies automatically and adaptively from input data. In a CNN, multiple layers are used to process the input, including convolutional and pooling layers, to extract and analyse meaningful features. The final fully connected layer, outputs final prediction based on the learned features. CNNs have achieved flagship results in many computer vision benchmarks and are vastly used in various applications such as object detection, image classification and segmentation.

Face Emotion Recognition (FER) is a field of study that focuses on detecting and categorising emotions expressed on a person's face. This is typically achieved through computer vision techniques and machine learning algorithms, which analyse images or videos of a person's face and use learned patterns to classify the emotions depicted. FER has applications in various industries, including psychology, marketing, human-computer interaction, and security.

Neural networks as a classifier: This paper (Amjad Rehman et al. 2022) claims that it combines two techniques. Face detection and classification involve feature extraction, neural networks, and two steps. Reprocessing images reduces processing time while improving image quality.

Adaboost as a classifier: According to (Akhand et al. 2021) the frontal face in a picture sequence is categorised into seven classifications including neutral, joy, sad, surprised, angry, fear, and disgust. Then, using techniques such as colour space transformation, connected component labelling, face region verification such as face segmentation, pupil detection such as referring to height and breadth of the face region, etc., facial feature detection is carried out. Thus, the landmark regions of the facial features such as the eyes, brows, and mouth are extracted. Finally, the author's technique utilising an AdaBoost-based classifier is used to distinguish between different facial emotions.

4. Conclusion and Future Work

Consumer sentiment analysis has been extensively studied, with most efforts focusing on analysing textual content in online reviews. However, previous attempts to analyse user image feedback in real-time have fallen short of success. Our three-perspective research approach, considering the product, picture quality, and consumer face expression, is expected to significantly improve the effectiveness of customer sentiment analysis. After reviewing various methods of collecting customer review datasets, sequential ML models, and Deep Neural Network functionality, it can be concluded that the CNN architecture is the most commonly used method for image analysis. CNNs can extract both low- and high-level features from images by applying filters or kernels to identify local features in the input image. As these filters are progressively applied through the layers, they can identify more complex features in the image. Furthermore, models predicting the sentiment polarity tend to have higher accuracy than those predicting specific emotions.

References

1. Biswajeet Pradhan et al. "Sentiment Analysis of Customer Reviews of Food Delivery Services Using Deep Learning and Explainable Artificial Intelligence: Systematic Review." Foods 11, no. 10 (2022): 1500.
2. Ajay Indian et al. "Sentiment Analysis from Images using VGG19 based Transfer Learning Approach." Procedia Computer Science 204 (2022): 411–418.
3. Thakre et al. "Customer emotions recognition using facial and textual review." INTERNATIONAL JOURNAL 5, no. 5 (2020).
4. Justin Zhan et al. "Sentiment analysis using product review data." Journal of Big Data 2, no. 1 (2015): 1–14.
5. Joan Santoso et al. "Multiview Sentiment Analysis with Image-Text-Concept Features of Indonesian Social Media Posts."(2021)
6. Zhang, Shitao. "Sentiment classification of news text data using intelligent model." Frontiers in Psychology 12 (2021).
7. Chafale et al. "Sentiment analysis on product reviews using Plutchik's wheel of emotions with fuzzy logic." International Journal of Engineering & Technology (AIJET) 1, no. 2 (2014): 1–8.
8. Hameed et al. "Content-based image retrieval: A review of recent trends." Cogent Engineering 8, no. 1 (2021): 1927469.
9. Lahariya, et al. "Real-time Emotion and Gender Classification using Ensemble CNN." arXiv preprint arXiv:2111.07746 (2021).
10. Bouzakraoui et al. "Customer satisfaction recognition based on facial expression and machine learning techniques." Advances in Science, Technology and Engineering Systems 5, no. 4 (2020): 594–599.
11. Ortis et al. "Visual sentiment analysis based on objective text description of images." In 2018 international conference on content-based multimedia indexing (CBMI), pp. 1–6. IEEE, 2018.
12. Hajek et al. "Opinion mining of consumer reviews using deep neural networks with word-sentiment associations." In IFIP international conference on artificial intelligence applications and innovations, pp. 419–429. Springer, Cham, 2020.
13. AlShariah et al. "Detecting fake images on social media using machine learning." Int. J. Adv. Comput. Sci. Appl.(IJACSA) 10, no. 12 (2019): 170–176.
14. Hong-Wei et al. "Deep learning for emotion recognition on small datasets using transfer learning." In Proceedings of the 2015 ACM on international conference on multimodal interaction, pp. 443–449. 2015.
15. Turgut et al. "Potential of image analysis based systems in food quality assessments and classifications." In 9th Baltic Conference of Food Science and Technology, Jelgava, Latvia, pp. 8–12. 2014.
16. Truong, et al "Visual sentiment analysis for review images with item-oriented and user-oriented CNN." In *Proceedings of the 25th ACM international conference on Multimedia*, pp. 1274–1282. 2017.

17. Lakshmi et al. "Segmentation algorithm for multiple face detection in color images with skin tone regions using color spaces and edge detection techniques." *International journal of computer theory and engineering* 2, no. 4 (2010): 552.
18. Kastrati et al. "A deep learning sentiment analyser for social media comments in low-resource languages." Electronics 10, no. 10 (2021): 1133.
19. Akhand et al. "Facial emotion recognition using transfer learning in the deep CNN." Electronics 10, no. 9 (2021): 1036.
20. Amjad Rehman et al. "Facial emotion recognition using conventional machine learning and deep learning methods: current achievements, analysis and remaining challenges." Information 13, no. 6 (2022): 268.

Note: All the figures in this chapter were made by the author.

Disruptive Technologies for Sustainable Development – Dr. G. Nagappan et al. (eds)
© 2024 Taylor & Francis Group, London, ISBN 978-1-032-55000-8

Intelligent Conversational AI based Net Banking Chatbot Using IBM Watson Assistant

45

V. Uma Rani[1]
Associate Professor, Department of Computer Science and Engineering,
Saveetha Engineering College, Chennai, Tamilnadu India

Sudarshana Raj[2]
PG Scholar, Department of Computer Science and Engineering,
Saveetha Engineering College, Chennai, Tamilnadu, India

Arun Kumar R.[3], Kanagavalli[4]
Assistant Professor, Department of Computer Science and Engineering,
Rajalakshmi Institute of Technology, Chennai, Tamilnadu, India

Abstract: Due to new digital era, Chatbots is a new kind of conversational interfaces which helps to manage customer queries in various fields like banking, automotive industries, travelling, healthcare and education. Due to demand of mobile banking and mobile communication, several artificial intelligences based chatbots are developed to assist banking services. This banking chatbots give assistant to bank executive as well as helping customers with financial advice, insurance policy details and automated notifications for remainders. In this research, Conversational banking chatbots are created by using IBM Watson assistant engine. The IBM Watson assistant is a simple and intuitive interface which helps to building, analyzing, testing and deploying Chatbots. This banking chatbot was evaluated by various kinds of peoples regarding their bank transactions and financial advice.

Keywords: Artificial intelligence, Chatbots, Banking services, IBM Watson assistant

1. Introduction

Due to advancement of mobile computing and Web 4.0, intelligent agents are designed and developed for various fields which provide interaction between human and machine. Machine Learning (ML) is one of the sub set of artificial intelligence (Hamidian, S,2020) helps to build machine that think, act like humanly and rationally. Conversational AI chatbots are intelligent assistants in digital form supported with an artificial intelligence-based computer program. These AI chatbots communicate with humans in a very natural form to resolve their queries and to assist them as required. Generally, chatbots are developed in two ways. They are rule based chatbot and AI based Chatbots (Narula, V.2020) Among these two types, AI based Chatbots are popular and it provide intuitive interface between system and humans. Several types of chatbot assistants are developed to manage customer queries in various fields like smart homes, banking, automotive industries, travelling, healthcare and education.

Banking Chatbots are digital assistants which helps to provide customer service related to bank transactions and insurance details. It improves the operational efficiency of the bank and its employees. Banking chatbots can process transactions in seconds instead of hours or days, allowing customers to complete their business faster. Quick resolution is something that

[1]umaranibharathy@gmail.com, umaraniv@saveetha.ac.in; [2]rsudharshanaraj@gmail.com; [3]arunkumar.r@ritchennai.edu.in, [4]kvalli.818@gmail.com

DOI: 10.1201/9781003428473-45

most customers love. Especially effective for large financial institutions that are notoriously known for long snaking queues. Banking chatbots can help banks reduce their employees' time on administrative tasks and mundane query resolution. This frees up time for bank personnel to focus on more sophisticated and impactful tasks. It's especially beneficial for customer service representatives who often spend a large chunk of their day answering phone calls, emails, and customer texts. Banking chatbots can help banks personalize their customer experience by incorporating information about each customer's preferences and history into the conversation.

The main objective of this research is to

- To assist both bank executives and bank customers
- To help customers with financial advice
- To automate notifications and reminders in net banking transactions

Rest of the paper is structured as follows. Section-2 reviews the existing literature in chatbot design. Section-3 describes the methodology for building conversational bank chatbot and deployment using IBM Watson Assistant. Section-4 explains the developed banking assistant and its system features. Section-5 discusses the conclusion and future work needed to improve the system.

2. Review on Chatbots

Several research works were carried out to developed an AI based chatbot for various fields. The main research work focused to build talkative systems that full of enthusiasm chat interplays with human inevitably. This in essence assistant making amusement for community, helping for solving the questions, appropriating driving guidance's, serving as human companion etc. Some of them are discussed below

Skrebeca (2021) have developed a academy chatbot which supplies an adept and accurate be accountable for any consumer questions about academy information. It uses Myanmar language and AIML for rule creation. Wang, Y (2022) have developed a AI based chat bot for psychological counseling. SF Suhel et al. (2020) have developed a chatbot for banking industry to provide awareness among privacy risk and awareness of the bank service. Yan, Z(2020) has created a smart chatbot which answers the users queries related to their daily activities.

Matheson, E.L(2021) has utilized a chatbot for recruitment process to reduces the manual effort, save time and cost needed for that process. This chat bot uses ML and cloud-based technology to reduce the recruitment process effort. Singh, R(2021) has created a personal finance chatbot using node.js and Microsoft bot framework for reducing the expenses , personalized recommendation and to provide financial advice in the time of emergency . Chen, Y (2021) has created a chatbot for virtual health resistant using Rasa and google cloud it helped many during the pandemic situations and provide medical assistance in the emergency cases and to reduce the risk of death. Zhi, Y (2019) have created an e-commerce chatbot using NLP and ML for easy product personalized recommendations to avoid the direct contact with sellers. It reduces the time and effort to buy or sell the products. Pradhan, A (2022) has created a chatbot for travelers using Dialog flow and cloud technology. It provides travel recommendations, reduce stress during travelling by providing travel trips.

Several chatbots (Gajbhiye, 2020) are created using AI technology and rule-based experts. From this analysis, Chatbot is necessary for various fields to assist peoples to aware about new things and enhance the usage of service.

3. Proposed Methodology

The banking AI chatbot is created using IBM Watson Assistant. The process of building and deployment of banking AI chatbot is shown in Fig. 45.1. Initially the login the "Watson Assistant" page and integrate the actions and response for the creation of chatbot. The following actions are created for integrating skills on customer service.

1. Greetings for start and end session
2. Creation of Index actions
3. Creation of bank account services
4. General queries, reminder and notification services

3.1 Greetings for Start and End Session

It determines a time to create a forceful first glance that will hopefully yield recurrent consumers. Initially the action skill name (Bot) is created and click on to create skills. By creating the end action is used to end the conversation, so that the user can reinitiate the session again.

3.2 Creation of Index Actions

The purpose of index is to build a aid procedure and further engage customers accompanying the new news sets.

3.3 Creation of Bank Account Services

The bank account services are categorized into current account section, saving account section, deposit section and loan account sections. The current account section helps to give details about customer account details like balance amount in savings amount and transaction history details. The saving account section helps to open an access plus account or a super value account. In deposit section, If the customer chooses time deposits for an account. Here the chatbot gives two options to choose from- Term (Fixed) Deposits or Recurring Deposits. Loan account section helps to give details about loan details taken by customer.

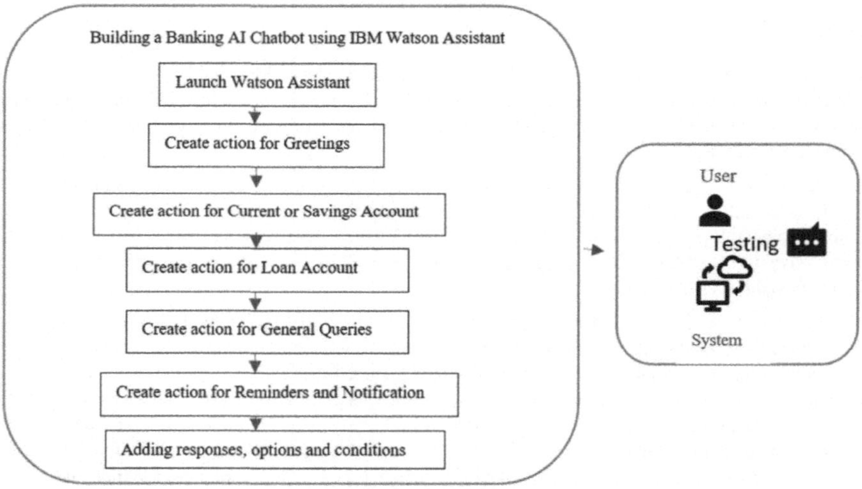

Fig. 45.1 Building and deployment of banking AI chatbot

3.4 General Queries, Reminder and Notification Services

It helps the customer general queries and please beneficial reminders, like for bill fee time limits or to lead particular documents for finance processes (e.g., loan requests). They can still please main notifications, like investment information and changes in credit scores. These push announcements may even form clients informed about latest trends belongings they would never previously have existed, accordingly increase clients' custom and understanding of banking duties. Finally, the responses for AI chatbots for each query, options and conditions are added and deployed in a website which demonstrate the working of customers.

4. Results

The AI based discourse for Banking Chat bots was developed by using an IBM Service called Watson Assistant. Initially the "Dallas" is selected as a region click on create. Upon successful login to the cloud, Click on Launch Watson Assistant Page. The skills of banking chatbot is a collection of actions and steps. Steps are the subset of actions where conversations are built. Banking Assistant is used to integrating skills. Totally 150 actions are created for start and end index session along with necessary services bank account services, general queries, reminder and notification services. The Fig. 45.2 shows the final set of actions created for banking chatbot.

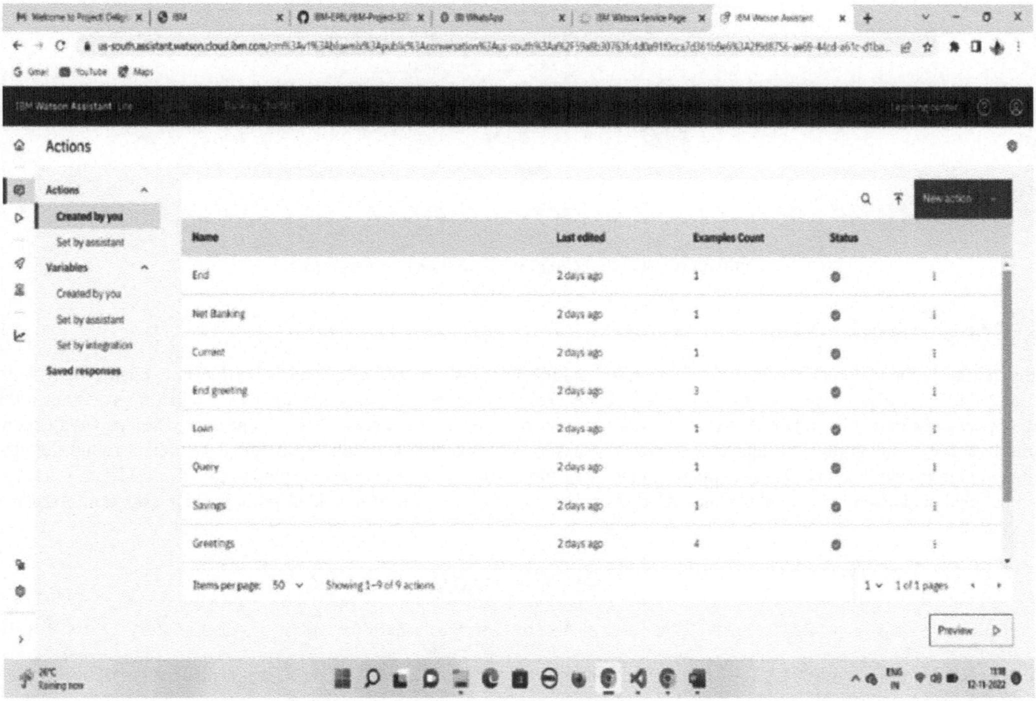

Fig. 45.2 Actions generated by Watson assistant

The website is developed for deployment. The Fig. 45.3 shows the bank webpage for chatbot deployment. Finally, the chatbot is deployed and tested by user queries. The Fig. 45.4 shows the chatbot questions and answers for customers.

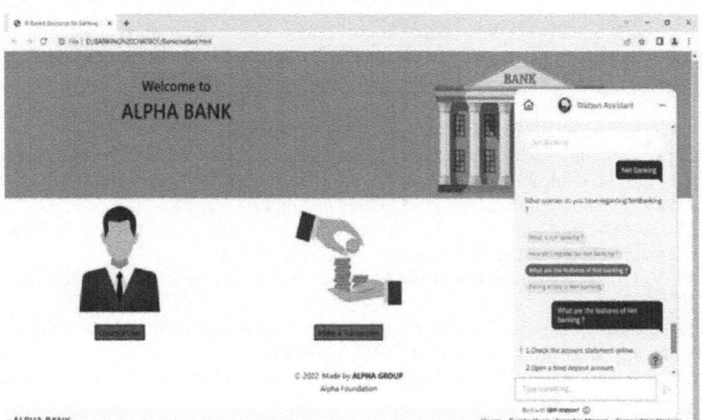

Fig. 45.3 Banking web page for Chatbot

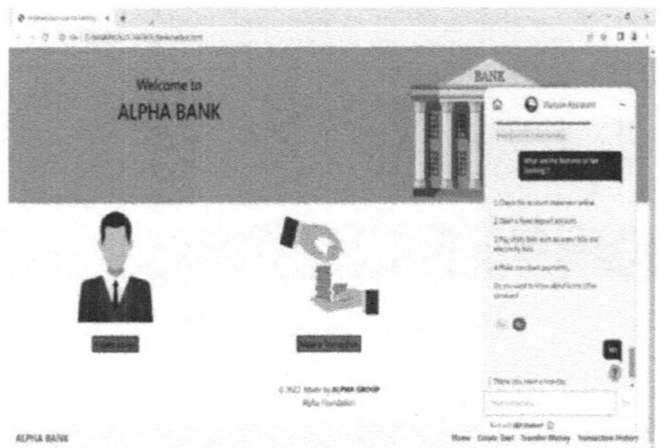

Fig. 45.4 Banking Chatbot for queries and responses

5. Conclusion

In this research, the chatbots was developed to assist banking services. The banking chatbot was designed using IBM Watson assistant Chatbots was adopted in the banking website used for answering customers questions and for providing a wide range of services which include bill payment, fund transfer, view recent transactions and much more. The designed Chatbots is intuitive due to integration of natural language processing and machine learning. By helping customers round the clock, they help banking staff focus on other important tasks. In future, this research work is extended to voice assistant for helping blind people and adding more services and will be evaluated by various kinds of peoples regarding their bank transactions and financial advice.

References

1. Hamidian, S. and Yunus, R. M. (2020). An AI-based chatbot to assist the diagnosis of Parkinson's disease. IEEE Access, 8: 15105–15117.
2. Narula, V. and Kohli, A. (2021). Design and Development of an AI-based Chatbot for Educational Institutes. Journal of Educational Technology & Society, 24(2): 296–308.
3. Skrebeca, J., Kalniete, P., Goldbergs, J., Pitkevica, L., Tihomirova, D. and Romanovs, A., (2021), October. Modern development trends of chatbots using artificial intelligence (ai). In 2021 62nd International Scientific Conference on Information Technology and Management Science of Riga Technical University (ITMS) (pp. 1–6). IEEE.
4. Wang, Y. and Luo, X. (2022). Development of an AI-based chatbot for psychological counseling. Journal of Medical Systems, 46(2): 1–12.
5. Suhel, S.F., Shukla, V.K., Vyas, S. and Mishra, V.P., (2020), June. Conversation to automation in banking through chatbot using artificial machine intelligence language. in 8th International Conference on Reliability, Infocom Technologies and Optimization (Trends and Future Directions) (ICRITO) 611–618, IEEE.
6. Yan, Z. and Huang, Y. (2020). An AI-based Chatbot for Supporting Elderly People in Daily Life. In Proceedings of the International Conference on Information and Communication Technology Convergence (ICTC) (pp. 322–326).
7. Matheson, E.L., Smith, H.G., Amaral, A.C.S., Meireles, J.F.F., Almeida, M.C., Mora, G., Leon, C., Gertner, G., Ferrario, N., Suarez Battan, L. and Linardon, J., (2021). Improving body image at scale among Brazilian adolescents: study protocol for the co-creation and randomised trial evaluation of a chatbot intervention. *BMC public health*, *21*(1), 1–14.
8. Singh, R. and Mishra, S. (2021). An AI-based chatbot for stock market analysis and trading. In Proceedings of the International Conference on Computational Intelligence and Data Science (ICCIDS) (pp. 1–6).
9. Todorov, D. and Toshev, V. (2019). Development of an AI-based chatbot for legal services. Journal of Information Science and Engineering, 35(6): 1339–1350.
10. Gajbhiye, S. and Sharma, A. (2020). AI-Based Chatbot for Automatic Diagnosis of Plant Diseases. In Proceedings of the International Conference on Smart Computing and Communication (ICSCC), 124–130.
11. Chen, Y. and Chen, W. (2021). Development of an AI-based chatbot for mental health counseling. Journal of Mental Health Counseling, 43(3), 213–224.
12. Zhi, Y. and Chen, Y. (2019). Design and Implementation of an AI-based Chatbot for English Learning. Journal of Language Teaching and Research, 10(5): 1059–1070.
13. Pradhan, A. and Gupta, N. (2022). Development of an AI-based chatbot for travel assistance. In Proceedings of the International Conference on Emerging Trends in Technology and Management (ICETTM), 25–30).

Note: All the figures in this chapter were made by the author.

Disruptive Technologies for Sustainable Development – Dr. G. Nagappan et al. (eds)
© 2024 Taylor & Francis Group, London, ISBN 978-1-032-55000-8

Wildfire Detection by Video Surveillance System Using Dark CNN

46

V. Umarani[1]
Associate Professor, Department of Computer Science Engineering,
Saveetha Engineering College, Chennai, Tamilnadu

Monica Manimaran[2], Pawan Kumar E.[3]
PG Scholar, Department of Computer Science Engineering,
Saveetha Engineering College, Chennai, Tamilnadu

A. Ramathilagam[4]
Professor, Department of Computer Science and Engineering,
P.S.R Engineering College, Sivakasi

Abstract: Wildfires are natural disasters that occur when a fire ignites and spreads through wild land vegetation, such as forests, grasslands, and shrub lands. Wildfire detection and prevention are critical components of efforts to reduce the impact of wildfires. Detecting wildfires early is key to preventing them from spreading and causing significant damage. There are several technologies and strategies that can be used for wildfire detection and prevention. In recent years, vision-based fire detection systems have become increasingly popular, as they can provide real-time data on the location and size of wildfires. To mitigate the impact of wildfires, the use of Dark Convolutional Neural Networks (CNN) has emerged as a promising technology in image processing and video surveillance. Our solution involves introducing a novel, computationally efficient architecture for Dark CNN networks that facilitates the identification, exact placement, and meaningful interpretation of the location of the fire.

Keywords: Dark convolutional neural network, Wildfire detection, Prevention of wild fire

1. Introduction

Forests are essential for supporting life on earth. However, when forest fires go undetected in their early stages, they can quickly spiral out of control, causing immense damage to all species. The damage caused by forest fires is irrecoverable and can account for up to 35% of carbon dioxide in the atmosphere. Each year, a significant number of living organisms and large areas of land are devastated, with the impact of forest fires on carbon monoxide emissions exceeding that of vehicular traffic.

However, in some cases, flames can ignite first, making early detection of both smoke and fire crucial to extinguishing the fire before it spreads. Various methods have been studied to detect smoke and fire early and minimize the damage caused by fires. Smoke sensors, flame sensors, and heat detection sensors that use infrared and ultraviolet beams are typically employed in early fire detection systems. However, these systems have the disadvantage of being challenging to notify the external world. To address the impediments of sensor-based detection systems, researchers have investigated several image-based methods of detecting smoke and fire using camera sensors. Video fire detectors have several advantages over sensor-based fire detectors, including quick response times, long-distance detection, and broad coverage areas.

[1]umaranibharathy@gmail.com, [2]monicamanimaran1601@gmail.com, [3]epawankumar12@gmail.com, [4]thilla2012@gmail.com

DOI: 10.1201/9781003428473-46

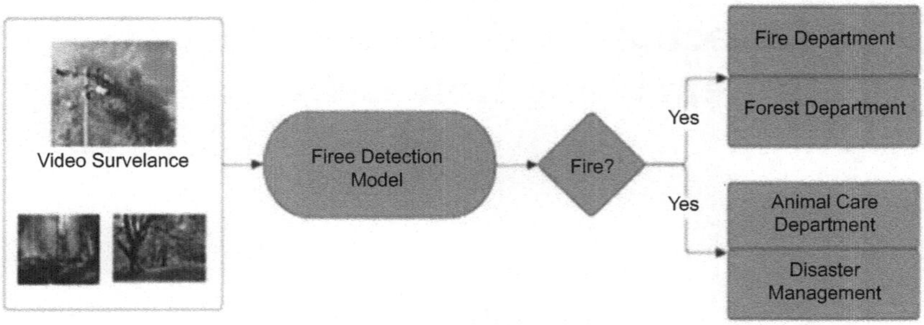

Fig. 46.1 Structure of proposed system

The objective of our paper is to introduce a fire detection system that utilizes the Super pixel inception algorithm and Dark-CNN to detect flames with higher accuracy, as depicted in Fig. 46.1. Our system aims to accurately identify wildfires without triggering false alarms that may occur with sensor-based detection methods. Once a fire is detected, our system promptly alerts both the Forest Department and the Disaster Management System.

2. Literature Review

Initially, wildfire monitoring systems relied on techniques such as satellite monitoring and manual patrolling. However, these methods have been revolutionized by various approaches that have improved their efficiency, accuracy, and other key objectives.

Manual forecast predictions were unreliable for detecting wildfires due to environmental factors and resulted in many false alarms. Satellites are not cost-effective for detecting fires. Sensor technology can detect physical and chemical parameters, but it has some drawbacks, such as high energy consumption and false alarms. Current optical sensor network systems also have limitations due to weather conditions and human activity. Today, two types of sensor systems are available for wildfire detection: camera observation and remote sensor networks. Image processing-based methods for detecting fires have several advantages, such as improved picture quality and lower costs. Several fire detection algorithms have been developed, including RCNN, which has shown to increase accuracy while reducing false alarms.

3. Methodology and Model Specifications

In the past decade, conventional methods for extracting features have been used to detect wildfires. However, false alarms often occur during surveillance due to shadows, light variations, and dark red-coloured objects, which can be mistaken for fire. To address these issues, we have proposed an architecture based on deep convolutional neural networks for early prediction of fires.

3.1 Model Specifications

Dark Convolutional Neural Networks (CNNs) are a type of neural network architecture that is more complex and deeper than traditional CNNs. They have demonstrated state-of-the-art performance on various computer vision tasks, such as image classification and object detection, by utilizing advanced techniques like skip connections, residual blocks, and dilated convolutions. The term "dark" refers to the fact that the inner workings of these networks are not easily interpretable. However, their high computational requirements and complexity can make them challenging to train and deploy in real-world scenarios.

The Dark-CNN framework uses deep convolutional neural networks to apply various filters of different sizes to input data, resulting in featured outcomes. These outcomes are then used as input for subsequent processes, including subsample operations and the tube-connected layer, which models top-level abstractions. During training, the neurons in the CNN and tube layers are learned and modified to improve the representation of the input data.

The Dark-CNN system is used to train both fire and normal datasets. The video input is segmented into frames and the proposed Superpixelization Algorithm is used to detect the severity of fire based on mean activation mapping. Although deep-CNN has a powerful ability to learn deep operations from the original data, it requires training various models with different properties to achieve the best solution for a specific problem.

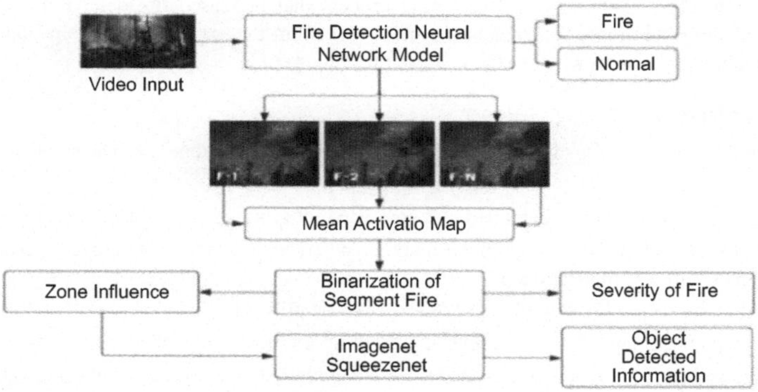

Fig. 46.2 Workflow of wildfire detection model

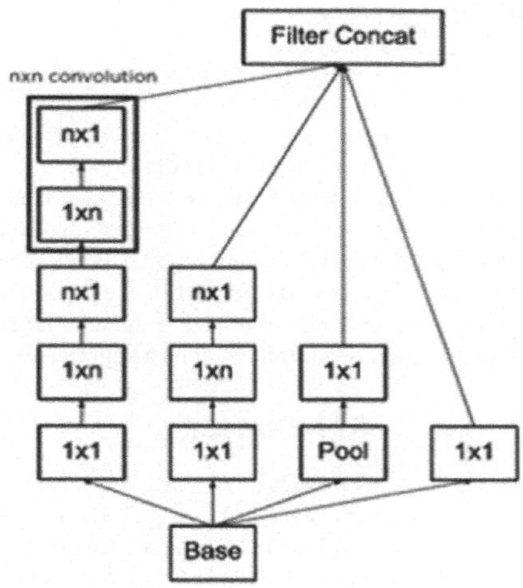

Fig. 46.3 Concatenation for N frames for Image convolution

To generate a featured outcome, the input data undergoes deep CNN convolution with various essential sizes. The featured outcomes are utilized as input for subsequent processes that involve using substitute sampling to extract activations from the neighborhood. These operations are essential in reducing the dimensionality of the featured operation and maintaining translation non-variation up to a certain degree. The tube connected layer is another critical layer in deep CNNs that models the highest-level abstractions from the input data.

Mean activation maps are a straightforward technique for identifying the image regions used by a CNN to recognize a particular class in the image. In essence, a mean activation map (MAM) enables us to visualize which regions of the image are pertinent to this class. This paper demonstrate that this approach facilitates reusing classifiers to obtain accurate localization results, even when training without bounding box coordinate data. This also highlights how deep learning networks possess a built-in attention mechanism of sorts.

Image binarization is the process of converting a grayscale image into a binary image by reducing its information to only two shades: black and white. This involves segmenting the image into its constituent objects, making it a type of segmentation.

We enhanced the efficiency and accuracy of fire detection by training several models with varying properties, taking into consideration their respective training datasets and the problem's nature. We also implemented a natural sharing strategy to leverage past learned knowledge, resulting in a fine-tuning process that increased fire detection accuracy from 84.53% to 90.36%. After several training iterations, We created a precise algorithm that can detect fires in both constrained and open environments, with the output indicating whether there is fire present or not.

3.2 Algorithm Super Pixelization

The algorithm is designed to detect fire in a given video file and output fire detection data. The steps involved in this algorithm are as follows:

1. Firstly, the frames of the video file are extracted and converted into individual frames, t1, t2, t3, ..., tn.
2. Next, each frame is transformed into a super pixel image, ti = {p1, p2, ..., pni}, where every pixel Pk is assigned a value between 1 and N, as illustrated in Figure 1.3.
3. A set of non-overlapping super pixels k, S = {S1, S2, ..., Sk}, is utilized.
4. Each pixel is then converted into a 32x32x3 input, representing a visual image of 32 pixels and 3 color channels (RGB).
5. The Dark Convolution layers are employed, along with a few gathered maximum subsampling values, through a series of Super pixelization models.
6. Finally, the neural network's output layer is a log-odds layer, which integrates an Aux-Logits component into the loss function immediately following the mixed output.

4. Conclusion

The outbreak of a fire that starts from a small spark can cause devastating consequences, posing a threat to both human and wildlife. To address this issue, we have developed an innovative wildfire detection system utilizing a new method that effectively eliminates false positive detections. Our approach involves the use of a Super pixelization Video Surveillance method coupled with a Dark Convolutional Neural Network (CNN) to intelligently retrieve data from surveillance cameras. We addressed the issue of false alarms in traditional deep information retrieval systems, caused by relying on the object's shape, by incorporating activity recognition and Mean Activation Mapping across the frames of the surveillance video. Our method can precisely determine the severity of the fire with high precision, while minimizing the computational input required for the dark CNN network. Compared to existing systems, our proposed algorithm exhibits a significantly lower probability of false alarms.

References

1. Aljarrah and D. Mohammad, "Video content analysis using convolutional neural networks," *2018 9th International Conference on Information and Communication Systems (ICICS)*, Irbid, Jordan, 2018, pp. 122–126, doi: 10.1109/IACS.2018.8355453.
2. S. Frizzi, R. Kaabi, M. Bouchouicha, J. -M. Ginoux, E. Moreau and F. Fnaiech, "Convolutional neural network for video fire and smoke detection," *IECON 2016 - 42nd Annual Conference of the IEEE Industrial Electronics Society*, Florence, Italy, 2016, pp. 877–882, doi: 10.1109/IECON.2016.7793196.
3. Muhammad, Khan & Ahmad, Jamil & Mehmood, Irfan & Rho, Seungmin & Baik, Sung. (2018). Convolutional Neural Networks Based Fire Detection in Surveillance Videos. IEEE Access. PP. 1–1. 10.1109/ACCESS.2018.2812835.
4. P M, Ashok Kumar & HarikaPalivela, Lakshmi & Krishna, V.V.. (2022). Fire Detection Using Deep Convolution Neural Network in Video Streams. 10.3233/APC220055.
5. Aljarrah, Inad & Mohammad, Duaa. (2018). Video content analysis using convolutional neural networks. 122–126. 10.1109/IACS.2018.8355453.
6. An Analysis Of Convolutional Neural Networks For Image Classification- Neha Sharma, Vibhor Jain, Anju Mishra Amity University Uttar Pradesh, Noida, India Available online 8 June 2018, Version of Record 8 June 2018
7. VIDEO ANALYSIS BY DEEP LEARNING - Mona Ramadan BS in Electronic Engineering, Sebha University, 2005 MS in Electrical Engineering, University of Pittsburgh, 2010

Note: All the figures in this chapter were made by the author.

Disruptive Technologies for Sustainable Development – Dr. G. Nagappan et al. (eds)
© 2024 Taylor & Francis Group, London, ISBN 978-1-032-55000-8

Malicious URL Detection Using CNN-LSTM Algorithm in Machine Learning

47

K. S. Rekha[1]
Assistant Professor, Department of Computer Science and Engineering,
Saveetha Engineering College, Chennai

Pungunuru Yuvaraj[2]
Department of Computer Science and Engineering,
Saveetha Engineering College, Chennai

Abstract: Hackers can embed malicious code (such as Trojans, Worms, etc.) into websites to collect customer information and utilize the money to make unauthorized transactions. Considerably more than one-third of all websites are known to be potentially dangerous. It makes sense to quickly identify fraudulent URLs online as a result. Malicious URLs are detrimental to computer users in all ways. It's crucial to identify dangerous URLs. At the moment, machine learning classification methods and black-list and white-list methodologies are employed to detect harmful websites. The black-list and white-list technologies are useless if a specific URL is not included. A CNN-LSTM model-based algorithm is created in the suggested system to solve this issue.

Keywords: CNN, CNN - LSTM

1. Introduction

Using a neural network and synthetic minority over-sampling, a potent model for identifying spam URLs is provided in this study. Using four metrics—accuracy, recall, precision, and G-mean—the model is applied to two benchmark email corpora and assessed against four other established classifiers. The discovered features are categorized using the fully connected neural network. To jointly improve the detection model, we employ a gradient descent approach.

Therefore, the project's primary contributions are: Data Analysis, Data Pre-processing, Training the Model, Testing of Dataset. An artificial neural network that mimics the brain's neuronal network is known as deep learning. Deep neural networks are a component of machine learning that is referred to as "deep learning."

Connected layers are used to develop deep learning systems. All the layers between the first and last are considered Hidden Layers. The top layer is referred to as the Input Layer. Deep networks are those that connect neurons over or more two layers.

2. Literature Review

In the data space, an oversampling strategy is suggested. To gather freshly found phishy and benign URLs, they crawled a crowd-sourced URL repository. Our tests show that the suggested oversampling strategy significantly improves performance (Ankesh Anand et al. 2018). For the purpose of intelligently identifying dangerous URLs, this paper developed a machine

[1]rekhaks@saveetha.ac.in, [2]yuvarajpungunuru600@gmail.com

DOI: 10.1201/9781003428473-47

learning framework. 41 harmful URL attributes are obtained from hostname, Amazon, and obscurity data sources. The 17 most crucial traits are determined using the XGBoost (eXtreme Gradient Boosting) method and the ANOVA (Analysis of Variance) test. The XGBoost classifier, which detects with greater than 99% efficiency, is learned using the dataset (Yu-Chen Chen et al.2020). They developed a programme that would be installed as a Chrome extension to address the issue of malicious URLs preying on people (Shivangi S et al. 2018). Many sample studies demonstrate that Random Forest's detection performance has greatly increased as a result of feature contribution and hyperparameter adjustment. The AUC value is 96.51%, while the detection precision reaches 94.85%. It has a significant application value for detecting rogue URLs (Shen He et al. 2021). By utilising URL-based characteristics in a multiclass classification context, they helped with the identification of dangerous URLs in this work. Phishing, spam, and malware were the three prevalent forms of URL attacks that they concentrated on (Tariro Manyumwa et al. 2020). In this study, they discuss the binary classification problem of detecting malicious URLs and assess the performance of many popular machine learning classifiers (Shantanu et al. 2021)

3. Methodology and Model Specifications

A suggested method next sends the dataset matrixes to an embedding layer with a batch size of 128. Each character is translated onto a real-valued vector of 128 lengths by an embedding layer. This can be thought of as one hyperparameter; we selected 128 to give deep learning architectures more flexibility. During backpropagation, this collaborates with other deep network layers. As a result, similar characters can cluster together in URLs more easily. Other layers can more easily identify domain names' semantics and contextual similarity structures thanks to this type of character clustering.

CNN and a hybrid network, like CNN-LSTM, are used in the proposed system to represent features. CNN aids in extracting spatial information between the characters, while recurrent layer LSTM extracts sequential information. The layers of deep learning models output a continuous number that indicates how well the pattern matched, and they encapsulate the ideal feature representation.

CNN is used to discover the spatial correlations between the characters. It comprises two layers; the first layer uses repeated convolution 1D operations to extract local features, which are then sent to the max pooling 1D layer. By gliding through a succession of character embedding vectors using 128 filters with 5-filter lengths, At each step, the 1D layer generates a continuous value. This properly displays how effectively the sub-sequence and pattern matching of the characters encoding vectors worked. This pool has a 4-foot length. The maxpooling1D function result is subsequently vectorized and transmitted to the network's repeated layer.

3.1 Data Pre-processing Module

Every machine learning approach that involves data preparation modifies or encrypts the data to make it simpler for the computer to parse. In other words, the algorithm is now able to swiftly recognize the characteristics of the input.A dataset may be considered of as a set of information items, sometimes referred to as recordings, pointers, arrays, trends, incidences, cases, trials, views, or groups.

The first pre-processing method is to remove @, which involves scanning the whole input dataset document and comparing it with @ before deleting @ from any comments that include @. The following phase in the pre-processing procedure is called remove URL, where the whole input document is read and checked with HTTP: and any comments that include a URL are removed. The next phase in data pre-processing is stopped word removal, so we go on. Stop words are precisely those words that are removed from the whole statement after scanning, leaving only the noun and adjective. Then, tokenization and normalization are performed. The Porter-Stemmer Algorithm is then used. The Porter stemming algorithm, sometimes known as the "Porter stemmer," is a method for eliminating the more typical morphological and inflexional ends from English words. Its primary use is in the term normalization stage of the setup of information retrieval systems.

3.2 Analysis of Email Module

Emails are only accessible for a brief period, usually seven days, after being posted. Dynamic elements, which are easily accessible, provide a better possibility to differentiate between spam and legitimate communications. Features of pairwise engagement are divided into Engage-with Features (ETF) are features that users may control or decide how to change that define user behaviours on Email. Friends count, statuses count, email kind, email production time, frequency, and other features are included in this group. Features in the Engaged-by Features (Eb F) group resemble ETF characteristics. These features are count, favourites count, amount of emails forwarded, etc.

3.3 Feature Selection Module

Features that fall within a certain category have values that are determined by that category. Another illustration is the Boolean set: True or False

Numerical: Components with continuous or integer-valued values. They have many of the characteristics of numbers and are represented by numbers. The number of steps you take each day or the pace at which you drive your automobile are two examples.

The email platform makes it easier for different people to engage and connect globally. Our feature extraction is based on the pertinent properties that allow people to connect. Features can also be categorized as derived features or fundamental features. Moreover, characteristics can be characterised as either passive or active. Time of sign up and access management are static attributes that cannot be modified once an account is established, however, statuses count and other dynamic features are constantly changing according to the user's degree of email involvement.

3.4 Spam Prediction Module

Beyond image recognition, convolutional neural networks are used in many other contexts. Neural networks, for instance, can anticipate time series and parse spoken language (NLP). A matrix serves as a neural network's input. Each picture's pixels are coded as numerical values that reflect the intensity of color for each pixel in image recognition.

We'll train a Word Neural Network and concentrate on the CNNs' use in NLP. Rows denoting words in a phrase and columns denoting word embeddings of n dimensions make up the input matrix of a word neural network.

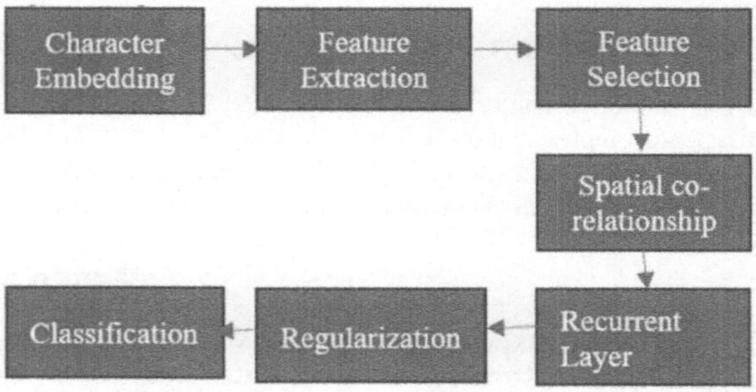

Fig. 47.1 Architecture diagram

Just using very few lines of programming, Keras makes it simple to construct a Word Neural Network. We create embeddings for this model using the Keras "embedding" layer within our corpus. As you can see, the convolutional layer requires a matrix as an input, which is the embedding layer's output.

4. Result and Analysis

Below is the result as per above trained model

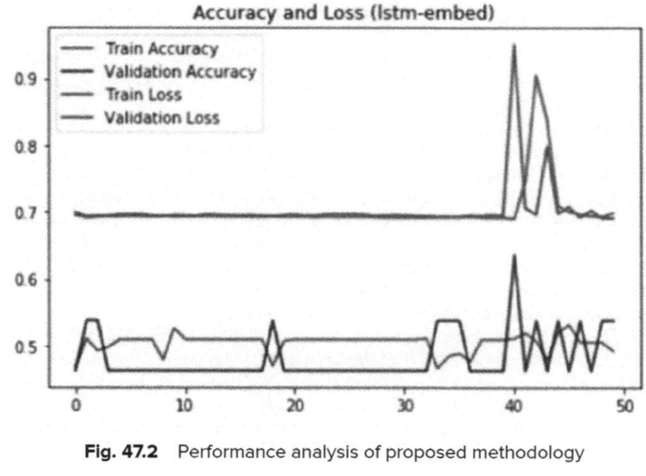

Fig. 47.2 Performance analysis of proposed methodology

```
In [13]: url_data.head()
Out[13]:
```

	text	label
0	http://facebook.com	0
1	http://youtube.com	0
2	http://yahoo.com	0
3	http://baidu.com	0
4	http://wikipedia.org	0

Fig. 47.3 Five rows of dataset and its classification

5. Conclusion

Deep learning techniques like CNN and CNNLSTM are preferred to machine learning techniques because they can produce the best feature representation on their own using only the URLs as input. Based on the findings, we can state that detection systems created using regular expression and blacklisting techniques cannot be defeated by machine learning and deep learning-based harmful URL detection. We randomly split the data set that we obtained from various sources to train the model.

References

1. Ankesh Anand, Kshitij Gorde, Joel Ruben Antony Moniz, Noseong Park, Tanmoy Chakraborty, Bei-Tseng Chu. (2018) .Phishing URL Detection with Oversampling based on Text Generative Adversarial Networks. 2018 IEEE International Conference on Big Data (Big Data). Seattle, WA, USA.
2. Akshay Sushena Manjeri, Kaushik R, Ajay MNV and Priyanka C. Nair. (2019). A Machine Learning Approach for Detecting Malicious Websites using URL Features. 2019 3rd International conference on Electronics, Communication and Aerospace Technology (ICECA). Coimbatore, India.
3. Fatimah Alkhudair, Mada Alassaf, Rehan Ullah Khan and Shouq Alfarraj. (2020) Detecting Malicious URL. 2020 International Conference on Computing and Information Technology (ICCIT-1441). Tabuk, Saudi Arabia.
4. Fatimah ALKHUDHAYR, Shouq ALFARRAJ, Buthina ALJAMEEL and Salim ELKHDIRI. (2019). Information Security:A review of information security issues and techniques. 2019 2nd International Conference on Computer Applications & Information Security (ICCAIS). Riyadh, Saudi Arabia.
5. Guolin Tan, Peng Zhang, Qingyun Liu, Xinran Liu, Chunge Zhu and Li Guo. (2018) .MalFilter: A Lightweight Real-time Malicious URL Filtering System in Large-scale Networks. 2018 IEEE Intl Conf on Parallel & Distributed Processing with Applications. Melbourne, VIC, Australia.
6. Hafiz Mohammd Junaid Khan1, Quamar Niyaz1,Vijay K Devabhaktuni1, Site Guo and Umair Shaikh. (2019). Identifying Generic Features for Malicious URL Detection System. 2019 IEEE 10th Annual Ubiquitous Computing, Electronics & Mobile Communication Conference (UEMCON). New York, NY, USA.
7. Muyang Du1, Yanni Han1 and Li Zhao. (2018). A Heuristic Approach for Website Classification with Mixed Feature Extractors. 2018 IEEE 24th International Conference on Parallel and Distributed Systems (ICPADS). Singapore.
8. Shantanu, Janet B and Joshua Arul Kumar R. (2021). Malicious URL Detection: A Comparative Study. 2018 9th International Conference on Computing, Communication and Networking Technologies (ICCCNT). Bengaluru, India.
9. Shen He, Jun Xin, Huaxi Peng and Erpeng Zhang. (2021). Research on Malicious URL Detection Based on Feature Contribution Tendency. 2021 IEEE 6th International Conference on Cloud Computing and Big Data Analytics (ICCCBDA). Chengdu, China.
10. S. Shivangi, Pratyush Debnath, K. Sajeevan and D. Annapurna. Chrome Extension For Malicious URLs detection in Social Media Applications Using Artificial Neural Networks And Long Short Term Memory Networks. 2018 International Conference on Advances in Computing, Communications and Informatics (ICACCI). Bangalore, India.
11. Tariro Manyumwa, Phillip Francis Chapita, Hanlu Wu and Shouling Ji. (2020) .Towards Fighting Cybercrime: Malicious URL Attack Type Detection using Multiclass Classification. 2020 IEEE International Conference on Big Data (Big Data).
12. Yu-Chen Chen, Yi-Wei Ma and Jiann-Liang Chen. (2020). Intelligent Malicious URL Detection with Feature Analysis. 2020 IEEE Symposium on Computers and Communications (ISCC). Rennes, France.
13. Wenchuan Yang, Wen Zuo and Baojiang Cui. (2019). Detecting Malicious URLs via a Keyword-Based Convolutional Gated-Recurrent-Unit Neural Network. IEEE Access (2169-3536).

Note: All the figures in this chapter were made by the author.

Disruptive Technologies for Sustainable Development – Dr. G. Nagappan et al. (eds)
© 2024 Taylor & Francis Group, London, ISBN 978-1-032-55000-8

Business Intelligence using Data Management, Data Analytic and Data Reporting

48

M. Vijayanand[1]
Professor, Department of Computer Science and Engineering,
Saveetha Engineering College, Chennai, Tamil Nadu

Deep Shankar K.[2]
Department of Computer Science and Engineering,
Saveetha Engineering College, Chennai, Tamil Nadu

Abstract: Data has substantially increased in recent years. One of the useful tools for assessing an organization's success or failure has been business intelligence (BI). The technology indicated as being suited for this use in business intelligence and analytics (BIA). For many firms, obtaining user requirements continues to be a challenge. To address this need, we established the data analytics method in business intelligence. This solution makes it easier to organize, manage, and communicate data throughout your whole business, empowering decision-makers to spot and seize strategic opportunities. It increases data availability and enables self-service analytics to give business users more flexibility and proactivity. Connectors assemble real-time data from multiple sources and automate reporting in a matter of minutes. New technologies and promising efforts benefit from high user adoption when user interactions are simple and appealing.

Keywords: BI, Data Reporting, Data Management

1. Introduction

Imagine a world without data storage, where every piece of information pertaining to the business, such as significant transactions and all other dependable features, is crucial and must be recorded, but is then permanently deleted after use. It would make it impossible for the firms to obtain relevant information and data in order to conduct in-depth studies and offer new prospects and advantages. Currently, information on client names and addresses, product availability, purchases made, personnel hired, and many other things are necessary for daily continuity. The ability of an organization to use data efficiently is essential to its success. Now think about the variety of species and the explosion of data and information that is now made available because of the technology age that we are currently living in. Now that massive amounts of data are easily available, storage technology and data collection methods have improved. Every second, more data is created, and in order to extract values, it needs to be analyzed and stored. Because it is now less expensive to store data, an organization must optimize the value of the massive volumes of data that are already present.

1.1 Visualizing Data for Easier Consumption

This application helps us to get a better idea by organizing the data into an understandable format and showing the trends and outliners, data visualization aids in the telling of stories. A strong visualization highlights important information while reducing data noises. Overall, by presenting the data in form of charts. Once all the members of the organization that have access to the

[1]Vijayananad@saveetha.ac.in, [2]deepshankar2@gmail.com

DOI: 10.1201/9781003428473-48

application are registered, it gives custom data analysis results to each and every one of them. For e.g., if the CEO access the data, he might want to know about the overall performance of all the departments and if the manager of the sales department accesses it, he might want to know about the performance of just the sales department alone. This information is presented to them in form of graphs or charts in which they like to view the data.

1.2 Getting Alerts when Business Needs Attention and Combining Data Sources

This application allows you to set an alert whenever something significant happens, via an email or a text. Sending personalized push notifications can establish your authority and build your reputation

2. Literature Survey

With many definitions, there is a second problem: throughout time, as people's perspectives on the world change, these meanings tend to alter as well. BI, for example, is involved in this circumstance. At first, private insight rather than public or shared knowledge was understood as business intelligence (BI) when it came to software companies. (Solberg Silen, 2015) Engineers and programmers continue to use BI even after many years have passed. BI is defined as frameworks that collect, transform, and present organized information from a variety of sources, reducing the time needed to obtain important business data and enabling their efficient use in management decision-making (Den Hamer, 2004), which enables the dynamic lookup, recovery, analysis, and explanation of corporate information for the purposes of administrative decisions (Nofal and Yusof, 2013). According to Tyson (1986), BI focuses on obtaining, processing, and presenting information on clients, competitors, industry sectors, technology, and goods. According to Pirttimäki (2007), BI is a process that entails a number of operations that are motivated by the specific data requirements of decision-makers and the desire to gain a competitive. The framework of business intelligence (BI) improves a company's fundamental decision-making process by turning information into data and then into learning (Singh and Samalia, 2014). BI is a tool and solution that aids decision-makers in understanding the financial situation of the company (Nofal et al. 2013). BI is described as a group of numerical and methodological methods for analysis used to extract data and useful information from unprocessed data for using unclear basic leadership preparation (Vercellis, 2013)." We may improve the knowledge provided by BI applications, especially by using information mining techniques, simulating and modeling the real world using a "systems thinking" approach, improving projections, and contributing to a better understanding of the business development of any firm (Raisinghani, 2004).

3. Implementation

3.1 Access Module

Increase self-service analytics and data access to provide business users more freedom to be more proactive and flexible. The executive of a company can keep track of and add extra users to the login screen thanks to the access module. Only the domains that he believes they need access to may be given to them, according to him. Due to the provider's control over who has access to the data, the company's data may be managed and kept secure in this way. On this basis, the CEO can guarantee both the safety of his company and the competence of the hands who are handling it.

3.2 Connection Module

The data source connection information for databases, the cloud, third-party APIs, and files (CSV and Excel) can be found in the connector module, which is used to obtain dataset specifics during integration. This module allows us to establish a connection to a certain database, such as MySQL, Oracle, or Mango DB, through connectors. Information is then verified against provided credentials, and there is a listing page where we can view the list of saved connections with the ability to delete them. Expand data access and self-service analytics to provide business users greater initiative and flexibility. We can connect to any kind of database connection and import the CSV file and Excel to import the data which is required by the provider or the person who has access.

3.3 Integration Module

In this integration module, we combine data from many data sources, including databases, the cloud, and files. Data can be merged from multiple sources, but there must be some kind of relationship between them in order for this to be possible. The combined data is referred to as a data set, and the resultant data is previewed before the dataset is stored. This combines real-time data from anywhere and automated reporting within minutes. This integration step plays a massive role in enhancing the data reporting feature because data cannot be reported until they all are available as a whole. This combines data from any number of files and creates a unified single file that can be used for further processing. A module called integrate is used to combine data from many data sources, including databases, the cloud, and files. Data can be merged from multiple sources, but there must be some kind of relationship between them in order for this to happen.

3.4 Publish Module

When combining data from many sources, such as files, databases, and the cloud, a module known as integration is employed. Several sources of data can be combined, but this requires that the sources have some sort of connection to one another. A data set is what is meant by the combined data, and the final data is shown before the dataset is saved. Simple and beautiful user interactions help new technologies and promising initiatives gain high user adoption. This publishes module literally publishes the data in a user-viewable and easily understandable format so as to make at-the-instant decisions on the data without going through the hassle of reviewing all the older documents and deciding from them.

4. Methodology

This paper presents a simple way of selecting the required fields based on 'the condition matching' strategy. While its peers such as Domo, SAS and so on used algorithm-based filtration to bring up the resultant set since it has to bring up suggestions to the user, but in our case, as there is no decision-making required to make, there is no need for any complicated algorithms. This not only increases the overall speed of the data retrieved and presented but also minimizes the complexity involved in doing so.

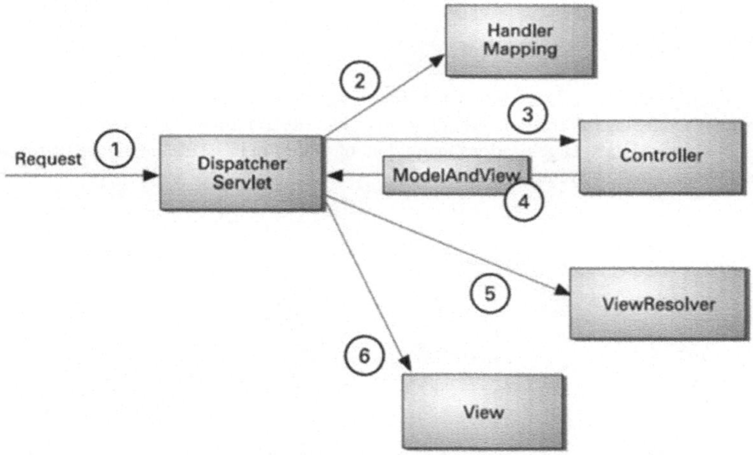

Fig. 48.1 Basic flow diagram

5. Experimental Setup and Result Analysis

This web app is developed on the Spring framework Spring is a container for inversion of control and an application framework for the Java platform. Any Java application can utilize the framework's fundamental components. This is accompanied by a tool called Maven. The Maven is a build automation tool, and it is primarily used for Java projects, this tool allows us to build and document the lifecycle framework. Maven's goal is to provide us with a thorough, manageable, reusable, and straightforward paradigm. A group of plug-ins and tools that can communicate with the declarative model. The Maven, once installed, is to be included in the spring to run or execute it. The maven creates a POM file in which any customization can be made. The repository link can be copied from the Maven central repository, and they are used inside the application to add the required functionalities.

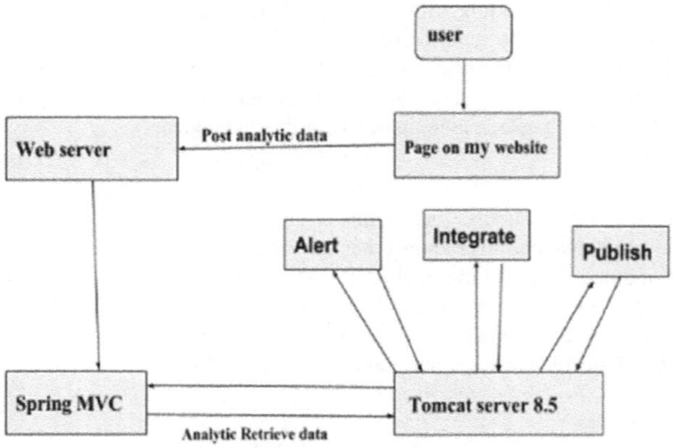

Fig. 48.2 Spring MVC flow diagram

Finally, the following are some observations on the most recent developments in business intelligence (BI). In terms of BI development, there are two key trends: BI in Business-to-Customer (B2C) apps and BI in enterprise management. We may infer that putting business intelligence into practice will have three key effects.

- Lower Costs
- Increased Revenue
- Satisfied Customers

The phrase "business intelligence" can end up being a fad. To manage today's global enterprises, it is crucial to understand the fundamental ideas behind leveraging information technology to provide decision-makers with useful information. BI makes use of semi-structured and structured data. Organizations invest a lot of money in business intelligence projects, but with the right management, they can yield significant returns.

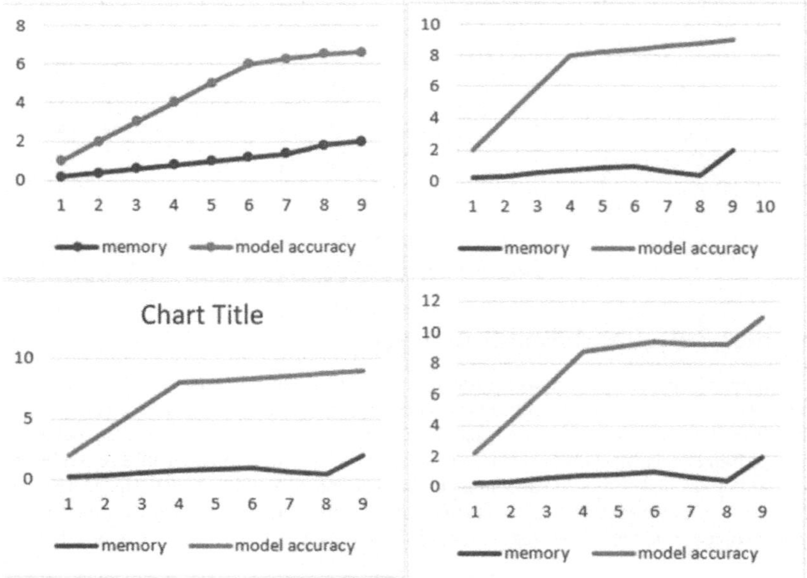

Fig. 48.3 Model validation

6. Conclusion

The use of business intelligence and analytics in organizational performance management is popular nowadays. Yet, obtaining user requirements for implementation continues to be a challenge for many organizations. The empirical study that was conducted to make it easier to manage an enterprise's total business activities is provided in this essay. Interviewing business intelligence professionals across a range of industries is part of the strategy. The suggested model can be used as a manual for integrating different company activities, processing data, and displaying it to users.

References

1. Dedić N. & Stanier C. (2016). Measuring the Success of Changes to Existing Business Intelligence Solutions to Improve Business Intelligence Reporting. Lecture Notes in Business Information Processing. 268. Springer International Publishing. 225–236.
2. Jump up (Rud, Olivia (2009). Business Intelligence Success Factors: Tools for Aligning Your Business in the Global Economy. Hoboken, N.J: Wiley & Sons. 39240-9.
3. N. H. Z. Abai, J. H. Yahaya, and A. Deraman (2014) "Requirements engineering model for the integration of performance management system with business intelligence and analytic," Int. J. Adv. Comput. Sci. Its Appl., 4(3), 8–12.
4. N.H. Z. Abai, J. Yahaya, and A. Deraman (2013),"User requirement analysis in data warehouse design: A review," in Proc. International Conference on Electrical Engineering and Informatics, 8, 769–774.
5. A. Chakraborty, M.K. Baowaly, A. Arefin, and A. N. Bahar, (2012) "The role of requirement engineering in software development life cycle," J. Emerg. Trends Comput. Inf. Sci., 3(5), 723–729.
6. L. Van Dyk,(2008) "A Data Warehouse Model for MicroLevel Decision Making in Higher Education," Electron. J. e-Learning, 6(3), 235–244.

Note: All the figures in this chapter were made by the author.

Disruptive Technologies for Sustainable Development – Dr. G. Nagappan et al. (eds)
© 2024 Taylor & Francis Group, London, ISBN 978-1-032-55000-8

Temperature Monitoring and Control System Using LabVIEW

49

M. Arivalagan[1]
Associate Professor, Department of Electronics and Instrumentation Engineering,
Saveetha Engineering college, Saveetha Nagar, Thandalam, Chennai

Abishek G.[2] **, Hemanth S.**[3]
Department of Electronics and Instrumentation Engineering,
Saveetha Engineering College, Saveetha Nagar, Thandalam, Chennai

Kishore K.[4]
Department of Electronics and Instrumentation Engineering,
Saveetha Engineering College, Saveetha Nagar, Thandalam, Chennai

Abstract: This Paper represents a Temperature Monitoring Control System Using LabVIEW. The primary goal of this system is to design and implement a LabVIEW-based remote monitoring system for temperature monitoring and Control. In this Modern World it will be very helpful to monitor and control the Temperature in some Places such as Industries and Hospitals etc.

Keywords: PID controller, LabVIEW, Temperature control, Heating tank, PID algorithm

1. Introduction

In this Modern World, there are so many places where we must have to control or monitor the temperature such as in Hospitals, Educational Institute, Power Plants and Industries with this project we can control and monitor such things. We have designed a system that is called a Remote Monitoring System Using LabVIEW with the help of PID Controller, and Heating Tank. This Project is achieved by using a Heating Tank and that can be accessed by the PID Controller through switching the Heater On and Off. The PID controller gives a Steady State Error in the Response outputs given by it in this Process of Temperature Monitoring System. It plays an Important Role in the Heat generation industry. This paper explains how the PID algorithm can be applied to the temperature process in real time. This PID Implementation was Tested and simulated by Using LabVIEW. High accuracy and Good performance has been Produced in the Control Technique. This paper illustrates a PID-based controller for an electromagnetic oven, which can be used to keep the oven at a constant temperature. Various tuning methods have been used and the controller parameters are adjusted and tuned in the Process. All of the answers from the Tuning Methods are analyzed, and then the device is created based on that data.

2. Controllers

PID Controllers

1. Proportional Control

[1]arivalagan@saveetha.ac.in, [2]Abishekgovind2001@gmail.com, [3]Hemanthsaravanan672@gmail.com, [4]Kishoreyuvan01@gmail.com

DOI: 10.1201/9781003428473-49

2. Integral Control
3. Derivative Control

Each Mode will have a different Reaction to the Error Obtained in the Process and there are totally three Modes. Adjusting the amount of reaction generated by the Control Mode is done by adjusting the Controller's setting Settings. The primary tool in the Process is a monitor known as a PID. In order to regulate variables in a process, such as temperature, flow, pressure, and speed, a PID controller is typically employed. In PID Controller, "P"- is Referred as Proportional, "I"- is Integral, "D"- Derivative and These three Controllers combined together and it uses a Control Loop Feedback Mechanism to control the Process various. This Controller is Very Accurate and Stable in the Working Process.

The Input and Output Relationship is Formulated for PID Controller:-

$$U(t) = K_p e(t) + K_I \int e(t)dt + K_D de(t)/dt \tag{1}$$

The Transfer Function in the Laplace Domain,

$$C(S) = K_P + K_I/s + K_D s \tag{2}$$

Algorithm sampling of incoming data and discretization of integral and derivative action are necessary for PID controller implementation in LabVIEW.

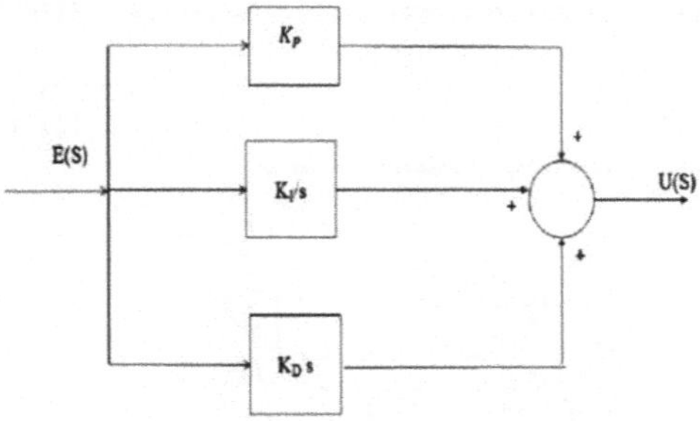

Fig. 49.1 Block diagram for PID

3. LabVIEW

LabVIEW is defined as Laboratory Virtual Instrument Engineering Workbench. A systems engineering tool, it facilitates the testing, measuring, and controlling of software while providing instantaneous access to hardware and data insights. The Block Diagram acquires some data that is displayed by indicators simulate output mechanisms.

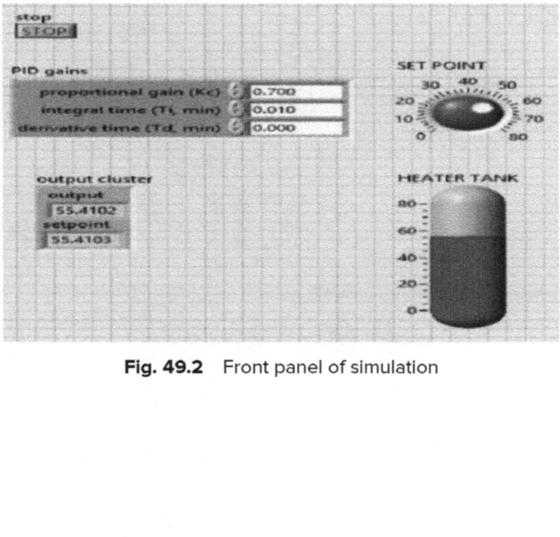

Fig. 49.2 Front panel of simulation

During a manufacturing process, the fix position is adjusted using a front-panel button. A System's Set point is the desired or predetermined value of an input or output quantity. The gauge will show you how hot or cold the water is inside the reservoir. On the Process map, you can see the P, I, and D numbers from the PID gain block. You can cancel the temperature programme by pressing the Stop button, and the output Cluster will show you the current process variable and the value you entered for the set point Process.

4. Block Diagram

The Graphical of the Functions to control Front Panel Objects in the Process is defined as the Block Diagram. The icon terminals Appear on the block are the Front panel objects. To express is, the Wires are connecting the Controls and Indicators and this is known as data Flow Programming. The starting point and the gain values are input into a PID controller. The Set point will be displayed with the help of Bundle block. The Actual Value and the Set point value is compared using the Feedback Node. The programme can be terminated by pressing the Stop Button, a Boolean input. For 100 ms, the timer synchronize the process and computer speed and the values will be entered into the process. The actual requirement of the input is the Set point. The Power Allocation, Frequency and Memory usage is reduced using the Timer. By means of the Feedback, the Outcome is fed back into the Primary Process Variable. In VI, the lower temperature is considered as 0°C and 80°C is considered as the Maximum temperature. The Output Response is represented by the Waveform Chart.

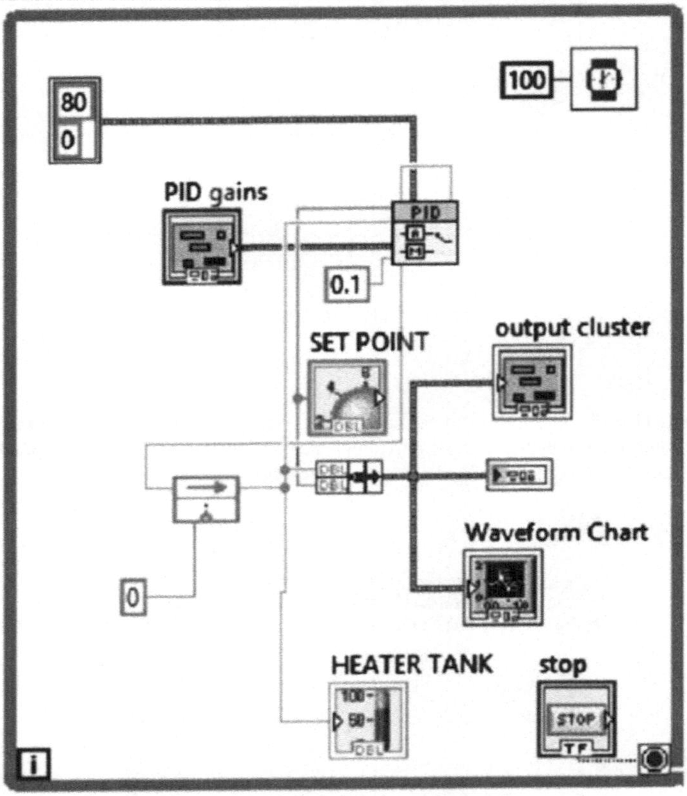

Fig. 49.3 Block diagram

5. Controller Tuning

The controller is a piece of equipment that controls the process or Operation of the Industrial Applications. The controller's parameters can be adjusted to determine P, I, and D values. The Process Reaction Curve in the empirical Tuning is the most popular method used in the Process. Values of P, I, and D can be calculated using a number of tuning techniques, some of which are the Cohen and Coon, Ziegler, and Nicholas Methods.

6. Schematic Diagram

This is a diagrammatic depiction of the mechanism used to regulate the temperature in a building. The drain is connected to a compressor that transports the water in it to a receptacle on the upper floor. The electrical compressor is powered by an alternating current (AC) supply of 230 volts at 50 hertz. A rotameter measures the liquid's movement into the receptacle in litres per hour. Using data acquisition (DAQ), a temperature monitor like an RTD or thermostat can be read. In order to prevent the process from going into Overshoot, the Controller will activate a switch.

7. Results

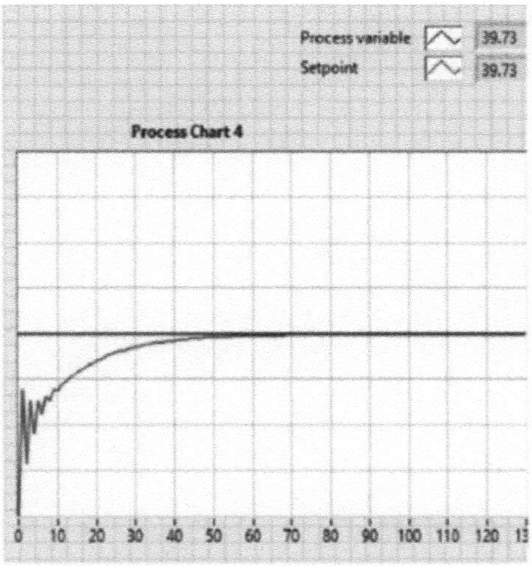

Fig. 49.4 Response of the heating tank

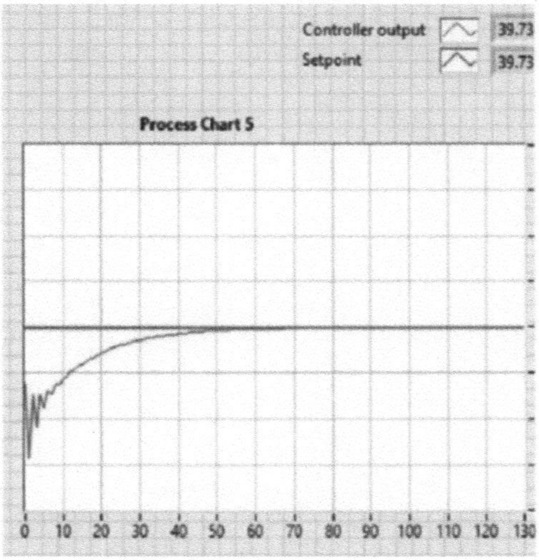

Fig. 49.5 Response of controller output

The Temperature Control System Simulation is Performed using PID controller in LabVIEW.

8. Conclusion

The Temperature Process is very non-linear in its Value. Therefore, it becomes more challenging to keep the system at its desired fixed position. Simulation can also be done using SCADA, NI LabVIEW and MATLAB. Here in this Process we are using LabVIEW due to its Effectiveness and Control Ability. However, the Expected response given from PID controller by Different tuning Methods. Keeping an eye on your equipment is a serious job, and so is keeping an eye on the temperature and humidity levels that can have a negative impact on your crucial equipment at your off-site locations.

REFERENCES

1. Salim, Sunil Kumar, Jyoti ohri, (2013). LabVIEW based DC motor and Temperature Control using PID Controller, International Journal of advanced research in Computer Science.
2. S Kohout, J Roos and H Keller, (2020). Automated Operation of Ahomemade torque magnetometer using LabVIEW.
3. Arnold Kohlendko ke long lee, Maya shendon , Efrain qualies, (2007). Temperature Control System for semiconductor process chamber, United States patent.
4. Pranav Kumar S., Suriya Prakaash D., Sathish Kumar M., Kabileshwaran R., (2022). Temperature Monitoring System Using LabVIEW, International Journal of Multidisciplinary Research Transactions.
5. Dhanraj, J.A., Ramanathan, K.C., Priyadharsini, S., Jayaraman, P. and Sankar, M.J., (2020). Boiler temperature and pressure monitoring system for thermal power plant through LabVIEW, IOP Conference Series: Materials Science and Engineering Vol. 988, No. 1, doi: 012079.
6. Dewi, A.K., Sahaya, A.A.B.A. and Sugiman, W., (2020). Level and temperature monitoring system in blending process using ZigBee wireless sensor network, 1st Borobudur International Symposium on Humanities, Economics and Social Sciences (BIS-HESS 2019) doi: 372–375.
7. Nanda, I., Shukla, V.K., Dhanekula, V.J., Gadipudi, M., Penugonda, V. and Maloji, S., (2021). Design and real time implementation of SmartWater management using LabVIEW and IoT. International Conference on Computational Intelligence and Knowledge Economy (ICCIKE) (IEEE: 524–528).
8. Zhang, B., Lin, F., Zhang, C., Liao, R. and Wang, Y.X., (2020). Design and implementation of model predictive control for an open-cathode fuel cell thermal management system. Renewable Energy, 154, doi: 1014–1024.

Note: All the figures in this chapter were made by the author.

Disruptive Technologies for Sustainable Development – Dr. G. Nagappan et al. (eds)
© 2024 Taylor & Francis Group, London, ISBN 978-1-032-55000-8

Comparative Analysis of Ranking Machine Learning Classifier Models for Parkinson's Disease (PD) Prediction

50

Akram Pasha[1]
REVA University, Bengaluru, India
Syed Zia ur Rahman[2]
Majan University College, Muscat, Oman
Syed Tauheed[3], **Syed Muzamil Basha**[4], **Anwar Basha H.***
REVA University, Bengaluru, India

Abstract: PD is one of the most common diseases among many men and women globally, whereas the early detection may well lead to a longer survival or even full recovery. With the development of clinical technologies, massive PD feature data become available to be collected and meanwhile many ML techniques have been introduced to support doctors in diagnostic decision-making process. The comprehensive studies conducted in past prove that the ML models typically exhibit improved results on a given classification problem when the key parameters involved in a model are effectively tuned. Accurately determining the most suitable combination of values for these key parameters in a ML model realistically is a search problem that exhaustively evaluates each combination of values of the key parameters that have typically produced the best values. Therefore, in this paper, an effort is made to carefully tune the parameters of four ML classification models: Decision Tree (DT), Random Forest (RF), Support Vector Machine (SVM) and Multilayer Perceptron (MLP), using grid and random search techniques, to ultimately determine the most efficient classification models. The models are assessed based on the distinct features they extract from the dataset and the amount of time to predict the PD accurately. The results obtained from all the cases of feature engineering are critically compared, and the most efficient parameter tuned top-ranked ML models are reported.

Keywords: Machine learning, Classification, Parameter tuning, Feature selection, Feature extraction, Parkinson's disease

1. Introduction

PD caused due to degeneration of the nervous system of a human brain is severely affected around ten million of world's population. The diagnosis of this disease is generally performed based on the features collected by the patients during the treatment (Jankovic J, 2008). Many research studies conducted in the past focus on the vocal instability of patients. These studies are typically driven by ML algorithms that are fed with various features corresponding to the symptoms of patients for initial detection of the disease (Peker, 2016; Erdogdu Sakar, 2017). These studies have assisted the clinicians in pursuing accurate decisions to provide rapid development to the patients at ease. Machine learning classification algorithms include parameters that are configured to produce superior efficiency. The parameters in such algorithms play a vital role in producing efficient results for an application at hand. These parameters, required to be tuned, are referred as hyper-parameters. The current study uses the grid and random search from the work of (Komer 2019) and uses the PD data set that is available in the UTI ML repository (Little 2007). The primary contributions of the current study are that 1) it implements two types of feature

*Corresponding Author: anwar.mtech@gmail.com
[1]akram.pasha@reva.edu.in, [2]syed.rahman@majancollege.edu.om, [3]syedthouheed.ahmed@reva.edu.in, [4]muzamilbasha.s@reva.edu.in

DOI: 10.1201/9781003428473-50

scaling techniques and studies hyper-parameter tuning separately for all the classification models. 2) It determines the best hyper-parameters and the feature scaling setting for the most efficient ML model to diagnose the PD. In conclusion, the study finds the four best ranked ML classifiers giving the best values for the hyperparameters. The remainder of this article is divided into the following sections. The work relating to the current investigation is summarised in Section 2. The in-depth analysis of the methods used in this study is presented in Section 3. The experimental findings are presented in Section 4 together with explanations. The summary of the work completed and the prospects for this effort are presented in Section 5.

2. Related Work

Many studies have used the two primary techniques of hyper-parameter tuning in the past. In this section, we present the summary of the significant contribution in this area that have been successful in using the hyper-parameter tuning techniques. In the work of (Baboota, 2019) the football results of English Premier League were analysed to predict the results using ML classification models. The authors used grid search and random search to tune the parameters of radial basis function based Support Vector Machine and Random Forest classification models to improve the accuracy of the model. Model-based optimization techniques have been used to study the performance of multi-objective ML algorithms in the work of (Tackman, 2019). The authors have compared the results of two optimizers using grid search and random search to tune SVMs. Similar kind of work can be seen in the work published in (Peker, 2016) where the authors use grid and random search for tuning RBF SVM. The parameters C and γ of SVM for classifying the patients having PD, were tuned using grid and random search techniques in the work of (Li, 2018). The 10-fold CV on training data with grid search mechanism were used to tune the parameters having various ranges for RBF kernel functions, were experimented in the work of (Peker, 2016). Some works in the field of parameter tuning have convinced to show how random search is better compared to grid search techniques. Such study conducted can be seen in (Bergstra, 2012) to compute the relevance of the hyper-parameters of neural networks and the authors conclude that some parameters are important on all datasets, while others are only important on some datasets. There have also been number of studies conducted in the literature (Pasha. A, 2020; Pasha. A, 2021) employing ML algorithms for disease prediction utilizing the datasets collected from the patients without focusing at the tuning of hyper parameters of the ML models incorporated. The literature shows that the hyper-parameter tuning is an important contribution to improve the efficiency of any ML model through optimizing the resources utilized by an algorithm and reducing the overall computational complexity of a model. The subsequent section discusses the Methodology used in this study.

3. Methodology

This section presents a framework for training and testing ML classification models on a PD patient data set. Two scaling techniques are used to standardize the independent variables, and two search methods are employed to find the optimal parameters for the ML algorithms that produce the highest accuracy. The top-ranked ML algorithms and their optimal values are reported.

3.1 Model Specification

Let (D_A) be the mathematical representation of the input data set with each attribute notation as $(\forall D_A \Rightarrow A_X)$ where (A_X) is the peak size or universal attribute set. The referential parameter of each of the attributes in the dataset is represented as in equations 1 and 2, where, weight of each attribute is associated and fragmented into the collective sum formulation:

$$A_X = \begin{cases} \{A_1, A_2, \ldots\ldots, A_n\} \\ \{W_1A_1, W_2A_2, \ldots\ldots, W_nA_n\} \end{cases} \tag{1}$$

$$\Rightarrow \quad A_X = \sum (W_nA_n).(D_A)_0^n \tag{2}$$

The formative attributes are extracted and customized with the framework of $(A_X \Rightarrow U_X)$ with each (U_X), the universal attribute $(U_X \cup A_X \Rightarrow D_X)$ with n^{th} attribute representation. Thus, equation 3 depicts the parameter used in building the ML model with the training datasets.

$$\Delta T = Lim_{n\to\infty}\left[\frac{\partial}{\partial t}(A_X)_U \leftrightarrow \frac{\partial}{\partial t}(D_A)\right] \tag{3}$$

According to the equation 4 the formulation of each of the subsequent bits of multi order w_n weight is co-ordinately aligned with ΔT with $n \to \infty$ order. The formative attribute paradigm is sequentially represented with multi order training datasets as $\Delta T = \{T_1, T_2, \ldots, T_n\}$ with changing interval of time as demonstrated in the equations 5 and 6.

$$\Delta T = Lim_{n \to \infty}\left[\sum_{i=0}^{\infty}\prod_{j=i}^{n}\sum_{u}(Au)_{(i,j)} \in \sum w_{AX}\right] \tag{5}$$

$$\Delta T = Lim_{n \to \infty}\left[\bigcup_{n}^{\infty}\left\{\sum_{n}^{0}[Au \cup w]_{(i,j)}\right\}\right] \tag{6}$$

Based on the application, the inclusion of the data formats is appended with training datasets (ΔT). The trained data set is then focused with interdependency evaluation to extract the feature (f) from (A_X) as shown in equation 7.

$$\sum f = \left[\sum_{i=0}^{\infty}\prod_{j=i}^{k}\left(\Delta T_0^k \cup AX_{i,j} \in \sum X\right)\right] \tag{7}$$

Where, each training dataset is characterized into multiple interdependency attributes. According to equation 7, the attribute ($AXi, j \Rightarrow DA$) and has fractional correlation matrix generated as shown in equation 8.

$$M = \begin{cases} \sum f \in (\Delta T)_0^n \\ \sum f \notin (\Delta T)_0^n \end{cases} \tag{8}$$

Based on the matrix (m), the individual matrix set of classifier algorithms are generated to supply the dataset to the classifiers such as DT, RF, SVM and MLP. Hybrid summarization is used as shown in equation 9.

$$\sum(C)_{(q,U)} = \lim_{n \to \infty}\left[\frac{\partial(m)_n}{\partial t} \cup \sum(f)_0^n\right] \tag{9}$$

Where, C is the defined ratio of classifiers used and implemented for performance upgrading.

$$\sum(C) = \lim_{n \to \infty}\left[\sum_{i=0}^{n}\prod_{j}^{n}\frac{\partial(m_n)_i}{\partial t} \cup \frac{\partial\sum\{(f)_0^n\}_j}{\partial t}\right] \tag{10}$$

Thus, a customized variant of classified algorithm is selected and appended for PD classification.

4. Experimentation and Discussion of Results

This section discusses the use of ML algorithms to classify PD patients based on a data set of 755 attributes. Four algorithms (Decision Tree, Random Forest, Support Vector Machine and Multilayer Perceptron) were used and their performance was measured using accuracy, precision, recall, F1-score, and kappa scores. Hyperparameter tuning was done using grid search and random search strategies to find the best combination of parameters for the algorithms. The best performance was found with the four algorithms.

4.1 Performance Evaluation of Classifiers

This research article explored the performance of various ML models using different feature selection and feature extraction techniques. It was found that the Support Vector Machine (SVM) produced the highest accuracy of 88.55% when all the features of the data set were selected. The Random Forest (RF) and Multilayer Perceptron (MLP) models produced the highest accuracy of 85.09% and 77.09% when the features were selected using Random Forest ensemble technique and transformed to a new search space using Principal Component Analysis (PCA). The RF model was found to produce the highest accuracy of 86.78% when all the features were transformed to a new search space using Kernel Principal Component Analysis (kPCA).

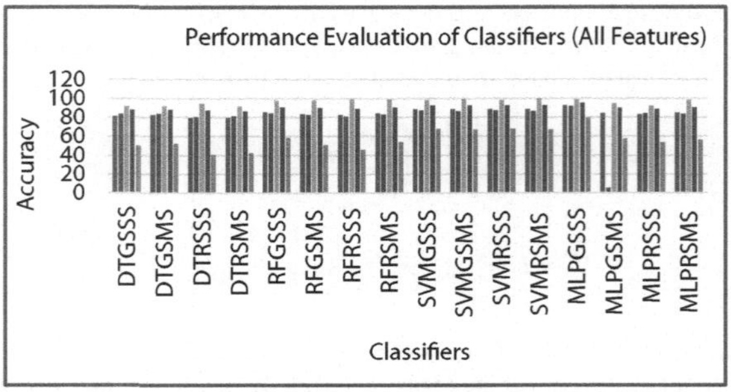

Fig. 50.1 Performance evaluation of the classifiers (all features)

4.2 Empirical Analysis of the Classifiers

The Fig. 50.2 shows the empirical analysis of DT when employed using grid and random search techniques. The performance of DT is best (taking only 2.38 seconds), when random search and standard scaler is employed by selecting 5 best features from the data set. Similar experiments conducted showed that the performance of RF is best (taking only 13.57 seconds), when random search is employed with kPCA (k=2), with standard scaler on the data set. The performance of SVM classification model is found to be best (taking only 0.1 seconds), when random search is employed with kPCA (k=2), with both standard scaler and maximum absolute scaler on the data set. The performance of MLP is best (taking only 3.4 seconds), when random search is employed with kPCA (k=2), with maximum absolute scaler on the data set.

Fig. 50.2 Elapsed time of DT

The Table 50.1 shows the parameters and their values of a top ranked DT (trained with All Features) selected based on the accuracy form many other DT classifiers. Similar experiments were conducted to also find the top ranked RF, SVM and MLP selected based on the accuracy form many other RF, SVM and MLP classifiers.

Table 50.1 DT (Rank-1)

Parameters	criterion	max_depth	max_leaf_nodes	min_samples_leaf	min_samples_split
All Features					
GSSS	'gini'	5	-	1	2
GSMS	'gini'	5	20	1	2
RSSS	'entropy'	3	14	3	15
RSMS	'entropy'	14	11	6	7

5. Conclusion and Future Enhancement

The study performed the grid search and random search to tune the hyper-parameters of ML models like Decision Tree, Random Forest, Support Vector Machine and Multilayer Perceptron. The two types of feature scaling techniques are implemented and the hyper-parameter tuning is studied separately for all the selected classifiers. The best hyper-parameters and the feature scaling setting for the most efficient ML classifiers to diagnose the PD are found. The study can be further extended to incorporate various Auto ML and /or bio-inspired computing algorithms to optimize the parameter values of the ML models trained to perform a task of regression, classification or clustering on any data set chosen from any domain.

References

1. Baboota, R., & Kaur, H. (2019). Predictive analysis and modelling football results using machine learning approach for English Premier League. *International Journal of Forecasting*, *35*(2), 741–755.
2. Bergstra, J., & Bengio, Y. (2012). Random search for hyper-parameter optimization. *Journal of machine learning research*, *13*(2).
3. Erdogdu Sakar, B., Serbes, G., & Sakar, C. O. (2017). Analyzing the effectiveness of vocal features in early telediagnosis of Parkinson's disease. *PloS one*, *12*(8), e0182428.
4. Feurer, M., & Hutter, F. (2019). Hyperparameter optimization. In *Automated machine learning* (pp. 3-33). Springer, Cham.
5. Jankovic, J. (2008). Parkinson's disease: clinical features and diagnosis. *Journal of neurology, neurosurgery & psychiatry*, *79*(4), 368–376.
6. Komer, B., Bergstra, J., & Eliasmith, C. (2019). Hyperopt-sklearn. In *Automated Machine Learning* (pp. 97–111). Springer, Cham.
7. Li, C., Hou, L., Sharma, B. Y., Li, H., Chen, C., Li, Y., ... & Chen, H. (2018). Developing a new intelligent system for the diagnosis of tuberculous pleural effusion. *Computer methods and programs in biomedicine*, *153*, 211–225.
8. Linmei, H., Yang, T., Shi, C., Ji, H., & Li, X. (2019, November). Heterogeneous graph attention networks for semi-supervised short text classification. In *Proceedings of the 2019 Conference on Empirical Methods in Natural Language Processing and the 9th International Joint Conference on Natural Language Processing (EMNLP-IJCNLP)* (pp. 4821–4830).
9. Little, M., Mcsharry, P., Roberts, S., Costello, D., & Moroz, I. (2007). Exploiting nonlinear recurrence and fractal scaling properties for voice disorder detection. *Nature Precedings*, 1–1.
10. Mantovani, A., Rinaldi, E., Zusi, C., Beatrice, G., Saccomani, M. D., & Dalbeni, A. (2021). Coronavirus disease 2019 (COVID-19) in children and/or adolescents: a meta-analysis. *Pediatric research*, *89*(4), 733–737.
11. Pasha, A., & Latha, P. H. (2020). Bio-inspired dimensionality reduction for Parkinson's disease (PD) classification. *Health information science and systems*, *8*(1), 1–22.
12. Pasha, A., & Latha, P. H. (2021). Well-Calibrated Probabilistic Machine Learning Classifiers for Multivariate Healthcare Data. *International Journal of Advanced Research in Computer Science*, *12*(2).
13. Pedregosa, F., Varoquaux, G., Gramfort, A., Michel, V., Thirion, B., Grisel, O., ... & Duchesnay, E. (2011). Scikit-learn: Machine learning in Python. *the Journal of machine Learning research*, *12*, 2825–2830.
14. Peker, M. (2016). A decision support system to improve medical diagnosis using a combination of k-medoids clustering based attribute weighting and SVM. *Journal of medical systems*, *40*(5), 1–16.
15. Rojas-Domínguez, A., Padierna, L. C., Valadez, J. M. C., Puga-Soberanes, H. J., & Fraire, H. J. (2017). Optimal hyper-parameter tuning of SVM classifiers with application to medical diagnosis. *Ieee Access*, *6*, 7164–7176.
16. Tackman, A. M., Sbarra, D. A., Carey, A. L., Donnellan, M. B., Horn, A. B., Holtzman, N. S., ... & Mehl, M. R. (2019). Depression, negative emotionality, and self-referential language: A multi-lab, multi-measure, and multi-language-task research synthesis. *Journal of personality and social psychology*, *116*(5), 817.

Note: All the figures and table in this chapter were made by the author.

Disruptive Technologies for Sustainable Development – Dr. G. Nagappan et al. (eds)
© 2024 Taylor & Francis Group, London, ISBN 978-1-032-55000-8

Design and Development of IoT Testbed with DDOS Attack for Cyber Security Research

51

E. Sujatha[1]
2Professor, Department of Computer Science and Engineering,
Saveetha Engineering College (Autonomous), Chennai, Tamil Nadu, India

M. Sangeetha Priya[2]
Department of Computer Science and Engineering,
Saveetha Engineering College (Autonomous), Chennai, Tamil Nadu, India

Abstract: The emerging growth of smart devices improved the quality of Living through various adaptations in services accessibility to reduce the time. Internet connecting devices offers great service outcome hens the number of users getting increased stopper this device are connected through common system. This devices are connected through a compromised network that leads to common server to communicate with all internet connecting devices. The influence of Denial of services and distributed Denial of services impact the normal functionality of the device. The IoT devices with security constraints are helpful to make resilience of changes over the network. The distributed system is thoroughly monitored to detect the software defined network problems effectively. The proposed system is focused on developing dynamic software defined in network models to detect the abnormal attacks coming on the massive network and predict the abnormal behaviours as early as possible. Early detection of a mitigating attack are helpful to secure the data over the cloud.

Keywords: Software defined network, Wireless communication, Smart devices, Dynamic systems, Internet of things

1. Introduction

Cloud Computing is the process of offering various services and resources over the internet. Instead of keeping the data in the physical storage devices people started storing the data into the cloud where the internet plays a major role in securing the data over the network. Preserving the data stored in the cloud is important.

Internet of things enabled services provides commercial and industrial applications offering with best outcome since the data get tuned through Reliable study of existing problems in the cloud. The attack have unique impact on the network and create heterogeneous result on generating output which are not feasible to accept. The major implications in IoT devices are unstructured data format and Forbidden occurrence of mitigating attacks over the network. With advancements of a software defined network (SDN) dynamic allocation of system and adaptive remote monitoring is feasible. The prime motive of software defined and network is to provide separate control for accessing the data. The challenging problem of software defined network is due to the different devices and patterns that frequently occur and attack the network.

Attacks are inevitable in massive network connectivity. Anomaly detection system provides Initial support to the IoT networks to detect and mitigate the problems occurring in the network. Various resources are saved in the cloud atmosphere any leakage are problem with in the network can happen to be sensitive and impact the device security. The mechanism created for software

[1]sujathae@saveetha.ac.in, [2]rksangeethapriya@gmail.com

DOI: 10.1201/9781003428473-51

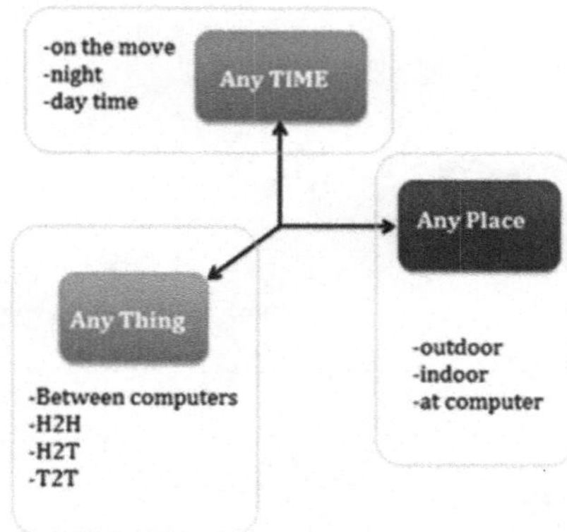

Fig. 51.1 IoT cloud utilities

defined network to detect the mitigating Attacks distributor Denial of services (DoS) attack secure the data communication and confidential data present in the network.

The rest of the paper is formulated as describing the detailed literature background on similar concept in section II. Followed by System tools selection, existing system development and challenges available in existing system are discussed in section III. The design methodology adopted for DDoS attack detection is explored in Section IV. Further the results and discussions are made in section V. The paper is concluded with obtained results and simulated models and further future exploration of the work is defined.

2. Background Study

Sood et al. (2019) Software defined in network is a Programmable network protocol that considered challenges in the existing wireless sensor Network and create adoptive network that can adjust the problems occurring over the network and dynamically change the configurations without disturbing the existing configurations. Because of high speed scheduling data transfer and traffic issues the physical objects may not responsible for the changes happening over the network. The author presented as system discussing on various constraints available in software defined and network and mitigating problems.

3. System Design

The presented system address the problem of security lags in networks. Software defined network (SDN) are developed to prevent anomaly attacks over the network instead of physical devices. The prime objective of the system is to create IoT devices to be protected completely from mitigating attacks. The attacks may be generated through various third party software devices, third party connecting equipment and service providers. The fast identification of anomaly attacks in the network and mitigation attacks provides best outcome of system performance in software defined network. The presented system on removing the existing problems in software defined in network model such as DDoS attacks, cyber-attacks, and distributed denial of services (DDoS) attack, anomaly attack, Probing and flooding attack. The security mechanism is planned to work with device independence model. Here mitigation of attacks rapidly in the early stages is focused. Detection of attacks with high precision rate and recall rate is focused.

The proposed system depicted here shows the system architecture of distributed Denial of services attack (DDoS) in the massive cloud platform. It compromises various development constraints in terms of designs specification, data transfer and data preparation.

4. Methodology

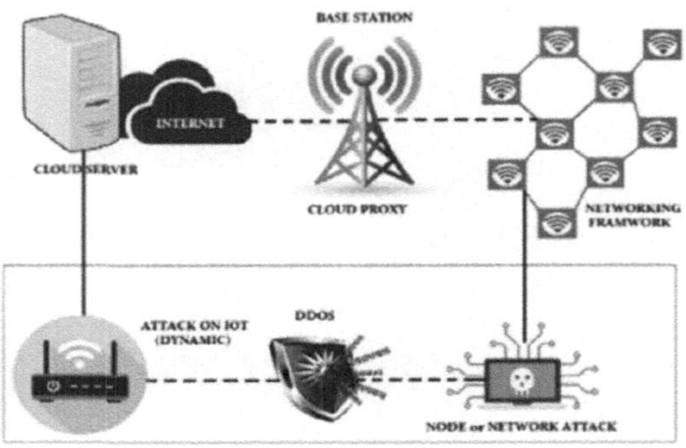

Fig. 51.2 System architecture of proposed model

Figure 51.2 Shows the system architecture of proposed attack detection model. The architecture comprised of connecting the cloud server with cloud proxy or the base station. Network model is created with various dynamic nodes consuming dynamic distance and weightage.

The distributed Deniel of services attack are the network attack created at the dynamic modems are detected by the presented approach by continuous analysis and detection model implemented over here. The proposed system considered them back scatter algorithm and attack origin detection algorithm for early detection of mitigating attack. The network topology consists of number of nodes connected with Local area network (LAN) and distribute the data to nearest node with the help of routing protocol. The core routing protocol consider the data package and transverse to the other nodes with shortest path sorting algorithm. The parts selection is based on reliability and accessibility of nearby nodes based on the demand. Once the node is getting connected with another node the data communication is established. Using the local area network, Data packets from one node is connected to the other node using packet marking algorithm. The packet marking and login create multiple hash values in the table. The proposed system here enabled to provide adaptable network topology. The pattern construction is the process of finding the new node with the same source and destination neglecting the attack in scenario. If dynamic attack available at the router, then the attack information transferred to other connecting devices in the distributed network. Authentication is a process of confirming the valid user to access the network. The distributor Denial of services attack is controlled through software defined in network computer with robust authentication protocol

4.1 Hybrid Multi-modality Screening Process

The hybrid multi-modality screening model (HMSP) considers trace back attack, back scatter attack and usage of root cause attack screening algorithms combined together.

4.2 Trace Back Algorithm

Here the routing of each packets are navigated to trace back algorithm. For the given time period the data packets are screened through Trace back algorithm. After the user login to the system the trace back server create a log that consider all the packets entering into the network and examine the health of the packet. If the trace back server identify any anomaly packet appear in the queue then immediate neighbor routing can reroute the process into another node as temporary node to avoid the negative packet can enter into the network.

Backscatter algorithm considered the technique of delayed tree method where the back scatter container phase shift key implementation with switch is provided.

The resilient problem is very important to detect.

The distributor denial of services attack need to be completely cleared if the root cause of the problem is being defined. Here the original cause of attack generation and enhance the route within short duration of time. The resilient problems need to be considered as an important criteria that should be removed from the deployment as early as possible.

5. Results and Discussions

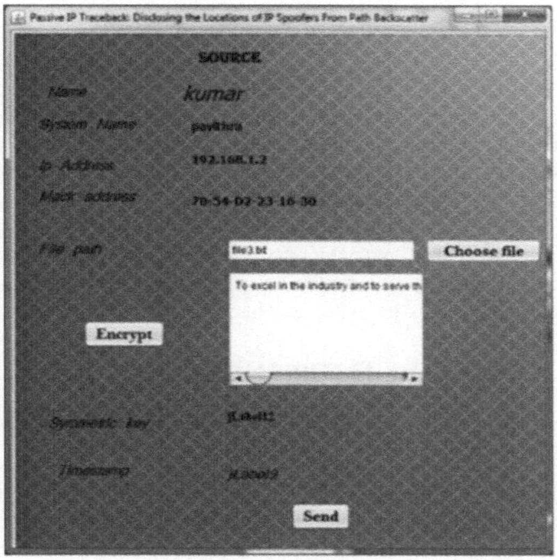

Fig. 51.3 Simulation result

The Figure 51.3 Shows the simulation result of user entering into the network to send the data packet. After implementing the HMSP algorithm a secret encryption need to be done. The encryption is a highly robust key and that consider the network parameter such as IP address, MAC address, system name and symmetric key. The high speed encryption process protects the data and the network before routing.

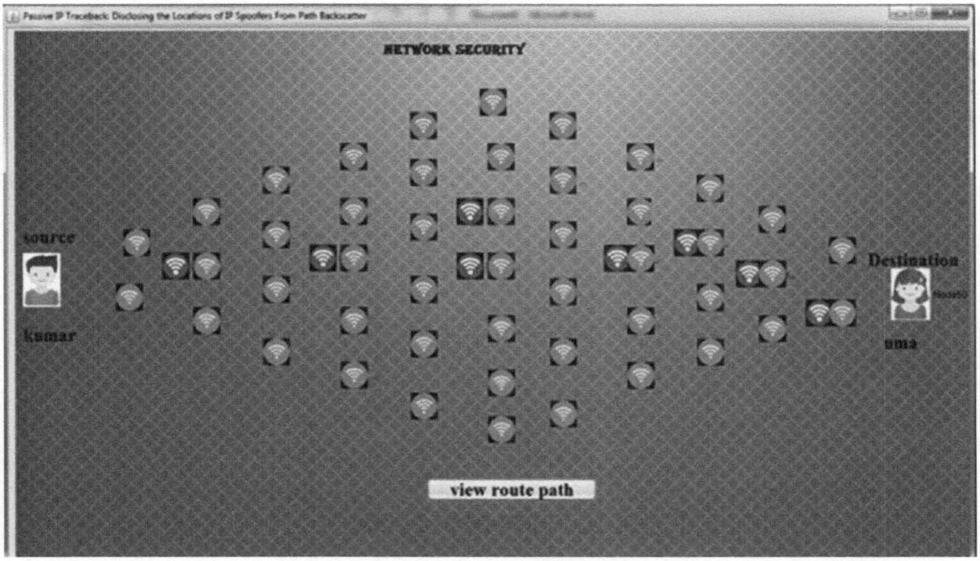

Fig. 51.4 Rerouted network

Figure 51.4 shows the rerouted hybrid multi-modal screening network (HMSP) based rerouted adjusted network. The network created here after the anomaly is detected and further the evaluation of trustworthiness in the network topology.

6. Conclusion

In the current development of 5G communication software defined a network plays a significant role in protecting the devices connected over the network. Different anomaly operations are happening in the network if there is no security constraints are mapped in the protocol. The main challenge of the developer is to create a software defined in network with highly robust protocol to protect the mitigating attacks at the node itself. Denial of services (DoS) is also one of the important issues considered. The presented system consider trace back algorithm, back scatter algorithm and root cause attack screening algorithm as a hybrid model. The three constructive method detect the complete architecture health and create a trusted network to effectively process the data. The simulation results detecting the distributor Denial of services attack and Readjustment of network route using the Hybrid multi-modal screening process is well studied. Further the presented system can be improved by implementing nature inspired algorithms to optimize the routing nodes and mitigating back the attacks as early as possible.

References

1. Mirai Malware (2016), http://blog.malwaremustdie.org/2016/08/mmd0056-2016-linuxmirai-just.html, (Online; accessed 18-07-2017).
2. Jayavardhana Gubbi, Rajkumar Buyya, Slaven Marusic, and Marimuthu Palaniswami (2013). "Internet of Things (IoT): A vision, architectural elements, and future directions," Elsevier FGCS, 29(7): 1645–1660.
3. Shaibal Chakrabarty and Daniel W. Engels (2020), Secure smart cities framework using IoT and AI. In 2020 IEEE Global Conference on Artificial Intelligence and Internet of Things (GCAIoT) (pp. 1–6). IEEE.
4. Keshav Sood., Kallol Krishna Karmakar, Shui Yu, Vijay Varadharajan, Shiva Raj Pokhrel and Yong Xiang (2019). Alleviating heterogeneity in SDN-IoT networks to maintain QoS and enhance security. IEEE Internet of Things Journal, 7(7): 5964–5975.
5. Waseem Iqbal, Haider Abbas, Pan Deng, Jiafu Wan, Yawar Abbas, Bilal Rauf and Imran Rashid (2020). ALAM: Anonymous lightweight authentication mechanism for SDN-enabled smart homes. IEEE Internet of Things Journal, 8(12): 9622–9633.
6. Moheeb Abu Rajab, Jay Zarfoss, Fabian Monrose and Andreas Terzis (2006). "A multifaceted approach to understanding the botnet phenomenon," 6th ACM SIGCOMM conference on Internet measurement, (pp. 41–52).
7. Yaser Jararweh, Mahmoud Al-Ayyoub, Ala' Darabseh, Elhadj Benkhelifa, Mladen Vouk and Andy Rindos (2015). "SDIoT: A Software Defined based Internet of Things framework," Journal of Ambient Intelligence and Humanized Computing, 6(4): 453–461.

Note: All the figures in this chapter were made by the author.

Disruptive Technologies for Sustainable Development – Dr. G. Nagappan et al. (eds)
© 2024 Taylor & Francis Group, London, ISBN 978-1-032-55000-8

Medical Analytical Decision Support System Investigative Study

52

Rekha K. S.[1], Jeevananthan R.[2], Gowtham G.[3], Sai Prahalatha Raghavan V.[4]
Department of Computer Science, Saveetha Engineering College, Chennai, India

Abstract: An appointment booking system with intelligent capabilities that enables users to easily schedule doctor's appointments online. This web-based application is designed to tackle the difficulty of scheduling and organizing appointments. based on users' preferences and demands. The manual assignment of appointments to patients can be a tedious task for compounders or doctors. Therefore, this system offers an effective solution where the platform allows users to access different available booking slots, select their preferred date and time, and book an appointment accordingly. Additionally, users have the option to cancel their booking at their convenience through the system. The disease predictor is the main component of this project, which enables users to identify their disease based on their symptoms and helps them to gather more information for the consultation of the appropriate doctors.

Keywords: Medical analysis, Disease predictor, Machine learning, Web user interface

1. Introduction

The Medical Analytical Decision Support System is an online consultation and end-user support project that aims to provide instant guidance on health issues using intelligent healthcare technology. This technological innovation offers numerous benefits for both doctors and patients, including reduction of charge, accessibility, ease, and improved security and thought interchange process. Currently, when an individual experiences a particular disease, they must visit a doctor, which can be both time-consuming and expensive. Moreover, if the user is located far away from a doctor or hospital, it may be challenging to identify the disease. Existing systems don't behave in a user-friendly way as most of them are built from the perspective of doctors and medical institutions. Additionally, current software systems are not advanced enough compared to the rest of the field. However, if an automated proposed work can predict the disease based on the user's symptoms, it could save the user time and money and make the connection between entities comfortable. The Disease Predictor is a web tool that lists possibilities of occurrences of the user's illness based on their symptoms. The Disease Prediction system collects datasets from various health-related websites to improve its accuracy.

The online appointment booking system and Disease Predictor can help users save time by allowing them to book appointments with appropriate specialists and predict the probability of a disease based on their symptoms. As the internet is thriving more than ever, people have greater access to medical intel. In many cases, people may not have immediate access to medical professionals, but they can access the Internet at any time. This system provides a helpful solution for individuals looking for guidance on their health issues. Users can learn more about their condition based on the results provided by the disease predictor and then book an appointment with the appropriate specialist.

[1]rekhaks@saveetha.ac.in, [2]jeeva.r.2311@gmail.com, [3]gowthamangowda7@gmail.com, [4]raghavanmarch2002@gmail.com

DOI: 10.1201/9781003428473-52

2. Literature Survey

2.1 Internet Use on Health-related Behaviors

The main objective of this study was to investigate the linkage possibility between the internet and health. Additionally, the study aimed to improve the association between people and health professionals via the Internet. The emergence of the internet has been a remarkable development that differs from other media. In addition, its impact has affected communication systems and interpersonal relationships. As a result, a vast majority of internet users now utilize the internet as a tool for accessing medical information. The younger generation of today often brings the knowledge they have gained from the internet to their medical consults with their doctors. To investigate this trend, Sandesh Kumar Sharma (2013) conducted cross-sectional study on patients. The participants answered a predetermined questionnaire that included information about their socioeconomic status, health behaviors, and relationship with their doctor.

2.2 Evaluation of Online Consultation

The personnel and various online medical consultation platforms share common themes and features, including the importance of obtaining a second medical opinion, Argeet Singh et al (2018) explains this deeply in his study. This is a key benefit of internet-based doctor services, as many individuals express a desire for an additional perspective on their health issues. In Sweden, second opinions are only granted for serious health conditions, making it challenging for patients with less severe medical issues or those struggling to communicate with their regular doctor to obtain a second evaluation.

2.3 Health System Research

Among the numerous expositions of a healthcare system and its functions along with its capabilities, there are still much more areas of the system that can be exploited for the care of the people. Bhandari TR (2013) concerns with making healthcare evidence based decisive process. In a broad aspect, the health system covers different aspects of society such as socioeconomic status, culture, religion, education, politics, public sector, and private sectors which are the major determinants of social epidemiology. Much more organized research will analyse ways to include the healthcare system into the public routine seamlessly. It also puts a doctor in the difficult situation of making a diagnosis without a proper understanding of the patient's medical history and social context, in addition to removing the opportunity to conduct a physical examination.

3. Proposed System

The architecture of the proposed healthcare system is based on various neural network algorithms and network-based operation handling for real-time maintenance of requests of doctors and patients. This system is aimed at developing a set of modules that can facilitate the diagnosis by analysing the various symptoms of patients. A database of doctors and hospitals available is also stored and maintained online in real time.

- The user interface system provides patients and doctors with a platform to manage their personal information, medical reports, and account details as shown in Fig. 52.1 and the dashboard of the system is shown in the Fig. 52.5.
- With the online appointment booking system, patients can schedule appointments during the available session and slot timings provided by the doctor. They also have the option to cancel their appointment at any time. Doctors can view their patients' appointments through their accounts. Mageshwari et al. (2012) cross examines appointment booking techniques in her study.
- Detailed Real-time results are used as training and testing datasets to predict the accurate results of the disease. To show the accuracy, many real testing datasets compare with trained datasets and then results are predicted.
- If a user is not aware of their disease, they can utilize the disease predictor system to identify their disease based on their symptoms. Additionally, the system can suggest an appropriate specialist for their specific illness.

3.1 Algorithm/Methodology

Naive bayes

Naïve bayes classifier is based upon the theorem of Bayes: q(b/a1,.., an)=q(b)q(a1,...,an/b) q(a1,...an) where, b is the class variable a1,a2,.., an are the dependent features. the value a(symptoms /disease) can be calculated by: $Q\ (symptomi/Disease) = Nbi + \alpha Nb + \alpha n$ Where,

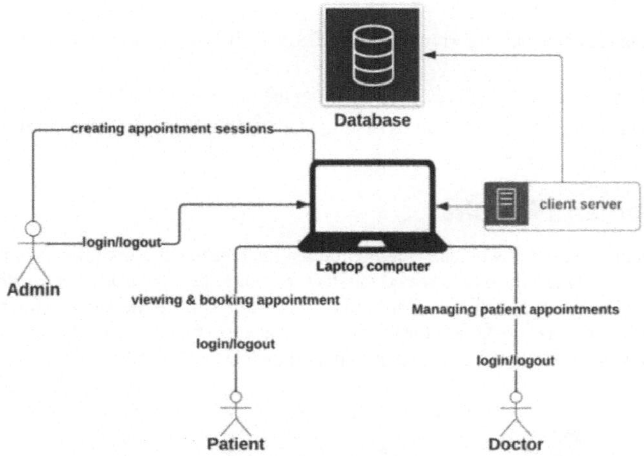

Fig. 52.1 User interface architecture

Nbi = Number of occurrences of the disease in the dataset

Nb = The addition of the symptoms that belong to that particular disease

n = The number that is outputted by adding all the symptoms in the dataset

α=1, known as Laplace Smoothing

The value of Q(Disease) can be estimated by: Q (Disease) = $N(Disease) + 1N + 2$ which is law of Success by laplace

Where, N (Disease) = The number of occurrences of the disease in the set of data

N= The added value of the diseases in the set of data.

Tomas Pranckevičius et al. (2017) also investigated these algorithmic methodologies in detail in thier studies.

Random forest algorithm (RFA)

RFA, which has been utilized in studies related to diseases such as diabetes, COVID-19 detection, septic shock early detection, and heart disease detection, is primarily based on DT as the main classifier. This way of decising also employs ensemble learning techniques to characterize information to categorize instances that are occurring presently by combining identifiers from various pre-trained classifiers. Random Forest, an algorithm composed of trained sapling-organized classifiers, is utilized in this process. An elected choice for the most tested class is rigorously contributed by each trained sapling. To create a decisive unit, a vital employment of exercising exam is done, and a random vector, which is not considerate of other factors of the previous like distribution arbitrary vectors with the, is utilized.

Working of RFA

The Random Forest algorithm is of dual main stages. The first stage involves the creation of the random forest, and the random forest classifier created in the first stage is used to make predictions on new, unseen data.

Random forest pseudo code

1. "L" features from total "n" features are selected randomly where l << n
2. Among the "L" attributes, reckon the node "e" using the best split
3. With the help of best split, split individual nodes into daughter nodes
4. Repeat the a to c steps until the reaching point of "m" number of nodes
5. Repeat the a to c steps until the reaching point of "m" number of nodes

Decision tree

1. There are many popular decision tree algorithms CART, ID3, C4.5, CHAID, and J48. From this J48 algorithm is used for this system.
2. In the J48 algorithm, each tree sampling is established using the pruning method
3. Pruning is a processing manner that makes trained sapling size reduction by eliminating overfitting data, which ultimately occurs in poor accuracy at prediction.

4. Implementation and Results

The system operates the prediction process by carrying the trained random forest algorithm test characteristics and the naive bayes algorithm via the edicts of each arbitrarily created tree that needed to be passed as shown with the help of Fig. 52.2 and Fig. 52.3. Suppose let's say one-tenth of a thousand arbitrary decision trees from the random forest are produced. Each tree in the algorithm will produce a different outcome for the same test dataset. Then by considering each outcome received final calculation of the predicted votes is done, the complete system is shown in the Fig. 52.4.

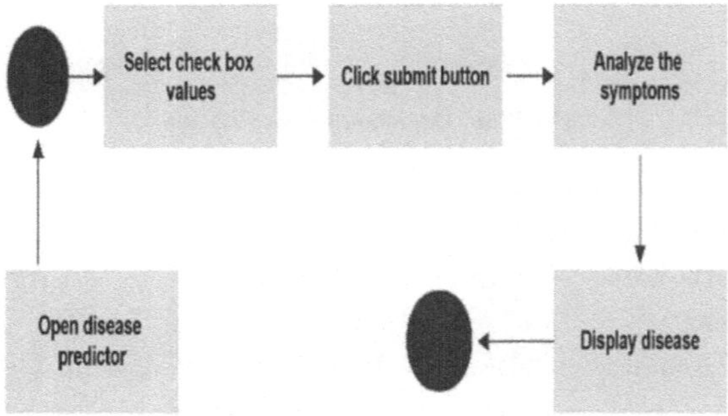

Fig. 52.2 Flow chart of the disease prediction system

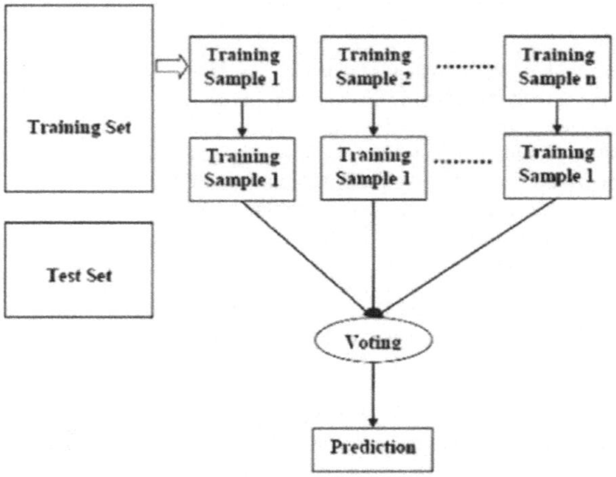

Fig. 52.3 Predicting the disease using RFA

Let's take the case that the one hundred random decision-trained saplings are predicting some 3 unique targets a, b, and c then the votes of (a) are out of one hundred random decision trees how many trees prediction is a? For the other targets in the set

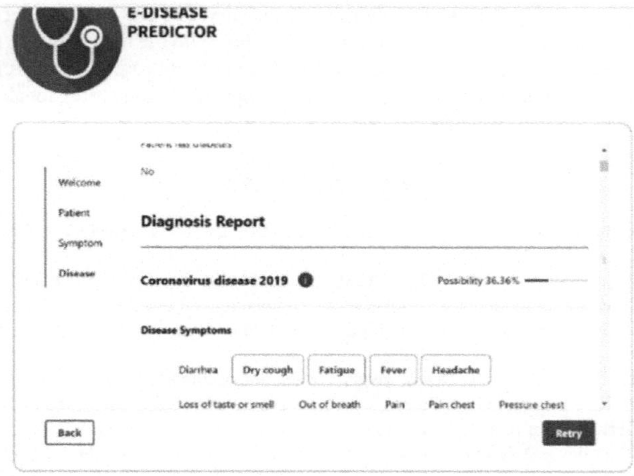

Fig. 52.4 Sample result of the prediction of COVID-19

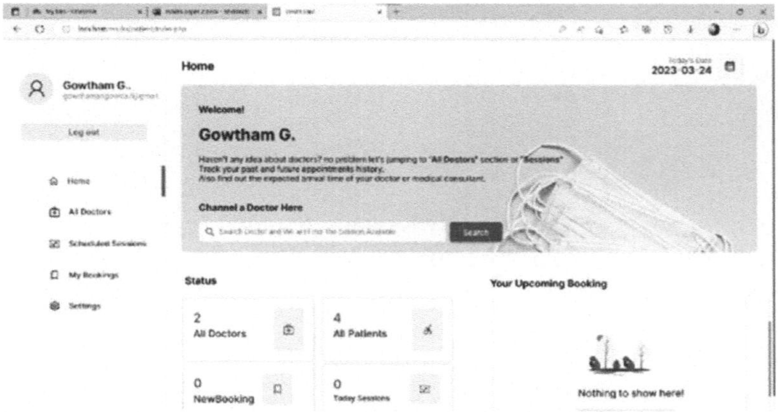

Fig. 52.5 Patient dashboard of the healthcare system

(b, c) are likewise. Whatever the target, which got the highest sum of votes in the forest will be produced as the selected result. This particular process described in this context is called majority voting.

5. Conclusion

The proposed work has been designed to provide feature sets that are versatile and can work efficiently in various situations, thereby increasing its potential for success. The user-oriented aspect of the proposed work includes a feature that enables users to easily locate doctors and specialists in hospitals. The disease predictor is another important feature that not only provides results based on symptoms but also indicates the probability of different diseases, which helps alleviate unnecessary anxiety for patients and ensures the reliability and safety of the medical system. This is especially important as it concerns the lives of patients. On the doctor's side, the proposed work helps manage a comprehensive database of patients, which would otherwise be a time-consuming task. Moreover, it simplifies the process of booking appointments, which reduces the risk of conflicting schedules for patients. This makes the process more convenient for both doctors and patients and optimizes the hospital's capacity. Overall, the proposed work is well-suited for various use cases and effectively interacts with all parties involved without any conflicts, making it an invaluable asset to the medical industry. The use of technology in the healthcare industry has revolutionized the way we approach and manage health-related issues. With the help of online tools like Disease Predictor and appointment booking systems, patients can receive quick and efficient healthcare services from the comfort of their own homes. This not only saves time and money for the patients but also helps healthcare professionals to efficiently manage their workload and provide better care for their patients.

REFERENCES

1. Ajeet Pal Singh and Hari Shanker Joshi (2018): "Online medical consultation: a review", International Journal of Community Medicine and Public Health; 5(4): 1230–1232.
2. Tomas Pranckevičius, Virginijus Marcinkevičius (2017): "Comparison of Naive Bayes, Random Forest, Decision Tree, Support Vector Machines, and Logistic Regression Classifiers for Text Reviews Classification", Baltic J. Modern Computing.
3. Carlisle George and Penny Duquenoy (2008): "Online Medical Consultations: Legal, Ethical and Social Perspectives" Article, IGI Global.
4. Bhandari TR (2013): "Health System Research: Development, Designs and Methods", Journal of Health and Allied Sciences, P: 68–72.
5. Dr. Sandesh Kumar Sharma, Dr. Sudhinder Singh Chowhan (2013): "Patient Survey to Measure the Quality of Care Provided by Health care Providers in OPD of Tertiary Care Hospitals", Indian journal of Research, Volume 2, Issue 2, ISSN: 2250–1991.
6. Anne Mills (2012): "Health policy and systems research: defining the terrain; identifying the methods",Health policy and planning, Volume 4, Issue 3, Pages 1–7.
7. G. Mageshwari, E. Grace Mary Kanaga (2012): "Literature Review on Patient Scheduling Techniques", International Journal on Computer Science and Engineering (IJCSE), Volume 4, No. 03, ISSN: 0975-3397.

Note: All the figures in this chapter were made by the author.

Disruptive Technologies for Sustainable Development – Dr. G. Nagappan et al. (eds)
© 2024 Taylor & Francis Group, London, ISBN 978-1-032-55000-8

Highly Efficient System of Spam Detection for IoT Gadgets with the Help of ML

53

E. Sujatha[1]
Professor, Department of Computer Science and Engineering,
Saveetha Engineering College, Chennai

P. Dhinesh Prabhu[2]
Department of Computer Science and Engineering,
Saveetha Engineering College, Chennai

Abstract: IoT stands for Internet of things. This is a special technology whereby the group comprising gadgets in large numbers with various sensors and the actuators. These sensors and actuators are linked for transmission of data over wired or wireless channels. In the past decade, internet of things or IoT has grown rapidly. It was expected that by the end of 2020, 25 billion devices would be connected to IoT additionally. From all these connected devices, the volume of data released is expected to increase in the future and coming times ahead. Since, volume increase is expected, the data released from these devices will be increased as well in the coming times. Along with it, there would be modalities of various kinds with variable quality of data that is well defined by speed, if we consider their dependency on various factors such as time and position. The ML algorithms and other machine learning algorithms in such an environment plays a key role in authorization and security ensuring on the basis of bio-technology, detection of anomalies for the usability improvement and IoT systems and their security. In this article, we suggest the IoT security to detect the possible spams with the help of techniques such as machine learning. This objective can be achieved by in the framework of machine learning for IoT, the spam detection is suggested.

Keywords: security of communication system, IoT (abbreviation of Internet of Things, ML (abbreviation of machine learning).

1. Introduction

The convergence and implementations are enabled along with IoT between the objects of real world regardless of where they are present at the globe. Such kind of networking management implementation and control are needed and for this purpose, the proper strategies are needed and this is one of the highly challenging aspect in such environments. The protection of data is a key feature for internet of things and their application and this protection is needed to fix the issues regarding security such as any kind of malicious or spoofy attack, networking interruptions, occurrences such as DoS attacks, malware, network overcrowding, network spamming etc.

The size and type of organization determines the safety measures of Internet of things whereby this technology is imposed. The user's behavior of this kind makes it essential to cooperate for the security gateways. It can also be said that the security measures are determined by the application of IoT devices, their location and nature. The various protocols and attacks appear on the layer of IOT is specified in Fig. 53.1.

[1]sujathae@saveetha.ac.in, [2]dhinesh.prabhu1998@gmail.com,

DOI: 10.1201/9781003428473-53

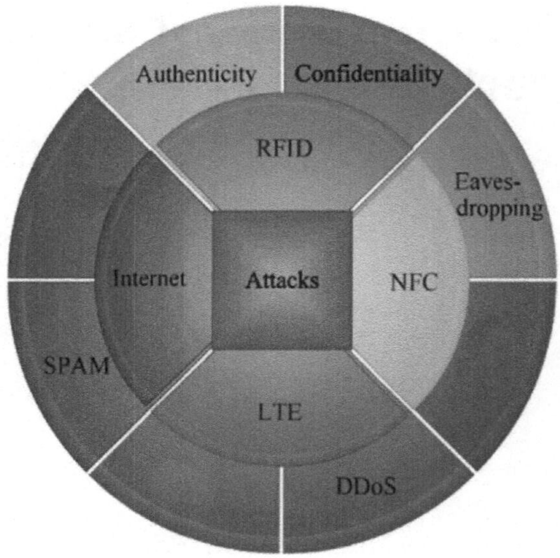

Fig. 53.1 Possible attacks and their protocols

2. Literature Review

The Denial of Service attack also abbreviated as DDOS attack: The targeted database can be flooded by the attackers. In doing so, they sent unwanted requests to withdraw Internet of Things gadgets to get access from various other nearby points and services. (Z.-K. Zhang et al 2014).RFID attacks: at the physical layers inside the internet of things gadgets, these kind of attacks are imposed. The integrity of the device is lost by this kind of attack (M. A. Rahman et al 18611–18621, 2019).At the sensor node, the common attack that is possible is the authenticity attack, confidentiality attack, availability attack and attack on cryptographic keys by the brute forces by (E. Bertino and N. Islam et al 2017).The attacks at internet: to access various resources, the internet connection is ensured by internet of things. The spamming techniques are there which are used by the spammers to steal the information from other system (W. Kim, O.-R. Jeong et al 2011)NFC (abbreviation of near field communication) if it is applicable, it is larger than NFC attacks. The electronic payment frauds are linked with these kind of attacks. The unencrypted tragic, eaves dropping, modification of tags are the possible attacks in this category (R. V. Kulkarni et al 2009).The protection of privacy and confidentiality is the solution to this problem. Here, the same profile cannot be created by the attacker even if the public key is used (A. L. Buczak et al 2015). Machine learning with supervised techniques: there are various SVMs (abbreviation of support vector machines), K nearest neighbors abbreviated as KNN, randomized forests and other kind of neural networks along with naïve bayes (A. Jamdagni et al 2013). All these techniques are used to label the given networks for the attack detection. The devices such as internet of things along with these kind of models are detected in a successful way by DOS and DDoS attacks, the malware attacks and other forms of intrusio (Y. Li, D. E. Quevedo et al 2016). The machine learning of unsupervised techniques: in the absence of labels, such techniques tend to outperform their given counterparts . The clusters are formed as they work. In the devices working with IoTs, the DoS attacks are detected often by multivariate correlation analysis(C. Giannella et al 2013).

3. Methodology and Sub Modules Specification

3.1 System Model

This can be retrieved as well in an efficient way. The occurrence of spam is reduced as a result of this suggested model for the known devices and this is explained with the help of equation given below:

$$\min P(s) = \aleph - \vec{s}. \quad (1)$$

the equation 1 suggests that $\aleph$ is an data collection. $\vec{s}$ is the information that is related to vector, it is also subtracted from $\aleph$ and as a result of it the spam information probability is reduced from the devices working with internet of things.

3.2 Proposed Methodology

From producing malicious information or data, the internet devices are supposed to be protected. In this suggestion, targets are set to detect the spam at web. In devices working with IoT various algorithmic machine learning techniques are considered. The issues in IoT devices are resolved with the help of suggested techniques. On the other hand, the suggested method, all the data engineering parameters are considered before machine learning models are validated with it. The machine learning algorithms work efficiently and reliably when appropriate attributes are given. From the smart objects from world, we gather the real world data examples and the process is known as feature extraction or selection as the engineering process and its core feature.

3.3 ML Models

The spam boundaries are supposedly needed for the suggested strategy and this is done by machine learning methods. For experiments, the models of ML are mentioned in Fig. 53.2.

There are some of the essential components that are needed. [18], [19].

1. The data at initial stage is consolidated. The data is determined quantitatively and the coefficient likelihood is calculated for the dispersion addresses.
2. The component of probability is also noticed as well. The outcomes are also analyzed by the address likelihood.

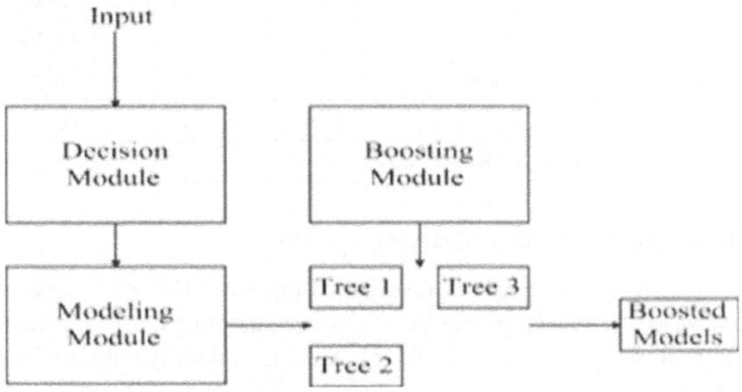

Fig. 53.2 Standard deviations of PCs

3. At third stage, the capability and likelihood also brings the coefficient values for the dissemination of results.
4. At fourth stage, from back dispersions, the recreations are created to create the populace boundary dissemination of the shaped values.
5. At fifth step, the measurable dispersion of reproduces is summarized from the opposite side. In this case, the insights from the basic are utilized.

Boosted Linear Model: For the components of data, trees of various choices are created. The isolation of information is made and the data is sent in series to the classes with major information. From the modular demonstration, the shaped models are portrayed in the following Fig. 53.2.

Xgboost (abbreviation of Boosting Extreme Gradient): It is a slope helping framework, which is proficient and versatile. There are various objective functions that are supported by it like.

Table 53.1 Filters based on entropy and their outcomes

Feature	attr importance
plug Id Ref	0.76342
Space Id Ref	0.12322
manufacturer	0.23432
model	0.20345

Feature	attr importance
Occupancy Type	0.10346
Built Form Type	0.20998
Wall Age Band	0.43219
Condition Type	0.76908
Room Type	0.03076
Wall Type	0.38151
Window Type	0.12602
Fuel Type	0.06642
Meter Type	0.47700
Healing	0.30522
Battery Life	0.61396

3.2 Implementation

Table 53.2 For the experimental models the performance based summary

Model	Precision	Recall	Accuracy
M1	0.650	1	79.81
M2	0.541	1	83.22
M3	0.567	1	84.35
M4	0.598	1	88.9
M5	0.513	1	91.8

3.3 Generalized Linear Model (GLM) With Stepwise Feature

GLM (abbreviation of generalized linear method) With Stepwise Component Determination: GLMs give a unique system to make sense of that how a reliant variable can be deciphered through various informative (indicator) factors. The boundary ward might be constant or distinct, and there are various kind of factors such as illustrative factors that are covariate or experimental in nature or wither factors or straight outs.

4. Results and Discussion

After the assessment of models of ML, we figured the spamcity score of the given apparatus. The trust value is shown by the score along with unwavering quality of the gadget. The following equation 2 defines it:

$$\sum_{j=1}^{n}(p_i - a_i)^2 n \tag{2}$$

$$\mathrm{S} \leftarrow \mathrm{RMSE}[i] * V_i \tag{3}$$

$e[i]$ in the conditions given above, is the rate of error is figured with the anticipated and real exhibits. S is the score of spamcity, that is registered with the backing of property significance score and rate of mistake or error. The complete process or the score of spamcity is checked with the help of given algorithm 1. In R the algorithms are implemented, the computed score is presented as well. The feature selection based on attribute importance is specified in Table 53.1 and the performance of machine learning model is given in Table 53.2.

5. Conclusion

With the help of machine learning models, the suggested framework detects the parameters of spam inside the Internet of Things. The IoT along with datasets that are utilized for the sake of experimentation as processed earlier as well with the help

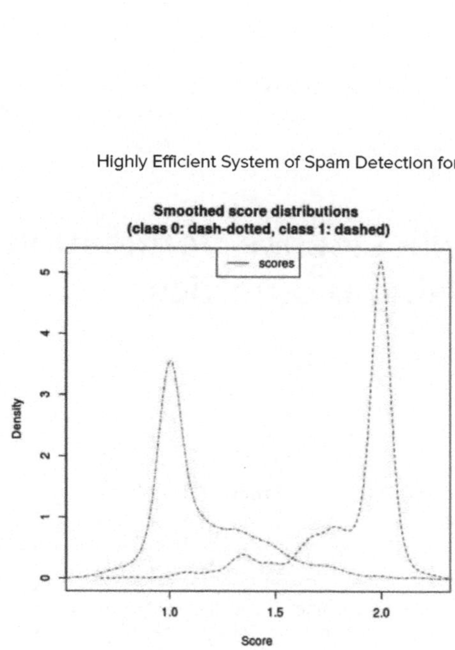

Fig. 53.3 Spam score distribution by extreme gradient boosting

of special engineering feature way. With machine learning models, when the frameworks are tested, same spam score was given to each appliance of IoT. For the successful working of internet of things in smart homes, the refined conditions are supposed to be taken. In the coming times ahead, the climatic and surrounding features shall be considered as well for internet of things. This action is carried out for their additional security and reliability.

References

1. Z.-K. Zhang, M. C. Y. Cho, C.-W. Wang, C.-W. Hsu, C.-K. Chen, and S. Shieh, Appl. (2014), pp. (230–234) "IoT security: Ongoing challenges and research opportunities," in Proc. IEEE 7th Int. Conf. Service-Oriented Comput.
2. M. A. Rahman (2019) et al., "Blockchain and IoT-based cognitive edge framework for sharing economy services in a smart city," IEEE Access, vol. 7, pp. (18611–18621).
3. Bertino and N. Islam, (2017) "Botnets and Internet of Things security," Com- puter, vol. 50, no. 2, pp. (76–79).
4. C. Zhang and R. Green, (2015) "Communication security in Internet of Thing: Preventive measure and avoid DDOS attack over IOT network," in Proc. 18th Symp. Commun. Netw., pp. (8–15).
5. W. Kim, O.-R. Jeong, C. Kim, and J.So, (2011) "The dark side of the internet: Attacks, costs and responses," Inf. Syst., vol. 36, no. 3, pp. (675–705).
6. H. Eun, H. Lee, and H. Oh, (Feb. 2013) "Conditional privacy preserving security protocol for NFC applications," IEEE Trans. Consum. Electron., vol. 59, no. 1, pp. (153–160).
7. R. V. Kulkarni and G. K. Venayagamoorthy, (2009) "Neural network based secure media access control protocol for wireless sensor networks," in Proc. Int. Joint Conf. Neural Netw., pp. (1680–1687).
8. M. A. Alsheikh, S. Lin, D. Niyato, and H.-P. Tan, (Oct.–Dec. 2014) "Machine learn- ing in wireless sensor networks: Algorithms, strategies, and applica- tions," IEEE Commun. Surveys Tuts., vol. 16, no. 4, pp. (1996–2018).
9. A. L. Buczak and E. Guven, (Apr.–Jun. 2015) "A survey of data mining and machine learning methods for cyber security intrusion detection," IEEE Commun. Surveys Tuts., vol. 18, no. 2, pp. (1153–1176).
10. F. A. Narudin, A. Feizollah, N. B. Anuar, and A. Gani, (2016) "Evaluation of machine learning classifiers for mobile malware detection," Soft Comput., vol. 20, no. 1, pp. (343–357).
11. Z. Tan, A. Jamdagni, X. He, P. Nanda, and R. P. Liu, (Feb. 2013) "A system for denial- of-service attack detection based on multivariate correlation analysis," IEEE Trans. Parallel Distrib. Syst., vol. 25, no. 2, pp. (447–456).
12. Y. Li, D. E. Quevedo, S. Dey, and L. Shi, (Sep. 2016) "SINR-based dos attack on remote state estimation: A game-theoretic approach," IEEE Trans. Control Netw. Syst., vol. 4, no. 3, pp. (632–642).
13. L. Xiao, Y. Li, X. Huang, and X. Du, (Oct. 2017) "Cloud-based malware detection game for mobile devices with offloading," IEEE Trans. Mobile Comput., vol. 16, no. 10, pp. (2742–2750).
14. J. W. Branch, C. Giannella, B. Szymanski, R. Wolff, and H. Kargupta, (2013) "In-network outlier detection in wireless sensor networks," Knowl. Inf. Syst., vol. 34, no. 1, pp. (23–54).

Note: All the figures and tables in this chapter were made by the author.

Disruptive Technologies for Sustainable Development – Dr. G. Nagappan et al. (eds)

Automobile License Verification Using Biometrics

54

Preethi P.[1]
Associate Professor, Department of Computer Science Engineering,
Saveetha Engineering College, Chennai, Tamil Nadu

Pawan Kumar E.[2]
PG Scholar, Department of Computer Science Engineering,
Saveetha Engineering College, Chennai, Tamil Nadu

Monica Manimaran[3]
PG Scholar, Department of Computer Science Engineering,
Saveetha Engineering College, Chennai, Tamil Nadu

Abstract: The process of verifying a match between two human fingerprints through automation is known as fingerprint authentication or recognition. This is a type of biometric identification used to verify the identity of individuals. To match fingerprints, various features of the print pattern are analyzed. Our project utilizes a non-imitable biometric authentication scheme called Fingerprint authentication, which can prevent unlicensed individuals from driving. Our proposed system involves a smart card capable of storing a specific person's fingerprint, which is recorded at the time of issuing their license and stored in a database. Whenever the person needs to verify their identity, they can place their finger on the fingerprint reader, and their information will be displayed. The Arduino is used in this license system to facilitate the biometric fingerprint authentication process.

Keywords: Arduino, License system, Biometric, Fingerprint

1. Introduction

In our country, many RTO offices lack a systematic driving license verification system. While obtaining a driving license from the RTO office has become relatively easy, maintaining the original license is a challenging task for vehicle users. Unfortunately, some vehicle users resort to cheating the police by possessing fake licenses, which is a criminal offense. To address these issues, we propose an authenticating driving license system for RTOs. Our project aims to provide a solution for proper maintenance of the driving license system. Firstly, we consolidate the databases of Aadhar card, license, insurance details, vehicle, and other relevant information into a single database. This merged database is then fed into the ARM9 processor. Whenever the license details of a particular person are required, biometric scanning techniques such as fingerprint or retina scanning can be used to retrieve the information.

2. Literature Review

Smart phones have been widely utilized in recent studies focusing on tracking the location of objects. With built-in GPS features, users frequently use smartphones to locate addresses, monitor traffic conditions, and plan travel time. The Android

[1]ppreethi@saveetha.ac.in, [2]pawankumarcse37@gmail.com, [3]monicamanimaran1601@gmail.com

DOI: 10.1201/9781003428473-54

operating system is a popular choice due to its straightforward programming, reliable data connection, and effectiveness in numerous studies. Based on a modified version of Linux, Android was utilized in one study where participants installed an APK file. During setup, users entered destination coordinates and the allowable area radius. The application was linked to the office's internet IP address, which allowed it to automatically record the date and time of an employee's entry into the office area. It's important to note that continuous monitoring of employee location can drain smart phone batteries; therefore, carefully selecting clustering on the map to define the radius is crucial.

3. Methodology and Model Specifications

We have introduced a novel approach that overcomes the limitations of the current system and provides precise outcomes. Our new project is called the Automobile License Verification Using Biometrics, which aims to identify counterfeit license IDs by utilizing a fingerprint reader. Initially, we obtain the user's fingerprints and input data such as their license identification number, Aadhar number and photo. We manage a database of driving licenses utilizing a PIC controller. When RTO officials need to verify a driver's license, they can authenticate the user by verifying their fingerprints. If the driver is verified, they will have a valid driving license. In case of authentication failure, the officer can detect that the driver is carrying an invalid license and proceed with necessary measures against them. Our module incorporates a fingerprint sensor, thermal printer, and PIC controller and utilizes MPLAB IDE for programming.

The system works by taking the user's fingerprint through the fingerprint reader, which is then matched with the respective details such as license ID, photo, and Aadhar number maintained in the driving license database through the ARM9 processor. The law enforcement officers will utilize this validation system to inspect the driving license of the vehicle operators.

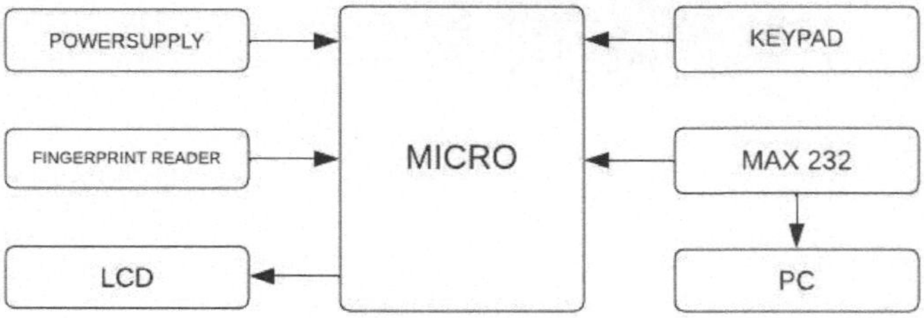

Fig. 54.1 Fingerprint sensor

The ID Module functions by capturing a unique image of the surface of the skin using the fingerprint sensor, which is then processed by the pattern recognition software on the ARM9 processor to identify the individual based on their fingerprint characteristics. The module compares the current reference with archive entries in Flash ROM to determine the authenticity of the identity, and a signal is sent to indicate acceptance or rejection of the verification result. The ID Module features several serial interfaces for control and verification, as well as module and file administration. It also has security advantages, as biometric data does not need to be communicated and can be secured with cryptographic mechanisms. The module is small, consumes low power in operation and standby modes, and can easily integrate with existing systems.

3.1 RS232 Connections, and Wiring up Serial Devices

To establish a connection with a PC, it is necessary to use a mating cable that features a DB25/F (female pin) connector, as the PC connector has male pins. However, if handshaking is looped, the CTS input will be activated immediately by the RTS output from the PC. Furthermore, the mating cable should have a DB9/F (female pin) connector at its termination.

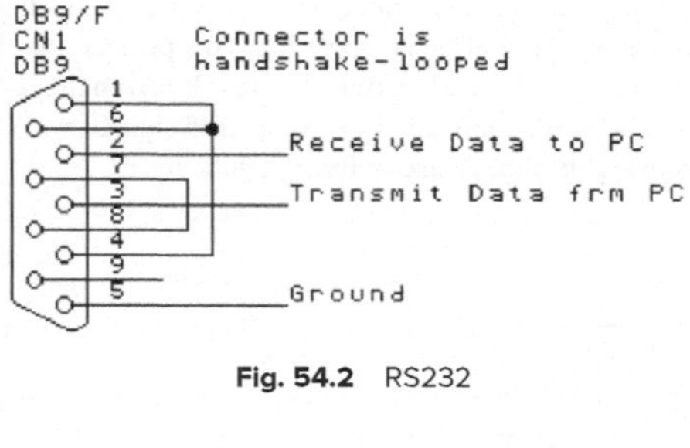

Fig. 54.2 RS232

3.2 Circuit Diagram

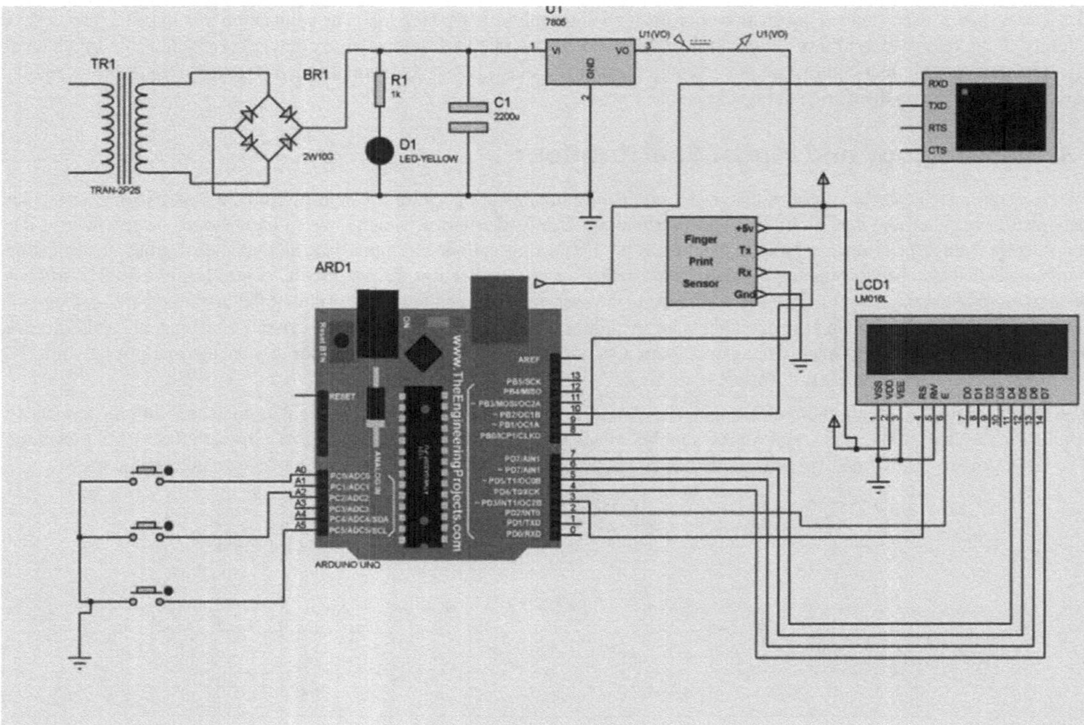

Fig. 54.3 Circuit connection diagram

4. Conclusion

By implementing a fingerprint-based authentication and licensing system, it becomes easy to identify individuals who are not authorized to drive and prevent them from doing so illegally. This authentication method relies on a unique and non-reproducible biometric factor: fingerprints. During the license issuance process, the individual's specific fingerprint ,along with their personal details, which are also recorded in a database. When the person places their finger on the fingerprint reader, their complete information is displayed. This project not only helps in catching unlicensed drivers and their vehicles but also detects fake users and unregistered vehicles. It also helps licensed individuals avoid fines in cases where individual do not have the liscense at the moment. To sum up, a straightforward fingerprint-based licensing authentication system has been implemented, ensuring its ease of use in future applications.

References

1. Ajay Kumar and Yingbo Zhou, Human Identification Using Finger Images" 2011, IEEE Transactions on image processing, pp. 2228–2244.
2. Anil Jain, Arun Ross and Salil Prabhakar, "Fingerprint Matching Using Minutiae and Texture Features" 2001 International Conference on image processing. Doi: 10.1109/ICIP.2001.958106.
3. Naoto Miura, Akio Nagasaka and Taka fumi Miyatake, "Extraction of Finger-Vein Patterns Using Maximum Curvature Points in Image Profiles " 2005, IAPR conference on Machine Vision Application, pp. 347–350.

4. I.A.Turkand , A.P.Pentland, "Face recognition using Eigenfaces"1991, IEEE Computer Society Conference on computer Vision and pattern Recognition.
5. J.Hashimo, "Finger vein authentication technology and its future" 2006 Symposium on VLSI Circuits, Doi: 10.1109/VLSCI.2006.1705285
6. Mulyona D, Jinn H.S, "A study of finger vein biometric for personal identification" 2008, International Symposium on biometrics and security Technologies, Doi: 10.1109/ISBAST.2008.4547655
7. R. K. Singh, "Crime in India2011 - Statistics", for National Crime Records Bureau, 2011.
8. Law Enforcement Guide to False Identification and Illegal ID Use Prepared by Pacific Institute for Research and Evaluation In support of the OJJDP Enforcing the Underage Drinking Laws Program. (2001)
9. S. Kranthi, K. Pranathi, A. Srisaila, "Automatic Number Plate Verification" 2011, International Journal of Advancements in Technology, 2(3): 408–422.

Note: All the figures in this chapter were made by the author.